W9-BZX-030

C01345

REVIEWING

The Living Environment: BIOLOGY

REVISED EDITION

With Sample Examinations

Rick Hallman

Principal, Benjamin N. Cardozo High School
Bayside, New York

AMSCO

AMSCO SCHOOL PUBLICATIONS, INC.
315 Hudson Street, New York, N.Y. 10013

The publisher wishes to acknowledge the helpful contributions of the following reviewers and consultants in the preparation of this book:

Bart Bookman
Assistant Principal, Science, Retired
Adlai E. Stevenson High School
Bronx, New York
Coordinating Biology Mentor, NYS

Elise Russo, Ed.D.
Administrator for Secondary Curriculum
Amityville Union Free School District, NY
Former Associate in Science Education
State Education Department of New York

Barbara Nicolato
Science Teacher, Biology
Newburgh Free Academy
Newburgh, New York

Lane Schwartz
Principal, Retired
Glen Cove High School
Glen Cove, New York

Christine Caputo
Science Editor / Educator
Ready 4 Print, Inc.
Tampa, Florida

Ira Rosenkrantz
Assistant Principal, Biology & Health, Retired
DeWitt Clinton High School
Bronx, New York

Edited by Carol Davidson Hagarty
Composition by Brad Walrod/High Text Graphics, Inc.

Cover and Text Design by Merrill Haber
Artwork by Hadel Studio

Photo Research by Tobi Zausner
Cover Photo: Gulf fritillary butterfly on purple daisy, © Royalty-Free/CORBIS

Please visit our Web site at: www.amscopub.com

When ordering this book, please specify: *either* **R 323 P** *or*
Reviewing The Living Environment: Biology, Revised Edition

ISBN: 978-1-56765-901-6
NYC Item: 56765-901-5

Copyright © 2005 by Amsco School Publications, Inc.

No part of this book may be reproduced in any form without
written permission from the publisher.

Printed in the United States of America
    4 5 6 7 8 9 10   10 09 08 07

CONTENTS

Student's Study Guide

This guide explains how to use this review book to prepare for the New York State Regents High School Examination: Living Environment.

I. WHAT IS THE REGENTS LIVING ENVIRONMENT EXAMINATION?

All students in New York State must pass one Regents examination in science in order to earn a high school diploma. The Regents Living Environment Examination can be used to meet this requirement. Students will usually take this examination after completing one year of high school biology.

The Regents Living Environment Examination is a three-hour test that consists of a variety of multiple-choice and open-ended questions that test comprehension of core topics (Parts A, B, and C) and mandated laboratory activities (Part D). Part A consists of multiple-choice questions. Parts B and C consist of graphing, short-answer, and short essay questions. Part D consists of multiple-choice and open-ended questions, which are based on three of the four required lab activities that are completed during the school year. Sample examinations are included at the back of this book. The format for the Living Environment Examination is as follows:

THREE-HOUR TEST:

Part A   Multiple-choice questions on key ideas and details of biology (approximately 35 questions, 35% of the exam grade)

Part B   Multiple-choice, short-answer, and short essay questions based on key ideas and details of biology, laboratory skills, and experimental problems (approximately 30% of the exam grade)

Part C   Short essay questions based on real-world problems and situations that require the application of knowledge of biology to the problem or situation (approximately 20% of the exam grade)

Part D   Multiple-choice, short-answer, and short essay questions based on the required laboratory activities; completed as part of the three-hour test (approximately 15% of the exam grade)

II. WHAT SHOULD I STUDY?

The following are the areas of study in which you will need to be prepared in order to be successful on the Regents Living Environment Examination.

A. Key Ideas and Details in Biology

It will be necessary to know the key ideas and details in high school biology in order to answer the Regents examination questions. The main concepts are the six themes of this review book:

♦ Evolution

♦ Energy, Matter, and Organization

♦ Maintaining a Dynamic Equilibrium

♦ Reproduction, Growth, and Development

♦ Genetics and Molecular Biology

♦ Interaction and Interdependence

The content in this review book and the questions in each chapter will help you study these ideas. Any words in **boldface** may appear on the examination without any explanation. You must know the meaning of these words. Other words in the review book are *italicized* for emphasis if they are important, but they do not need to be memorized. Chapter 4, *Human Evolution*, is an advanced topic that will NOT be tested on the Regents examination. Also, the details in the review book on each of the human organ systems (for example: digestive, circulatory, reproductive) are details that are taught before high school. They also will not be specifically tested on the Regents Living Environment Examination; but a knowledge of these systems is important as background information for concepts that may be tested on the Regents.

B. The Nature of Scientific Inquiry

What science is and how scientific research is conducted will also be tested. This topic is explained in more detail in the Appendix of this review book.

It is important to realize that science is both a body of knowledge and a way of learning about how the world works. Scientific explanations are developed using observations and experimental evidence to add to what has been learned previously. All scientific explanations may change if better ones are found. Good science involves asking questions, observing, experimenting, finding evidence, collecting and organizing data, drawing valid conclusions, and discussing results with other scientists.

In order to make useful observations to test proposed explanations, a research plan involving well-designed experiments must be prepared. Included in creating a research plan are the following:

♦ researching background information on the major concepts being investigated;

♦ developing proposals, which include hypotheses, to test the explanations; i.e., predict what should be observed under specific conditions if the explanation is true;

♦ designing experiments that include techniques to avoid errors or false conclusions; these techniques may include repeated trials, large sample size, objective data-collection techniques, and use of a control.

Observations made through research need to be analyzed to determine if they support or contradict the proposed explanations. Methods used to analyze data include the use of diagrams, tables, charts, and graphs as well as the use of statistical techniques. These statistical techniques should help determine the closeness of the predicted results in the hypothesis to the actual results, in order to reach a conclusion as to whether the explanation on which the prediction was based is supported.

It is important in science to make the results of an investigation public. It is assumed that different scientists will arrive at the same explanation if they are able to analyze similar evidence. Therefore, scientists tell each other how they did their experiments in order to allow others to repeat the investigations.

C. Laboratory Skills

As a biology student, you need to be able to successfully conduct laboratory investigations. Skills needed in the laboratory that may be tested on the Living Environment Regents Examination are listed below:

♦ Follows safety rules in the laboratory

♦ Selects and uses correct instruments

    ✔ uses graduated cylinders to measure volume

    ✔ uses metric ruler to measure length

    ✔ uses thermometer to measure temperature

    ✔ uses triple-beam or electronic balance to measure mass

♦ Uses a compound microscope/stereoscope effectively to see specimens clearly, using different magnifications

    ✔ identifies and compares parts of a variety of cells

    ✔ compares relative sizes of cells and organelles

    ✔ prepares wet-mount slides and uses appropriate staining techniques

♦ Designs and uses dichotomous keys to identify specimens

♦ Makes observations of biological processes

♦ Dissects plant and/or animal specimens to expose and identify internal structures

♦ Follows directions to correctly use and interpret chemical indicators

♦ Uses chromatography and/or electrophoresis to separate molecules

♦ Designs and carries out a controlled, scientific experiment based on biological processes

♦ States an appropriate hypothesis

♦ Differentiates between independent and dependent variables

♦ Identifies the control group and/or controlled variables

♦ Collects, organizes, and analyzes data, using a computer and/or other laboratory equipment

♦ Organizes data through the use of data tables and graphs

♦ Analyzes results from observations/expressed data

♦ Formulates an appropriate conclusion or generalization from the results of an experiment

♦ Recognizes assumptions and limitations of the experiment

III. HOW SHOULD I STUDY?

A. Basic Strategies for Study

♦ Make a regular time and place for study at home, and plan a daily schedule. As a guideline, on a regular school night, 30 minutes *per subject* in high school should be spent on homework and study. Success will come—**if you work hard!**

♦ Prepare your own notes of important ideas and terms as you are reading. This helps to keep your mind concentrated on the subject while you are reading.

♦ Write your own questions about what you read, close the book, and then try to answer your own questions. After completing this task, open the book again, and review and correct your answers.

♦ Study when you are feeling alert and try to avoid distractions (such as the Internet, telephone, radio, and television).

♦ Take five-minute breaks every hour to help you stay alert and retain what you have studied.

B. Test-Taking Strategies

Multiple-Choice Questions

For multiple-choice questions, **read** the statement or question carefully. If a diagram is included, study the diagram as well. Do not look at the answers until you have looked carefully at the questions first. Reread the question. It is helpful to underline key words as you read. Now look at each of the answers. Study all of the answers. Do not stop when you think you have found the right one! Look at all the answers. Be aware that you are looking for the word or expression that **best** completes the statement or answers the question. There will be only one answer that is the best choice.

As you study the answers for a multiple-choice question, eliminate the answers that you know are not correct. It is helpful to think about why answers are wrong as you eliminate them. This will assist you in selecting an answer. Refer back to the statement or question as often as necessary as you study the answers. After eliminating any answers you know are wrong, choose wisely, or in other words, guess from the remaining answers. **Be certain to make a selection. Do not leave a blank.** You will not be penalized for entering a wrong answer. Therefore, any attempt at an answer is better than none at all.

If you have additional time at the end of the test, and you are reviewing your answers for multiple-choice questions, be very careful before making any changes.

After following the thinking process described above, the answer you choose the first time is more likely to be the correct one than another choice made later. In other words, be careful not to "second-guess" yourself.

EXAMPLE:

The energy found in ATP molecules synthesized in animal cells comes directly from

1 sunlight
2 organic molecules
3 minerals
4 inorganic molecules

ANALYSIS:

Although "energy" and "ATP" may seem to be the key terms in the question, you should realize that they describe the same thing: ATP *is* a form of chemical energy. There is something else important in the question—the word "animal." Now we know that the question is about energy and animals. At this point, we look at the answers. We recognize that Answer 1—"sunlight"—is wrong because sunlight provides plants with energy, not animals. Also, when we look back at the question, we realize from this wrong answer that another key word in the question is "directly." Animals get energy indirectly from sunlight when they eat plants (or other animals), not directly. Answer 2—"organic molecules"—describes substances from other living things, i.e., organic. This fits what we think would answer the question. We continue by looking at Answer 3—"minerals." Minerals are from the earth and do not contain stored energy. The same is true for Answer 4—"inorganic molecules." These answers help make us sure that the word "organic" in Answer 2 makes that answer the correct choice.

This kind of thinking process should be done for each multiple-choice question. While this requires time, it is the only method that will allow you to make use of what you have learned during the year and avoid making incorrect selections.

Reading-Comprehension Questions

The Living Environment test may have questions based on one or more reading passages. You will usually be required to prepare written answers to questions based on the reading passages. Prepare yourself for the reading passage by giving the passage you are about to read your **full attention**:

♦ make a decision that the reading passage **will** contain information that will interest you, even if the topic is something you have never paid any attention to before; comprehension of reading material is much higher when people are interested in what they are reading;

♦ stop worrying about other things in your life;

♦ stop looking around at other people in the room or at things happening outside the classroom;

♦ stop looking at your watch to check the time.

The following are suggested strategies for reading-comprehension questions. You may wish to use one or more of these strategies, based upon your personal preference:

♦ look over the questions before reading the passage; this may give you an idea of what to look for while you are reading; however once you start reading, do not stop for individual answers because this will break your concentration;

♦ as you read each paragraph, ask yourself what the paragraph was about; if a reading passage has more than one paragraph, this is because it has more than one important idea; each paragraph has one important idea;

♦ look carefully at the first sentence of the first paragraph; it usually states the topic for the reading passage;

♦ underline the sentence that states the main idea of the reading passage and of each paragraph;

♦ circle specific facts;

♦ reread the entire passage before working on the questions.

Work on one question for the reading passage at a time. Note the number of points indicated after the question for the answer. If the question is worth two or more points, be certain that you write a complete answer to earn these points. Underline the **key words** in the question and **include these words in your written answer**. This will make it much more likely that the teacher who reads your answer and marks your test will understand what you are trying to say. Also, write very neatly. If this is not possible, then print your answer. A teacher marking the test cannot give points for the answers that cannot be read!

Data Interpretation and Graphing

The Regents Living Environment Examination may include questions that require you to read about an experiment, study a data table showing the results of the experiment, graph the data, and answer one or more questions about the data. The data for a typical experiment may be presented in a Data Table with at least two columns. The first column is usually the *independent variable*. This is the condition or variable that the researcher set. The numbers in this column show each level at which the variable was set. The other columns are for the *dependent variables*. The values in these columns are the data that was measured or collected at each level of the independent variable. These values depend upon the independent variable.

As you read the passage about the experiment, study the data table. Use the reading passage to understand the data table. If you are required to construct a line graph for the data, follow the directions carefully. The title for a graph should refer to **both** the independent and the dependent variables. The most important part of a graph is the marking of an **appropriate scale**. The scale begins at the lower left, with the values increasing going to the right or going up. The scale for the independent variable is on the horizontal axis and the scale for the dependent variable is on the vertical axis. The scale must be prepared to include the lowest and the highest value for the variable. Each space on the graph paper must represent the same quantity. To plot the data points, refer to the data table. Each point on the line graph is located by two pieces of data from the data table, i.e., the value for each level of the independent variable with the corresponding value for the dependent variable.

A bar graph is used to compare data rather than to show relationships between variables. A bar graph has no scale along the horizontal axis. Rather, labels for each set of data are placed on the horizontal axis, with a scale for the data on the vertical axis.

C. Final Test Preparation

During the review period at the end of the term, prior to taking the Regents examination, you should set aside time to take a complete practice exam all at once. Pretend that you are taking the real exam and do all the questions on the test. Record your answers according to the directions given in the practice test. Once you have taken the complete test, mark your answers. Note what topics were particularly difficult for you and return to the review book to study those topics further. Then take the other practice test provided in this review book. Finally, do not try to cram all the material the night before you take the actual test; it is important for you to be well rested when you take the Regents examination.

Remember: Hard work increases your abilities and will improve your test results.

The Process of Evolution

⬛ THE THEORY OF EVOLUTION

Discoveries in modern science have shown that, over many thousands of years, populations of living things change. In 1859, the English naturalist Charles Darwin proposed a theory to explain how organisms change over time. It is called the theory of **evolution**. (See Figure 1-1.)

The theory of evolution explains how the immense variety of living things on Earth has developed from ancestral forms during the past three billion years. This theory is considered to be the most important unifying idea in biology. It offers an explanation, based on fossil and other scientific evidence, of how Earth came to be populated by the millions of **species** (different kinds of organisms) now alive and of

Figure 1-1 Charles Darwin proposed the theory of evolution to explain how organisms change over time.

how these species are related to one another. As a result of the changes that occur in living things and in the environment over time, nearly all species that once lived on Earth are no longer living today.

⬛ A STRUGGLE FOR EXISTENCE

During its lifetime, a female elephant may produce six offspring. Darwin calculated that, over hundreds of years, millions of elephants could descend from one original pair of elephants. Similarly, if a plant produced only two seeds a year, in 20 years there could be a million new plants descended from the original parent plants. However, this does not happen. In fact, only a relatively small number of offspring of any species survive to produce their own offspring. Darwin realized that Earth cannot support huge increases in populations; thus, they do not increase that dramatically. He concluded that there is a "struggle for existence" in which only a few offspring of any type survive to maturity and reproduce.

In this struggle for existence, there is **competition** among organisms for various resources. Lack of food and space are two factors that can limit an organism's chances to survive and reproduce. Plants must have minerals, sunlight, and space in order to make their own food and grow. Animals also need resources such as food, water, space, and shelter in order to survive. Competition for available resources exists among individuals of the same species and between members of different species living in the same area.

GENETIC VARIATION

Through the process of *reproduction*, characteristics are passed on from parents to offspring. The resulting offspring resemble their parents. However, all the offspring produced by one pair of parents are not identical. The various characteristics that are passed on from one generation to the next are called **hereditary** traits. The differences among the offspring that inherit these characteristics are called *genetic variations*.

It is very important to recognize the difference between hereditary, or inherited, characteristics and changes that occur to an individual during its lifetime. For example, eye color is inherited. It is a hereditary trait. By contrast, becoming extremely muscular due to weight lifting is an *acquired* trait. It is not inherited. A characteristic that is hereditary can be passed along to offspring. An acquired characteristic cannot be passed on to offspring. Thus, only changes that are in the sex cells, or **gametes**, of the parents can become the basis for evolutionary change.

There are two types of reproduction. In **asexually** reproducing organisms, a single parent organism splits in two to produce two new organisms. (See Figure 1-2.) In **sexually** reproducing organisms, a male and a female mate to produce offspring. Sexual reproduction involves the combining of genetic material from two individuals. The traits in the offspring are the result of a new assortment, or **recombination**, of traits inherited from both parents. Thus, sexual reproduction produces greater genetic variability among offspring than does asexual reproduction.

In addition to the recombination of genetic information that occurs during sexual repro-

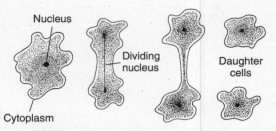

Figure 1-2 Asexual reproduction, as seen in the ameba, requires only one parent to produce two new organisms.

duction, genetic variation can arise from mutations. A **mutation** is a sudden change that occurs in the genetic material of an organism. These changes occur randomly and spontaneously, and may be caused by **radiation** or chemicals. A mutation in the gametes may produce a small change in the resulting offspring, a major effect in the offspring, or no noticeable effect in the offspring at all.

Darwin recognized that there is variation among individuals produced by the same parents. He wondered how the differences within a group of organisms could lead to differences between groups of organisms. Darwin asked this question about the species of small birds, called finches, that he had observed on the Galápagos Islands. He realized that there probably were small differences among the first finches that arrived on the islands from the South American mainland. (See Figure 1-3.) How did these minor variations within the ancestral finch species lead to the significant differences that now exist between the groups of finches? In other words, how did they develop into separate species? (See Figure 1-4.) Darwin recognized two important facts that play a role in the development of new species:

♦ There is a struggle for existence, which limits the number of offspring that survive.

♦ There are differences among offspring due to individual, inherited variations.

Perhaps more important, Darwin posed the question: What determines which individuals survive to reproduce and thus become the parents of the next generation of offspring? His answer to this question formed the basis for his theory of evolution and revolutionized our understanding of how various forms of life have come to be.

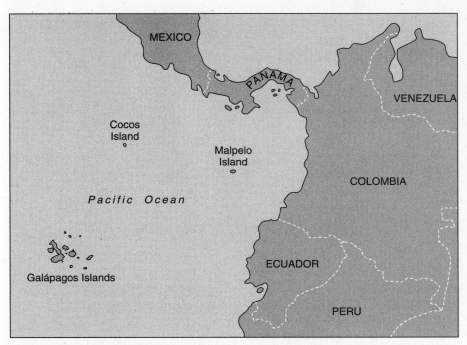

Figure 1-3 The first finches to arrive on the Galápagos Islands came from the South American mainland, hundreds of kilometers away.

NATURAL SELECTION

The special characteristics that make an organism well suited to a particular environment are called *adaptations*. How do organisms evolve the adaptations that enable them to survive so well in a particular environment?

Ground finch eats seeds

Catches insects with beak

Tool-using finch digs insects from under bark with thorn

Figure 1-4 The different beaks of these Galápagos finches show adaptions to different environments and different food sources.

Darwin attempted to answer this question. He developed an answer by combining what he knew about the inheritance of traits with what he observed about an organism's struggle for existence. He concluded that whatever slight variations an organism had that gave it an advantage over other individuals in that environment would make it more likely to survive. That is what is meant by "survival of the fittest." An organism that was more likely to survive also would be more likely to reproduce and pass on its genetic variations to future offspring. Those individuals that did not have such characteristics would be less likely to reproduce and pass on their characteristics. Darwin used the term **natural selection** to describe the way that environmental conditions determine which organisms survive to reproduce. Over time, the proportion of those more fit individuals would increase in a population. This is because of the increase in frequency of the genes responsible for those traits that give the surviving individuals the advantage.

ANTIBIOTIC RESISTANCE IN BACTERIA—A CASE STUDY IN NATURAL SELECTION

Penicillin can kill some of the bacteria that cause diseases in humans, leaving human cells unaffected. Penicillin was the first **antibiotic** discovered. Later, many more antibiotics with the ability to kill different kinds of bacteria were discovered.

Over time, scientists noticed that some strains of bacteria once killed by antibiotics

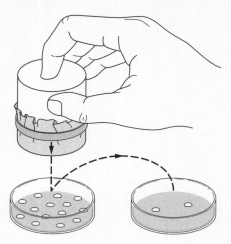

Figure 1-5 In this experiment, colonies of bacterial cells were transferred to a dish that contained an antibiotic. The only bacteria that survived were those that already had a natural resistance to antibiotics.

were no longer affected. They had developed a *resistance* to penicillin and some of the other antibiotics. How did the bacteria develop this resistance? Were there genetic variations that made some bacteria naturally resistant to the antibiotic, without having prior exposure to it? (See Figure 1-5.)

If some bacteria were resistant to antibiotics from the start, they would have a survival advantage when such chemicals were added to their environment. In fact, this is what happened. By killing off nonresistant bacteria, the antibiotics had decreased the competition for food that existed in the original population. Many more resistant bacterial cells could survive, grow, and reproduce. The result would be a strain of bacteria that has resistance to antibiotics. Similar results have been noted in some insects exposed to **pesticides**; those that are resistant can survive to reproduce. These are examples of natural selection at work—where external factors affect the survival of individuals within a population.

⚏ ARTIFICIAL SELECTION

People who raise dogs to perform certain tasks intentionally select, train, and breed those pups that have the characteristics best

Figure 1-6 The selective breeding, or artificial selection, by people of specific characteristics produces particular breeds of dogs, such as these.

suited to their intended function. This selection by people, of organisms with specific characteristics, is known as **selective breeding**, or *artificial selection*. It is a process similar to natural selection. However, humans—not the natural environment—select the organisms that have certain desirable traits and decide which ones will survive and pass on those traits to their offspring. Plant and animal breeders have practiced selective breeding for centuries, resulting in a variety of domestic livestock breeds and crops that are different from their wild ancestors. (See Figure 1-6.)

⚏ THE SCIENTIFIC METHOD

Experiments conducted by Darwin, as well as by other researchers, show how knowledge has been gained in science, namely by the use of the scientific method—an organized approach to problem solving. The main steps in this process are as follows:

♦ State the problem (in question form)

♦ Collect information

♦ Form a hypothesis

♦ Perform an experiment: Use experimental group with a variable; use control group without the variable

♦ Record observations and data

♦ Check results; redo experiment

♦ Draw your conclusions

♦ Communicate your results

Chapter 1 Review

Part A—Multiple Choice

1. How did Darwin explain the fact that only a small number of offspring of any species survive to reproduce?

   1 Each species acts to limit the size of its own population.
   2 Every species is limited to a certain number of offspring.
   3 The members of a species allow only specific offspring to reproduce.
   4 Offspring must compete for available resources in order to survive.

2. Which statement best describes competition? It exists

   1 only among individuals within the same species
   2 only between different species in the same area
   3 only between different species living in different areas
   4 among individuals of the same species and between different species living in the same area

3. Suppose two animals live in the same location and eat the same kind of food. What adaptation would decrease the competition between them?

   1 Both animals eat at the same time.
   2 Both animals breed at the same time.
   3 One animal has hair and the other has feathers.
   4 One eats during the day and the other eats at night.

4. Heredity is best described as

   1 behavioral differences among offspring
   2 the struggle for existence among living things
   3 traits that are passed from one generation to the next
   4 the slow and gradual change in organisms over millions of years

5. A couple had two children. One child has blue eyes and the other child has brown eyes. This difference is an example of

   1 natural selection
   2 artificial selection
   3 genetic variation
   4 acquired characteristics

6. Which description relates to an acquired characteristic?

   1 Jamal is tall and thin.
   2 Olivia has curly, blond hair.
   3 Brittney has a widow's peak like her father.
   4 Jose has large muscles from doing exercises.

7. What happens during asexual reproduction?

   1 Two organisms join together to become one new organism.
   2 A single parent organism splits to produce two organisms.
   3 Two organisms mate to produce a new single offspring.
   4 A single organism forms from the halves of two organisms.

8. In sexually reproducing organisms, the offspring inherit a combination of genetic traits from the

   1 mother only
   2 father only
   3 mother and father
   4 grandparents only

9. When compared to asexual reproduction, sexual reproduction produces

   1 less genetic variation among offspring
   2 greater genetic variation among offspring
   3 offspring that are identical to the parents
   4 offspring that are identical to one another

10. A mutation results from

    1 artificial selection by humans
    2 the fact that only the fittest organisms survive
    3 a sudden change in the genetic material of an organism
    4 competition for resources such as food and water

11. When mutations occur in body cells, they can be passed along to

    1 sex cells only
    2 other body cells only
    3 offspring only
    4 gametes only

12. Which statement best describes the current understanding of natural selection?

    1 Natural selection influences the frequency of adaptive traits in a population.
    2 Changes in gene frequencies due to natural selection have little effect on evolution.

3 Natural selection has been discarded as an important concept in evolution.
4 New mutations of genetic material are due to natural selection.

13. The Florida panther, a member of the cat family, has a population of fewer than 100 individuals and has limited genetic variation. Based on this information, which inference would be most valid?

1 The panthers will begin to evolve very rapidly.
2 The panthers can easily adapt to their environment.
3 Over time, the panthers will be less likely to survive in a changing environment.
4 Over time, the panthers will become more resistant to diseases.

14. Which statement represents the major concept of the biological theory of evolution?

1 A new species moves into a habitat when another species becomes extinct.
2 Present-day organisms on Earth developed from earlier, different organisms.
3 Every period of time in Earth's history has its own group of organisms.
4 Every location on Earth's surface has its own unique group of organisms.

15. Which concept is *not* a part of the theory of evolution?

1 Present-day species developed from earlier species.
2 Complex organisms develop from simple organisms over time.
3 Some species die out when environmental changes occur.
4 Change occurs according to the needs of an individual organism to survive.

16. Which statement best illustrates a rapid biological adaptation that has actually occurred?

1 Pesticide-resistant insects have developed in certain environments.
2 Paving large areas of land has decreased habitats for certain organisms.
3 Scientific evidence indicates that dinosaurs once lived on land.
4 The characteristics of sharks have remained unchanged over a long period of time.

17. When a breeder allows only the strongest and fastest horses to reproduce, she is practicing

1 artificial selection
2 natural selection

3 artificial mutation
4 asexual reproduction

18. Unlike in natural selection, in artificial selection

1 genetic information is passed from one generation to the next
2 humans, not the natural environment, decide which organisms will reproduce
3 the natural environment, not humans, decides which organisms will reproduce
4 mating is random and all organisms may pass their traits on to their offspring

19. People can develop new varieties of cultivated plants by carrying out

1 selective breeding for all traits
2 random breeding for all traits
3 selective breeding for specific traits
4 random breeding for specific traits

20. Selective breeding for particular traits can be used to

1 develop cultivated plants only
2 develop domesticated animals only
3 develop cultivated plants and domesticated animals
4 breed rare, wild animal species only

21. Which situation would most likely result in the highest rate of natural selection in a population?

1 reproduction of organisms by an asexual method in an unchanging environment
2 reproduction of organisms in an unchanging environment that has few predators
3 reproduction of organisms that have a very low mutation rate in a changing environment
4 reproduction of organisms that show genetic differences due to mutations in a changing environment

22. Some behaviors, such as mating and caring for the young, are genetically determined in most species of birds. The presence of these behaviors is most likely due to the fact that

1 birds do not have the ability to learn
2 these behaviors helped birds to survive in the past
3 individual birds need to learn to survive and reproduce
4 within their lifetimes, birds developed these behaviors

23. According to the theory of natural selection, why are some individuals more likely than others to survive and reproduce?

1 Some individuals pass on to their offspring new characteristics they have acquired during their lifetimes.

2 Some individuals do not pass on to their offspring new characteristics they have acquired during their lifetimes.

3 Some individuals are better adapted to exist in their environment than others are.

4 Some individuals tend to produce fewer offspring than others in the same environment.

24. According to modern evolutionary theory, genes responsible for new traits that help a species survive in a particular environment will usually

1 not change in frequency over time

2 decrease rapidly in frequency

3 decrease gradually in frequency

4 increase in frequency over time

25. The analysis of data gathered during a particular experiment is necessary in order to

1 formulate a hypothesis for that experiment

2 develop a research plan for that experiment

3 design a control for that experiment

4 draw a valid conclusion for that experiment

Part B—Analysis and Open Ended

Base your answer to question 26 on the information and data table below.

A student hypothesized that lettuce seeds would not sprout (germinate) unless they were exposed to darkness. The student planted 10 lettuce seeds under a layer of soil and scattered 10 lettuce seeds on top of the soil. The data collected are shown in the table below.

Data Table

Seed Treatment	Number of Seeds Germinated
Planted under soil	9
Scattered on top of soil	8

26. One way the student could improve the validity of these results would be to

1 conclude that darkness is necessary for lettuce seed germination

2 conclude that light is necessary for lettuce seed germination

3 revise the hypothesis

4 repeat the experiment

27. Briefly explain how the diversity of species alive today is related to the process of evolution.

28. How is the "struggle for existence" important to the study of evolution? Give an example.

29. In terms of evolution, why are the variations among individuals within a population more important than the similarities between them?

30. Explain the main difference between changes due to evolution and changes that are due to aging or that are acquired. Your answer must include the following:

♦ which type of change can be passed on to offspring

♦ why only that type of change can be passed on to offspring

♦ *one* example of a change due to evolution

♦ *one* example of a change due to aging or that is acquired

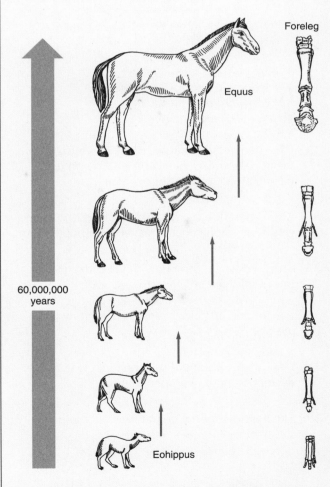

31. Which concept is best illustrated by the above diagram, which shows changes in the body size and form of horses over time?

1 acquired characteristics

2 artificial selection

3 partial inheritance
4 evolution by natural selection

32. Why does sexual reproduction produce greater variation among offspring than asexual reproduction? Why is this important for the process of evolution?

33. Briefly explain why mutations are important to evolutionary change.

34. The best title for the chart below would be

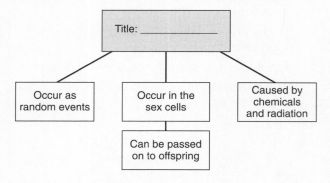

1 Types of Natural Selection
2 Characteristics of Mutations
3 Survival of the Fittest
4 Asexual Reproduction

35. Give two possible causes of genetic mutations. What cells would they have to occur in to be passed along to offspring? Explain.

36. Briefly describe four factors that lead to natural selection among organisms within a population.

37. The following terms relate to important factors in evolutionary change: struggle for existence; natural selection; environmental change; variation among offspring. Which concept includes the other three? Explain why.

38. Describe how natural selection can lead to the development of new species.

39. The best title for the chart below would be

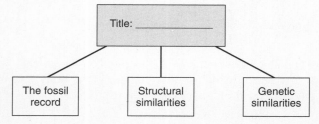

1 Evolutionary Pathways
2 Proof of Evolution
3 Natural Selection
4 Mutations in Evolution

40. A patient was given an antibiotic for an infection. The doctor told the patient to take it for 10 days; but the patient took it for 2 days, felt better, then stopped taking it. After several days, the patient became sick with the same infection again. Why?

41. Suppose there are two types of fur color, brown and white, in a species of rabbit that lives in an area with very little snow all year. Most of the rabbits have brown fur. Then the environment changes so that there is snow much of the year. Based on your knowledge of natural selection, you might predict that the proportion of white fur to brown fur in the new climate would change so that

1 equal numbers of rabbits would have brown fur and white fur
2 more rabbits would have white fur than brown fur
3 more rabbits would have brown fur than white fur
4 more rabbits would have white fur with brown splotches

42. Write an essay in which you identify the adaptive value of a particular trait in a population. In your answer, be sure to:

♦ define, and give an example of, an adaptation

♦ explain how the adaptation helps an organism survive

♦ describe how the adaptation becomes more widespread

43. Many pesticides have been used to kill insects that destroy crops or that spread diseases such as the West Nile Virus. Unfortunately, some of these pesticides are no longer as effective as they once were for getting rid of insect pests. Why?

44. A television advertisement claims that a certain brand of cough drop reduces coughing for 8 hours. Describe an investigation that could be used to determine if this claim is valid. In your answer, include at least a description of:

- the treatment to be given to the experimental group
- the treatment to be given to the control group
- the data to be collected

- when the data should be collected
- *one* observation that would lead to the conclusion that the claim is valid

Part C—Reading Comprehension

Base your answers to questions 45 to 47 on the information below and on your knowledge of biology. Use one or more complete sentences to answer each question.

The main purpose of science is to look at events, occurrences, and patterns in nature and develop explanations for them. These explanations can always be changed as new observations are made and new evidence is found. A possible explanation of a natural event or pattern is called a hypothesis. Charles Darwin, in his own words, showed why he was a true scientist:

"From my early youth I have had the strongest desire to understand or explain whatever I observed—that is, to group all facts under some general laws. I have steadily endeavored to keep my mind free, so as to give up any hypothesis, however much beloved (and I cannot resist forming one on every subject), as soon as facts are shown to be opposed to it. I followed a golden rule that whenever a published fact, a new observation or thought came across me, which was opposed to my general results, to make a memorandum of it without fail and at once. During some part of the day I wrote my Journal, and took much pains in describing carefully and vividly all that I had seen; and this was good practice . . . and this habit of mind was continued during the five years of the voyage. I feel sure that it was this training which has enabled me to do whatever I have done in science."

On the voyage of the *Beagle*, Darwin saw seeds that had washed ashore on a small island near Australia. He wondered whether seeds could travel long distances in the ocean and still be able to grow. Back in England, he enthusiastically filled his home with bottles of seawater. Darwin soaked many different kinds of seeds in the salty water, for short and long periods of time. He used a variety of crop seeds—such as cabbage seeds, lettuce seeds, and celery seeds—23 kinds in all. He then tried to grow them in soil. Darwin's experiments showed exactly what the scientific method is: State the problem; collect information; form a hypothesis; perform an experiment; record observations and data; draw a conclusion; and share your results.

45. When does a scientist find it necessary to change a hypothesis?

46. Why did Charles Darwin keep a journal?

47. Describe how Darwin followed each of the steps of the scientific method in his experiment with seeds and salt water.

Evidence for Evolution

No single idea explains the enormous diversity and complexity of life on Earth more powerfully than the theory of evolution by natural selection as proposed by Darwin. Evidence that supports this theory includes fossils, the shapes and structures of living organisms, the chemicals all living things are made of, and the distribution of species on Earth today.

❖ CREATING A FAMILY TREE

Simple diagrams can be used to represent the evolutionary relationships among different species. Suppose that A, B, C, and D represent four living species. The letters E, F, and G represent ancestral forms of the species that are most likely extinct. In this case, organisms B and C are more closely related because they evolved from their common ancestor E most recently. B and C are both equally related to A, with the more distant common ancestor F. Organism D is the least closely related to the others because it evolved from their common ancestor G the longest time ago. (See Figure 2-1.)

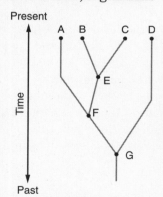

Figure 2-1 A diagram can be used to represent the evolutionary relationships among different living and ancestral species.

❖ EVIDENCE FROM FOSSILS

Fossils are traces or remains of dead organisms that have been preserved by natural processes. Usually only the hard parts of or-

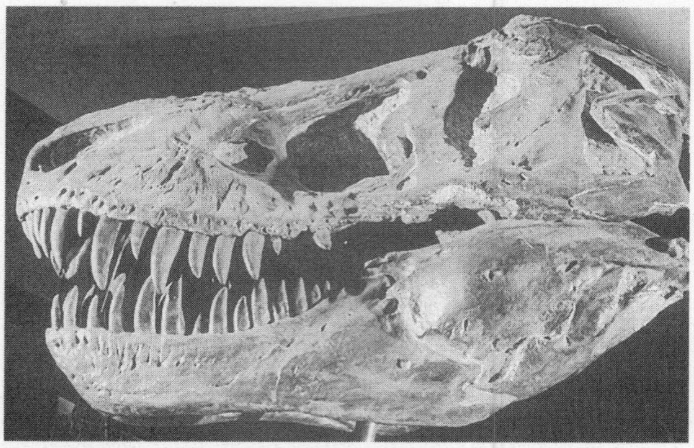

Figure 2-2 Dinosaur skull—a type of fossil that is formed when minerals slowly replace hard parts such as bones and teeth.

ganisms—that is, the bones, shells, or teeth— become fossilized. A common way fossils are formed is through the gradual replacement of an organism's remains by other substances. In this process of fossil formation, the organism is usually buried in sediments, its hard tissues being slowly replaced by minerals dissolved in underground water. Over time, these minerals harden to form an exact copy of the original organism. Fossils can also be formed if the body of a plant or animal creates an impression in soft mud or clay. This process shows only the original external shape of the organism and not the internal structure, as do some other types of fossil formation. By studying fossils, scientists can see that species have changed over time and that most ancient life-forms no longer exist. (See Figure 2-2.)

❖ EVIDENCE FROM COMPARATIVE ANATOMY

One way to determine the evolutionary relationships among different organisms is to find

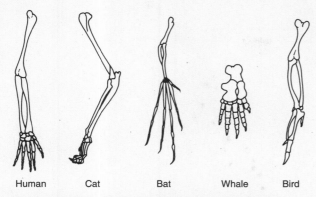

Figure 2-3 The similarities among the bones in each forelimb indicate that these animals shared a common ancestor.

Human   Cat   Bat   Whale   Bird

EVIDENCE FROM COMPARATIVE EMBRYOLOGY

Figure 2-4 shows five animals at very early stages in their development. Although all of these **embryos** resemble one another, they are actually the embryos of a salamander, a chicken, a pig, a monkey, and a human. These diagrams provide evidence that all vertebrates follow a common plan in the early stages of development. This is due to having similar sets of **genes**, and this similarity comes from their having had common ancestors.

some similar structures, or characteristics, that they inherited from a common ancestor. For example, similarities exist in the forelimb bones of some very different animals. The wing of a bat, flipper of a whale, front leg of a cat, arm of a human, and wing of a bird—although they appear to be quite different—are all made up of the same types of bones. These bones are attached to each other and to other bones in similar ways. The forelimbs indicate that, long ago, these five animals all evolved from a common ancestor. (See Figure 2-3.)

Sometimes a structure has little or no function in one organism, but is clearly related to a more fully developed structure that does function in another organism. For example, the appendix in humans is a small sac attached to the place where the small and large intestines meet. In appearance, the appendix is a smaller copy of the cecum, which is a large pouch found in plant-eating mammals such as rabbits. In a rabbit, the cecum contains microorganisms that digest plant materials that are ingested. The fact that a similar organ is still useful in another species is evidence that humans evolved from an organism that also had this larger, functional structure.

EVIDENCE FROM COMPARATIVE BIOCHEMISTRY

The similar chemistry of living things provides some of the strongest evidence that organisms evolved from common ancestors long ago. All organisms store their genetic information, which is passed from one generation to the next in DNA molecules, in almost exactly the same manner. This genetic code shows that all organisms are related in fundamental ways.

Proteins, a type of molecule in all living things, are made up of smaller units called **amino acids**. The same protein in two different species may be made up of similar but not identical amino acids. Biologists now know that a small number of amino acid differences in the same protein means that the two species are closely related in evolutionary terms. On the other hand, a large number of amino acid differences means that the two species are more distantly related. (See Figure 2-5.)

In addition, by matching a DNA sequence from one organism with a DNA sequence from

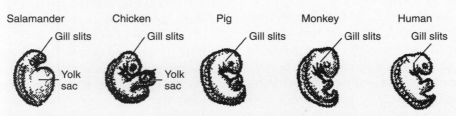

Salamander   Chicken   Pig   Monkey   Human

Figure 2-4 The very early embryos of these five animals show many similarities, such as gill slits and a tail—indications that they shared a common ancestor.

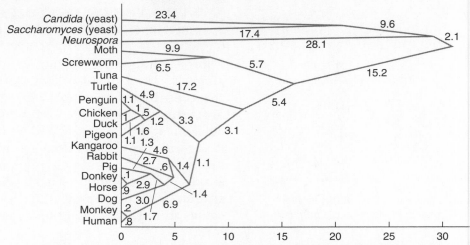

Figure 2-5 Similarities and differences in the same protein among different species can be used to determine their evolutionary closeness.

another organism, scientists can determine if the sequences belong to organisms of the same, closely related, or distantly related species. Again, the greater the similarity, the closer the species are related.

◼ EVOLUTION WITHIN POPULATIONS

Evolution occurs within a population as frequencies of inheritable traits change. This process occurs due to natural selection from one generation to the next. The peppered moth, carefully studied in England for more than a century, provides one of the best-known examples. When the peppered moth was first studied, most of its population was light colored. The moths were well camouflaged when they rested on trees and rocks covered with light-colored lichens.

In 1845, a dark-colored peppered moth was observed for the first time. At that time, soot and smoke produced by coal-burning factories had begun to pollute the air. The trees and rocks became dark with soot and the lichens began to die. As a result, the light-colored moths were easily seen against the darker backgrounds and became the easy prey of insect-eating birds. By 1900, most of the peppered moth population was dark colored. Why did this happen? The darker moths were better camouflaged when resting on the tree trunks and rocks blackened by soot. It is im-

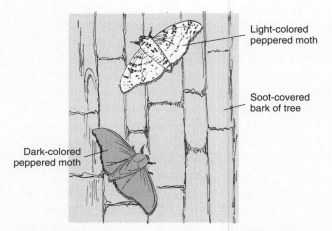

Figure 2-6 The changing frequencies of the light trait and dark trait in the peppered moth population are due to natural selection.

portant to understand that the light-colored moths did not change color; they were replaced by increasing numbers of dark-colored moths through natural selection. The dark-colored moths were less likely to be preyed on, and so were more likely to survive to pass on their genetic traits. Interestingly, as air pollution decreased, more light-colored moths were once again seen in the study area. (See Figure 2-6.)

Sometimes evolution leads to the development of new adaptive features, new species, and groups of species. Examples of new adaptive features are the legs of an amphibian (such as a salamander), the shell-encased eggs of a reptile (such as a turtle), and the large brain of a primate (such as an ape). Examples of entirely new groups of species that arose are the mammals and the flowering plants. Another example of large evolutionary developments is both the appearance and the extinction of all dinosaur species.

Chapter 2 Review

Part A—Multiple Choice

1. All of the following can be used as evidence to support Darwin's theory of evolution *except* the
   1. similarity of chemicals in all living things
   2. distribution of species on Earth today
   3. shapes and structures of living organisms
   4. distribution of mountain ranges on Earth's surface

2. Which statement is best supported by the fossil record?
   1. Many organisms that lived in the past are now extinct.
   2. The struggle for existence between organisms results in changes in populations.
   3. Species occupying the same habitat have identical environmental needs.
   4. Structures such as leg bones and wing bones can originate from the same type of tissue found in embryos.

3. Over time, fossils can be formed when an organism is
   1. buried in sediment; then its hard tissues are replaced by dissolved minerals
   2. buried in sediment; and then an impression of its internal parts is formed
   3. entirely preserved in soft mud or clay, including its internal organs
   4. buried in mud; then its bones dissolve and its internal organs remain intact

Base your answer to question 4 on the diagrams below, which show the forelimb bones of three different mammals.

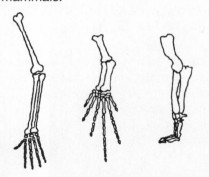

4. For these mammals, the number, position, and shape of the bones most likely indicates that they
   1. developed in the same environment
   2. have an identical genetic makeup
   3. developed from a common earlier species
   4. have identical methods of obtaining food

5. Suppose a scientist suggests that humans are related to rabbits because the human appendix resembles the cecum of a rabbit. The scientist is probably using evidence from
   1. fossil remains
   2. embryology
   3. comparative anatomy
   4. comparative biochemistry

6. The information below was printed on a calendar of important events in the field of biology.

> 1859
> Darwin Publishes
> *On the Origin of Species by Natural Selection*

The title is most closely associated with an explanation of
   1. the change in mineral types in an area due to ecological succession
   2. the structural similarities observed among diverse living organisms
   3. the reasons for loss of biodiversity in various habitats on Earth
   4. the effect of carrying capacity on the size of animal populations

7. Different organisms store their genetic information in the form of DNA in
   1. their own unique way
   2. a very similar manner
   3. a way that differs between groups
   4. one way in plants and another way in animals

8. Two species that have only a small number of amino acid differences in the same protein probably
   1. are identical in their appearance

2 share the same parent organisms
3 are closely related in evolutionary terms
4 are distantly related in evolutionary terms

9. A scientist comparing organisms, in terms of biochemistry, might analyze

1 their DNA sequences
2 the stages of their embryos
3 the organs that have similar uses
4 the shapes of their fossilized footprints

10. After the Industrial Revolution in England, the number of light-colored moths decreased and the number of dark-colored moths increased. How can this be explained in terms of natural selection?

1 The dark-colored moths chased the light-colored moths away from the soot-covered trees.
2 The light-colored moths changed their colors in order to blend in with the darker trees.
3 Once the trees were dark, light-colored moths had a genetic variation that gave them an advantage over dark-colored moths.
4 Once the trees were dark, dark-colored moths had a genetic variation that gave them an advantage over light-colored moths.

11. Which is an example of an evolutionary change at the population level?

1 the development of legs on amphibians
2 the evolution of large brains in primates
3 the replacement of light-colored moths by dark-colored moths
4 the appearance of the flowering plants group

Part B—Analysis and Open Ended

Base your answer to question 12 on the diagram below, which shows the evolutionary relationships of several living and extinct mammals.

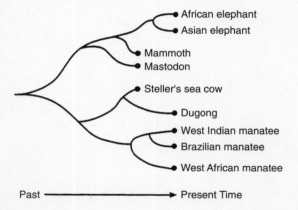

Past ————————→ Present Time

12. According to the diagram, which statement about the African elephant is correct?

1 It is more closely related to the mammoth than it is to the manatees.
2 It is not even remotely related to the Brazilian manatee or the mammoth.
3 It is more closely related to the West Indian manatee than it is to the mastodon.
4 It is the common ancestor of the Steller's sea cow and the dugong.

13. Explain how a family tree can be used to show evolutionary relationships among organisms.

Use the diagrams below, which illustrate the forelimb bones of three different mammals, to answer question 14.

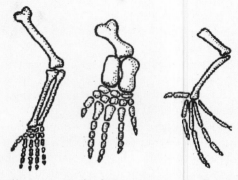

14. Differences in the bone arrangements support the hypothesis that these animals

1 are probably members of the same species
2 have adaptations for different environments
3 most likely have no ancestors in common
4 all contain the same genetic information

15. Why is an "evolutionary bush" considered a more accurate way to illustrate relationships among species than just a linear, or ladder-like, diagram?

16. The diagrams at right show the bones in the forelimbs of a cat and a human. The similarities between these appendages suggest that humans and cats

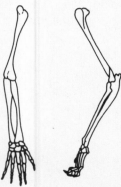

1 have identical genetic material
2 have the same direct ancestor
3 once shared a common ancestor
4 evolved in the same environment

17. Describe two ways fossils are formed, and tell which parts of an organism are usually fossilized. Your answer should explain the following:
    ♦ how fossils form through replacement by minerals
    ♦ how fossil imprints, molds, or casts are formed

18. The diagram at right represents a series of undisturbed sedimentary rock layers in a given area. Several layers show representative fossils of organisms. Relative to the other layers, the fossils of older, more primitive organisms would be found in the

    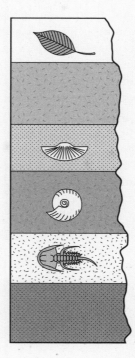

    1 top layers only
    2 bottom layers
    3 middle layers
    4 top and bottom layers

19. Explain why the presence of a body structure with no current function can provide evidence of an evolutionary relationship. Give an example.

20. The diagrams below show the embryos of three different vertebrate species. It is thought that they provide proof of evolution based on their similar

    Turtle   Chicken   Pig

    1 sizes
    2 fossils
    3 structures
    4 molecules

21. Explain how studies of similarities in the biochemistry of proteins can be useful in determining evolutionary relationships among organisms.

22. The three species shown below have similar enzymes, hormones, and proteins; this supports the idea that they share a common ancestor, based on their similar

    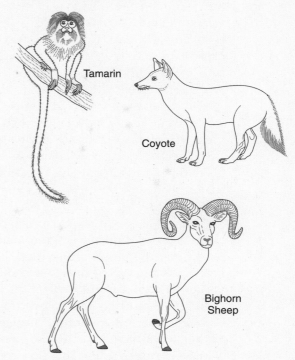

    1 external structures
    2 biochemistry
    3 convergent evolution
    4 behavioral patterns

23. Refer to Figure 2-6, on page 12, which shows the light-colored and dark-colored peppered moths. Over time, depending on changes in the environment, the percentage of each color type in their population has varied. Are these changes in frequency due to natural selection, artificial selection, or acquired characteristics? Explain.

24. As stated in the text, "sometimes evolution leads to the development of new adaptive features" within populations. Describe how such a change may result in the development of a new species. Give either a real or an imagined example.

Part C—Reading Comprehension

Base your answers to questions 25 to 27 on the information below and on your knowledge of biology. Use one or more complete sentences to answer each question.

In 1987, a group of scientists thought they could use a better method to study human evolution. Instead of studying bits and pieces of fossil remains, they decided to study the genes contained within the cells of living people. These genes, passed from generation to generation, have stored within them a history of our origins. The molecular biologists decided to examine the DNA located in our cells' mitochondria. Unlike ordinary DNA—the genetic material in the nuclei of our cells that we get from both parents—mitochondrial DNA (mtDNA) in our cells comes only from our mother. From one generation to the next, mtDNA never gets mixed with the DNA in the genes we get from our father.

The researchers collected mtDNA from women living in many parts of the world. By studying the similarities and differences in mtDNA in these women, the researchers were able to look back in time to study the origins of human history. Their startling conclusion was that the molecular evidence indicated that all humans alive today are the descendants of a single female who lived in Africa about 200,000 years ago. Some people began to call this person "Mitochondrial Eve."

Since 1987, scientists have disagreed widely on the results. Some scientists claim that the computer program the researchers used for their analysis was not used correctly. Others think that Mitochondrial Eve lived only 150,000 years ago. Still others point to evidence showing that modern humans evolved much earlier—and in several parts of the world, not only in Africa.

In 2001, scientists determined that mtDNA from a human fossil found years earlier in Mungo, Australia, showed no linkage to any humans living today. Therefore, "Mungo Man," as the fossil is called, could not have descended from Mitochondrial Eve. "Put the gloves on, Mitochondrial Eve, because Mungo Man has stepped into the ring," began a recent article on the topic, showing how the debate still continues. This kind of open discussion is what science is all about—questions are asked, answered, and then, when more evidence is found, even more questions arise. For now, the answer to this question about our ancestry remains undecided.

25. How is mitochondrial DNA different from the ordinary DNA found in cell nuclei?

26. Compare how these scientists used mitochondrial DNA with the way other scientists have used fossils.

27. Scientific research often produces some answers and then even more questions. How is this true about the Mungo Man and Mitochondrial Eve research?

The Origin and Extinction of Species

ADAPTATIONS TO THE ENVIRONMENT

Every species lives in a particular place. And every place on Earth has specific conditions, such as average air temperature, monthly rainfall, kinds of minerals in the soil, and wind speeds. Darwin's theory states that those organisms that are best suited to tolerate the conditions of their environment—that is, the ones that have beneficial traits—will be most likely to survive and pass their traits on to their offspring. Since environments on Earth are constantly changing, however slowly, evolution of living things is an ongoing process.

Different adaptations in living things occur by chance as a result of the genetic variations within a population. Sometimes an adaptation works well for an organism—that is, it helps it survive—and sometimes it does not. The *adaptive value* of a trait is determined by the specific conditions of the environment. For example, at first, the dark coloration of some peppered moths did not aid survival, because it made those moths more visible on lichen-covered trees and rocks. The dark moths tended to be eliminated by natural selection. Yet that same dark coloration gave some moths an advantage when pollution caused the trees and rocks to darken. In that environment, such a trait was "chosen" by natural selection, and more dark moths lived to pass on their genes. If the main predator of moths did not hunt by sight, the dark coloration of the moths would not have provided an advantage or a disadvantage.

TYPES OF ADAPTATIONS

Some of the most common types of adaptations are those that involve the shape and structure of organisms or the parts of organisms. Leaves, for example, are adapted in form and structure to the conditions of their environment, such as temperature, amount of sunlight, and water. These physical adaptations develop over time as plant populations adapt to changes in the environment. (See Figure 3-1.)

An adaptation to conditions in the environment may also involve the functions and behaviors of an organism. For example, to survive the long, cold winters during which food is scarce, some animals, such as the black bear, slow their metabolic functions. This adaptation leads to hibernation, a type of behavior in which the bear retreats to a den to sleep during the winter months.

SPECIATION

In order for a population to change, it must be physically separated from other populations of its kind, usually for a long time. As a result of evolution, such a population may continue to change until its members are no longer able to reproduce with members of any other population. This is known as *reproductive isolation*. The population would have undergone speciation, that is, it has become a new species.

It is also recognized that a new species may evolve to fill a niche in the environment that was available. A niche includes all the things

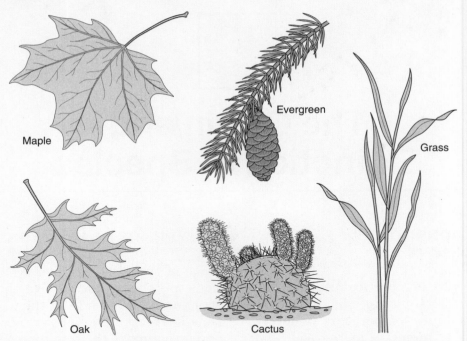

Figure 3-1 The different shapes and structures of leaves are physical adaptations of plants to their environment.

an organism does to survive, such as how it gets its food, reproduces, and avoids predators. A new species in a particular niche will have some advantages making it the organism to survive in that niche.

◼ GEOGRAPHIC ISOLATION

The most common type of separation that leads to the formation of new species is *geographic isolation*. An actual physical barrier, such as a river or a mountain, can prevent organisms from moving between related populations. For example, the Galápagos Islands, which were colonized by finches from mainland South America, each have different environmental conditions. Due to natural selection, the finches changed in different ways on each of the islands and evolved into several species. Those individuals that had advantageous traits survived to produce offspring. (See Figure 3-2.)

In addition to islands, there are other types of geographically isolated areas in which speciation can occur. Examples include mountaintops, lakes, and forests that are separated by different types of terrain. (See Figure 3-3.)

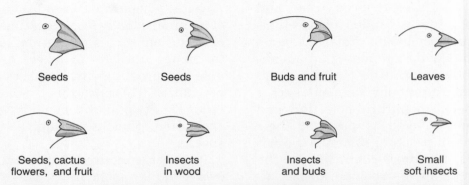

Figure 3-2 Geographic isolation of the ancestral finches led to the formation of several species of finches, each with a different type of beak for a different type of food.

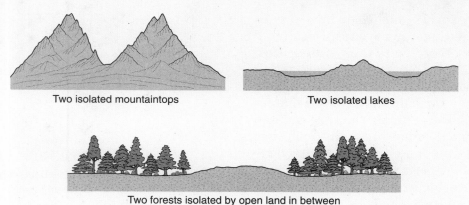

Two isolated mountaintops   Two isolated lakes

Two forests isolated by open land in between

Figure 3-3 Three examples of physical barriers that can cause geographic isolation.

■ EXTINCTION

Another natural part of the process of evolution is **extinction**, the disappearance of species from Earth. Extinction occurs when a species no longer produces any more offspring. This inability to reproduce may occur when members of the population cannot adapt to changes in the environment. For example, a significant decrease in rainfall or a drop in average temperature are two environmental changes that may affect an organism's ability to survive and reproduce. The extinction of a species may also arise from problems that occur within a population. Harmful genetic traits that become widespread in a population may cause a species' extinction. (See Figure 3-4.)

At particular times in Earth's history, *mass extinctions* have occurred. These extinctions are usually due to natural events that drastically change the planet's climate, such as the impact of an asteroid or large comet. In a mass extinction, a large number of species disappears forever. Largely due to these mass extinctions, about 99 percent of all species that have ever existed on Earth have become extinct. (See Figure 3-5.)

■ UNITY AND DIVERSITY

The Earth is estimated to be about 4.5 billion years old. Life on Earth is thought by many scientists to have begun about 3.5 billion years ago as simple, single-celled organisms. Through the process of evolution, about 1 million years ago, increasingly complex multi-cellular organisms began to appear.

During the constant process of change in living things, some characteristics are lost while other characteristics are gained. As a result, Earth is populated by many different kinds of organisms. Scientists estimate that there are millions of species alive today. To

Figure 3-4 The woolly mammoth went extinct about 10,000 years ago, most likely due to changes in its environment to which it could not adapt.

Pterodactyl

Figure 3-5 Flying reptiles, such as the pterodactyl, died out more than 60 million years ago during the mass extinction that killed off their relatives, the dinosaurs. Such mass extinctions result from major changes in Earth's climate.

study so many species, biologists have organized them into numerous groups, classifying the organisms according to their similarities.

The system of classification used today mainly groups organisms according to their evolutionary relationships and shared characteristics. Similar organisms that are capable of producing fertile offspring with each other are placed in the same *species*. Closely related species that most recently evolved from a common ancestor are placed in the same *genus*, and so on up to the level of **kingdom**. In the most commonly used system for grouping organisms, all living things are placed within the following five kingdoms: Monera,

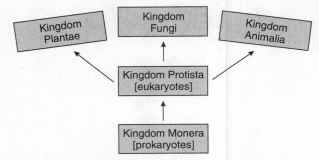

Figure 3-6 The most commonly used classification system groups all living things within five main kingdoms.

Protista, Fungi, Plantae, and Animalia. (See Figure 3-6.)

Chapter 3 Review

Part A—Multiple Choice

1. Which example describes an adaptation that aids survival?
   1 The height of a giraffe enables it to feed on the leaves of trees that other grazing animals cannot reach.
   2 A person's poor eyesight makes it difficult for him to see without glasses.
   3 A white peppered moth is clearly visible against the background of a dark-colored tree.
   4 The broad leaves of a maple tree shrivel up when placed in the hot climate of a desert.

Base your answer to question 2 on the diagram in the next column, which illustrates the change that occurred in the physical appearance of a rabbit population over a 10-year period.

2. Which circumstance would explain this change over time?
   1 a decrease in the mutation rate of the rabbits with black fur
   2 an increase in the advantage of having white fur
   3 a decrease in the advantage of having white fur
   4 an increase in the chromosome number of the rabbits with black fur

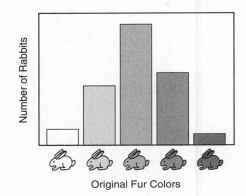

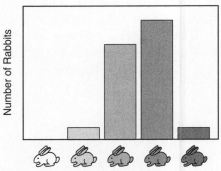

3. The physical adaptations of an organism involve its
   1 shape and behavior
   2 shape and structure
   3 structure and behavior
   4 behavior and ecology

4. In an area in Africa, temporary pools form where rivers flow during the rainy months. Some fish have developed the ability to use their ventral fins as "feet" to travel on land from one of these temporary pools to another. Other fish in these pools die when the pools dry up. What can be expected to happen in this area after many years?

   1 The fish using ventral fins as "feet" will be present in increasing numbers.
   2 The fish using ventral fins as "feet" will develop real feet.
   3 The "feet," in the form of ventral fins, will develop on all fish.
   4 All of the varieties of fish will survive and produce many offspring.

5. One explanation for the variety of organisms present on Earth today is that over time

   1 new species have adapted to fill available niches in the environment
   2 each niche has changed to support a certain variety of organism
   3 evolution has caused the appearance of organisms that are similar to each other
   4 the environment has remained unchanged, causing rapid evolution

6. The Galápagos finches eventually formed 13 new species as a result of

   1 the eruption of volcanoes on the islands
   2 their geographic isolation on different islands
   3 the extinction of other species on the islands
   4 artificial selection by researchers who visited the islands

7. According to modern evolutionary theory, genes responsible for new traits that help a species survive in a particular environment will usually

   1 not change in frequency
   2 decrease rapidly in frequency
   3 decrease gradually in frequency
   4 increase in frequency

Base your answer to question 8 on the following information.

As the Colorado River formed the Grand Canyon, a population of squirrels gradually became separated. The conditions on the northern portion of the canyon were different from those on the southern portion. The squirrels on the northern portion evolved into a different species of squirrel.

8. This scenario is an example of

   1 extinction, in which a species could no longer survive on Earth
   2 migration, in which a species traveled to a new environment
   3 distribution, in which members of the same species are distributed randomly
   4 isolation, in which a new species evolved as a result of a physical separation

9. The complete disappearance of a species from Earth is known as

   1 isolation
   2 extinction
   3 speciation
   4 classification

10. A species will become extinct when its individuals

    1 adapt to new environmental conditions
    2 can no longer adapt and reproduce
    3 move to a new environment and adapt
    4 have survived beyond a specific period of geologic time

Base your answer to question 11 on the diagrams below.

11. According to some scientists, patterns of evolution can be illustrated by diagrams such as those shown below. Which statement best explains the patterns seen in these diagrams?

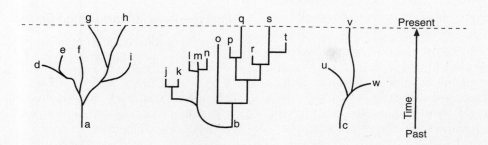

1 The organisms at the end of each branch can be found in the environment today.
2 Evolution involves changes that give rise to a variety of organisms, some of which continue to change through time while others die out.
3 The organisms that are living today have all evolved at the same rate and have all undergone the same kinds of changes.
4 These patterns cannot be used to illustrate the evolution of extinct organisms.

12. Organism X appeared on Earth much earlier than organism Y did. Many scientists think that organism X appeared between 3 and 4 billion years ago, and that organism Y appeared approximately 1 billion years ago. Which row in the chart below most likely describes both organisms X and Y?

Row	Organism X	Organism Y
(1)	Simple multicellular	Unicellular
(2)	Complex multicellular	Simple multicellular
(3)	Unicellular	Simple multicellular
(4)	Complex multicellular	Unicellular

1 (1)   3 (3)
2 (2)   4 (4)

13. Of all the species that have ever existed on Earth, approximately what percentage are now extinct?

1 5 percent
2 20 percent
3 50 percent
4 99 percent

14. The system of classification used today is based mainly on

1 what an organism typically eats
2 the relative size of an organism
3 where an organism lives on Earth
4 evolutionary relationships among organisms

15. Which statement about the rates of evolution for different species is in agreement with the theory of evolution?

1 They are identical, since all species live on the same planet.
2 They are identical, since each species is at risk of becoming extinct.
3 They are different, since each species has different adaptations for its environment.

4 They are different, since each species has access to unlimited resources.

16. The largest division of the classification system is a

1 species
2 class
3 kingdom
4 genus

17. Norway maples, sugar maples, and red maples are probably classified in

1 the same species and same genus
2 the same species but different genuses
3 different species but the same genus
4 different species and different genuses

18. Which process is correctly matched with its explanation?

	Process	Explanation
(1)	Extinction	Adaptive characteristics of a species are not adequate
(2)	Natural selection	The most complex organisms survive
(3)	Gene recombination	Genes are copied as a part of mitosis
(4)	Mutation	Overproduction of offspring takes place within a certain population

1 (1)   3 (3)
2 (2)   4 (4)

19. Organisms that are alike and capable of producing fertile offspring with each other are placed in the same

1 genus
2 family
3 species
4 kingdom

Part B—Analysis and Open Ended

20. Why is the evolution of living things considered to be an ongoing process?

21. What determines the "adaptive value" of a trait within a population? Give an example, either real or imagined.

Base your answer to question 22 on information in the table below and on your knowledge of the process of evolution.

Habitat	Number of Toes	Type of Horse Species
Plains	One toe (hoof)	Modern horse (*Equus*)
Forest	Four toes	Ancestral horse (*Eohippus*)

22. You could hypothesize that modern horses have fewer toes than ancestral horse species had because the

    1 changed habitat wore down their side toes as they ran fast over the plains
    2 ancestral horse species mated with several mutant one-toed horses
    3 people who first rode them preferred horses that had one large hoof per leg
    4 changed habitat favored survival of faster horses, which had reduced toes

23. In what way is the behavior of an organism related to the evolutionary process? Your answer should describe or explain the following:

    ♦ *one* type of behavioral adaptation of an organism

    ♦ how natural selection affects the development of behaviors

    ♦ why a population's behavior may change over time

Base your answer to question 24 on the information in the paragraph and map below.

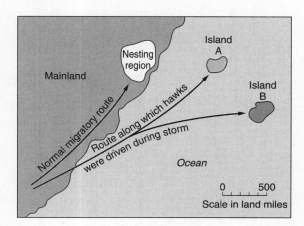

Thousands of years ago, a large flock of hawks was driven from its normal migratory route by a severe storm. The birds scattered and found shelter on two distant islands, shown on the map below. The environment of Island A is very similar to the hawk's original nesting region. The environment of Island B is very different from that of Island A. The hawks have survived on these islands to the present day with no migration between the populations.

24. Which statement most accurately predicts the present-day condition of these island hawk populations?

    1 The hawks that landed on Island B have evolved more than those on Island A.
    2 The populations on Islands A and B have undergone identical mutations.
    3 The hawks that landed on Island A have evolved more than those on Island B.
    4 The hawks on Island A have given rise to many new species of hawks.

25. The following diagrams indicate that the frequency of unspotted beetles is decreasing relative to the frequency of spotted beetles in this population of insects. Possible explanations for the changing frequencies of these traits include all of the following *except* that

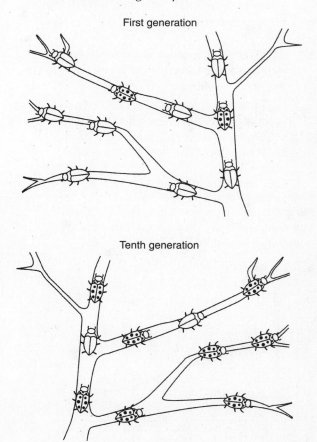

1 the beetles' environment has been changing over time
2 spotted beetles are better adapted to the changing habitat
3 unspotted beetles are better adapted to the changing habitat
4 natural selection is occurring, which affects survival rates

Answer questions 26 and 27 based on the information in the paragraph below.

The variation of organisms within a population increases the likelihood that at least some members of the species will survive changing environmental conditions. A large population of houseflies was sprayed with a newly developed, fast-acting insecticide. While most of the houseflies were killed off, some houseflies that were resistant to the new insecticide survived.

26. The changing environmental condition in this case was the

1 original population of houseflies
2 appearance of resistant houseflies
3 newly developed fast-acting insecticide
4 houseflies that were exposed to the spray

27. Which of the following items represents the variation that enabled some flies to survive the changing conditions?

1 The insecticide was new and fast acting.
2 Most of the flies were killed by the spray.
3 Some flies were resistant to the spray.
4 Only some of the flies were sprayed.

Base your answers to questions 28 and 29 on the graph below, which illustrates changing percentages of two varieties within a species' population over time.

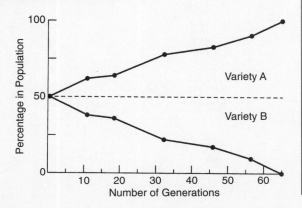

28. Which variety will most likely contribute to this population's traits in the future?

1 variety A only
2 variety B only
3 both variety A and B
4 neither variety A nor B

29. What is the probable reason that the percentage of variety A is increasing while the percentage of variety B is decreasing?

1 There is no opportunity for variety A to mate with variety B.
2 Variety A has some adaptive feature that variety B does not have.
3 Variety B has some adaptive feature that variety A does not have.
4 There is no genetic variation between variety A and variety B.

30. Distinguish between a situation in which variations within a population enable some organisms to survive changing conditions and a situation in which a mass extinction occurs. What is the main difference in each situation's cause and effect?

Answer questions 31 and 32 based on the data in the chart below.

Taxonomic Group	Tiger	Wolf
Kingdom:	Animal	Animal
Phylum:	Chordates	Chordates
Class:	Mammals	Mammals
Order:	Carnivores	Carnivores
Family:	Cats	Dogs
Genus:	Panthera	Canis
Species:	tigris	lupus

31. The tiger and the wolf are classified within the

1 same species but different genuses
2 same genus but different species
3 same order but different families
4 same family but different orders

32. According to the chart, the largest division used in this classification system is the

1 species
2 family
3 class
4 kingdom

Part C—Reading Comprehension

Base your answers to questions 33 to 35 on the information below and on your knowledge of biology. Use one or more complete sentences to answer each question.

Ferrets have become popular as pets in some places in recent years. Their clever antics can be a source of amusement. The unusual appearance of the small, long, slender, furry body of a pet ferret at the end of a leash always draws the attention of onlookers. However, there is a much more serious story about a ferret that is a native of North America—a story of life, death, and near-extinction.

The black-footed ferret is the only ferret species native to North America. These animals lived mostly in western parts of Canada and the United States. The black-footed ferret has a black face, a black tip on its tail, and—as its name suggests—black feet. It is a carnivore and survives mainly on a diet of prairie dogs. The only places where it has ever been found are near prairie dog tunnels. Prairie dogs are not well liked by farmers because their tunnels interfere with the planting of crops. Ranchers don't like them either, because they think their cattle can fall and injure themselves when they step in the openings of prairie dog tunnels. Farmers often put poison in the tunnels to kill prairie dogs. When prairie dogs are eliminated, black-footed ferrets also die. They were last seen in the wild in 1974, and by 1979 they were thought to be extinct.

However, to the great delight of wildlife biologists, a small population of black-footed ferrets was discovered living in a field near Meeteetse, Wyoming, in 1981. A species thought to be extinct was, in fact, still here! However, by the end of 1987, scientists counted only 18 survivors, so these animals were captured. In time, a captive-breeding program produced 400 individuals, which were released in seven different areas. Black-footed ferrets have now been reintroduced into the wild in Wyoming, Montana, South Dakota, and Nebraska. By 2002, they were also being released 100 miles south of the United States border, in Mexico, where many healthy prairie dogs live. The hope of biologists is to reach a goal of 1500 free-living black-footed ferrets by 2010.

The successful reintroduction of black-footed ferrets to the wild—along with the protection of prairie dogs and their grasslands habitat—has brought this interesting animal back from the brink of extinction. Still, we will never know what other species were unintentionally eliminated when the early settlers eradicated 90 percent of the North American prairie dog population.

33. Describe the connections and conflicts between farmers, prairie dogs, and black-footed ferrets.

34. Why did scientists in 1987 capture all the wild black-footed ferrets?

35. Why does the protection of prairie dogs provide protection to black-footed ferrets?

4

Human Evolution

✖ THE SEARCH FOR HUMAN ORIGINS

The theory of evolution is a wonderful example of what science does best: tests ideas against evidence and observations in the real world to determine if the ideas are correct. The study of how the human species evolved is a good example of how science has tested the idea of evolution.

For about 130 million years, reptiles such as dinosaurs were the dominant large animals on Earth. Then suddenly, in terms of **geologic time**, dinosaurs became extinct. The first ancestors of today's major groups of mammals appeared on Earth about 70 million years ago. (See Figure 4-1.) Faced with fewer reptile competitors, an enormous variety of mammals evolved. One group of mammals that evolved—called *primates*—was adapted to a life in the trees. The order of primates includes prosimians, monkeys, apes, and humans. The first monkeys evolved from their prosimian ancestors about 50 million years ago. (See Figure 4-2.) The higher primates include the apes and humans. Ape fossils indicate

Figure 4-1 The early mammals lived at the same time as the dinosaurs.

Figure 4-2 Primates, such as this New World monkey, show adaptations to a life in the trees.

that these primates first evolved from monkeys in Africa and Asia.

✖ HOMINIDS: THE EARLIEST HUMANS

Approximately 20 million years ago, large changes in climate and landforms caused forested areas to diminish in size. At that time, Asian and African apes became separated from one another. Between 14 and 8 million years ago, an African ape evolved that became the common ancestor of both chimpanzees and humans. This does *not* mean that our ancestors are chimpanzees. It *does* mean that we are most closely related to chimpanzees because we share a more recent common ancestor with them than with any other living animal. (See Figure 4-3.) All living and extinct human species are grouped together as *hominids*.

Figure 4-3 The chimpanzee, our closest living relative, can walk upright on two feet for limited periods of time.

✖ WHY DID EARLY HOMINIDS WALK ON TWO FEET?

Our earliest hominid ancestors lived in the trees. One of the first big steps on the path to becoming human occurred when early hominids started to walk—on two feet—on the ground. The shift to a life on the ground may have occurred in response to a changing environment, in which there were fewer forested

Figure 4-4 This reconstruction of a four-million-year-old skeleton shows that, early on in their evolution, hominids were already upright and bipedal—an adaptation to life on the ground, not in the trees.

areas and more open grasslands. The oldest hominid fossils that have been found so far are dated at 4.4 million years old. The fossils of these individuals, who lived in eastern Africa, show that the skull was balanced on top of the skeleton for walking upright. The ability to walk on two feet, called *bipedalism*, may have helped hominids survive by freeing their hands to gather food and to care for their young more efficiently. (See Figure 4-4.)

◆ OUR OWN GENUS

Hominids in the genus *Homo* are characterized by having a large brain, which sets them apart from earlier hominids as well as from all other primates. *Homo erectus* is considered to be the first ancestor within our genus. *H. erectus* existed for a long time, from about 1.8 million to 300,000 years ago. It had a body skeleton much like ours and walked erect just as modern humans do. The brain of *H. erectus*, at 700 to 1200 cc, was much larger than that of earlier hominid species and almost as large as that of modern humans. *H. erectus* was the first hominid species known to build fires, live in caves, wear clothes, and manufacture tools such as stone hand axes. With these skills, *H. erectus* was able to migrate north to colder regions beyond Africa and thus colonize the Eurasian continent. *H. erectus* was ancestral to our own species, *Homo sapiens*. (See Figure 4-5.)

Figure 4-5 Our ancestor *Homo erectus* made and used stone hand axes such as the one shown in this photograph.

Today, there is much debate about human evolution. There are two main scientific views on the subject. One group of scientists sees it as a ladder, with one hominid species at a time leading to the next species. The other group sees it as a tree with several branches of hominids: one leads to modern humans; the others lead to extinct hominid species.

Chapter 4 Review

Part A—Multiple Choice

1. For about 130 million years, the dominant large animals were
   1. sea mammals such as whales
   2. land mammals such as elephants
   3. reptiles such as dinosaurs
   4. primates such as apes

2. Based on the information about reptile and mammal evolution, it is correct to say that geologic time
   1. is a very short-term measure of time
   2. is a very long-term measure of time
   3. is a very brief moment in time
   4. really has nothing to do with time

3. Why did an enormous variety of mammals evolve after dinosaurs became extinct?
   1. Mammals had been afraid of the dinosaurs.
   2. The dinosaurs had been eating all the mammals.
   3. Dinosaurs and mammals did not exist at the same time.
   4. Without dinosaurs, there was less competition for resources.

4. Primates are mainly adapted for a life in the

1 water
2 trees
3 arctic
4 desert

5. Monkeys evolved from small primate ancestors about

1 1 million years ago
2 25 million years ago
3 50 million years ago
4 100 million years ago

6. Apes first evolved from monkeys in

1 Africa and Asia
2 South America
3 Europe
4 North America

7. What is the evolutionary connection between humans and chimpanzees?

1 Humans evolved from the first chimpanzees.
2 Humans have exactly the same DNA as chimpanzees.
3 Humans and chimpanzees both always walk on two feet.
4 Humans and chimpanzees share a common ancestor.

8. Walking on two feet helped hominids survive because it

1 made them look taller
2 made their brains grow larger
3 left their hands free for gathering food and carrying their young
4 let them swing from branch to branch more efficiently

9. The oldest hominid fossils indicate that our ancestors first evolved in

1 Asia
2 Africa
3 Eurasia
4 America

10. What characteristic separates hominids in the genus *Homo* from other primates?

1 large feet
2 an enlarged brain
3 large hands
4 a height of over 1.7 meters

11. *Homo erectus* were probably the first hominids to

1 build fires
2 walk on two feet
3 climb trees
4 eat meat

12. *Homo erectus* had skills that enabled them to

1 survive in a warm climate only
2 migrate to live in colder climates
3 keep a written history of their lives
4 grow grain and domesticate cattle

Part B—Analysis and Open Ended

13. Based on your reading, state whether modern humans are more closely related to modern Asian apes or modern African apes. Give your reason.

14. Which statement is more correct: "humans and gorillas probably share a common ancestor" or "humans probably evolved from gorillas"? Explain your answer.

15. Study the photograph below, which shows a chimpanzee using two sticks. Explain the importance of this chimp's actions in terms of the following ideas:

◆ the significance of the chimp's behavior in relation to hominid evolution

◆ how the chimp's behavior is quite similar to that of humans

◆ how the chimp's behavior is different from that of humans

16. Describe how the differences between a grasslands environment and a forest might affect an animal's adaptations and survival.

17. Briefly explain the importance of bipedalism to early hominids. Your answer should describe the following:

♦ environmental changes that might have led to bipedalism

♦ possible effect of bipedalism on methods of food gathering

♦ possible effect of bipedalism on hominid so-cial behavior

18. In what way do hominids in the genus *Homo* differ from earlier hominids and from other pri-mates? Why do you think this difference is so important?

19. List some traits *Homo erectus* had that were more advanced than those of earlier hominids and similar to those of modern *Homo sapiens*. Your answer must include at least:

♦ *one* physical characteristic different from that of earlier hominids

♦ *two* physical characteristics similar to those of modern humans

♦ *three* behavioral characteristics similar to modern humans

20. The figure below illustrates what an early *Homo sapiens* (Neanderthal) might have looked like. List four features you can observe that set it apart from most other primates and make it closest to modern humans. Tell why these traits and abilities were important for the survival of early humans.

Part C—Reading Comprehension

Base your answers to questions 21 to 23 on the information below and on your knowledge of biology. Use one or more complete sentences to answer each question.

In the 1920s, large amounts of limestone were being dug from the ground in Taung, an area of South Africa. Many human fossils were also being dug up in the limestone quarry. Raymond Dart, an Australian teaching at the medical school in Johannesburg, heard about the fossils. Dart was an expert on the anatomy of the human head and was anxious to examine the fossils—a natu-ral curiosity. Dart contacted the owner of the quarry and, in time, two large boxes of fossils arrived at his home.

When he examined the material in the boxes, Dart found a dome-shaped piece of stone and immediately recognized that it was shaped like a brain. In this fossil, Dart saw the folds of tissue that make up the brain and even the blood vessels on the surface. Dart realized what had happened many years be-fore. Long ago, someone had died in the vicinity of this present quarry. Sand and water that contained minerals had entered the skull; eventually these materi-als hardened into rock in the exact shape of the brain.

On close examination, Dart felt that the fossil brain looked like it had come from an ape, but he recognized that the fossil also had some similarities to a hu-man brain. The skull might provide some clues to the brain's origin. Dart looked again in the box that contained the fossilized brain. Much to his amazement and delight, he found pieces of the lower jaw and the skull.

However, the front of the fossil skull—the face—was covered by layers of rock. In a procedure that took several months, Dart chipped away at the rock layers. What he eventually revealed was the face of a young, humanlike creature, later dubbed the "Taung Child," which Dart believed was an early ancestor of the human species. His find turned out to be one of the most important fossil discoveries ever made, adding crucial details to our understanding of human evolution.

21. Why is it valuable for a scientist studying human evolution to have a background in medicine or human anatomy?

22. How did the bones of a humanlike skull come to be preserved as the fossils that were studied by Dart?

23. Imagine that you are Raymond Dart and you are writing a letter to a friend to describe your discovery. In the letter, explain to your friend why you think you have discovered evidence of an early ancestor of the human species.

Energy, Matter, and Organization

5

From Atoms to Cells

▚ ENERGY, MATTER, AND ORGANIZATION

Everything in the world tends to get more disorganized as time passes. Physicists recognize that natural events tend to increase disorder in the universe; or, put more simply, the universe is running down. It becomes more disordered over time.

Living things are highly organized. How do they keep themselves carefully arranged and functioning properly in a universe where things are constantly running down?

To maintain the state of organization necessary for life to exist, all living things require **energy**. Energy exists in different forms, such as heat, motion, light, and electricity. And organisms always need more energy—a continuous input of energy—to stay organized and remain alive.

▚ MATTER, ATOMS, AND LIFE

All matter is made up of **atoms**, particles far too small to be seen with the unaided eye or even through an ordinary microscope. Each atom has an extremely dense nucleus in its center. Distributed in the mostly empty space around the nucleus are *electrons*. Electrons have energy. The orbitals in which electrons move make up shells, or energy levels. The electron shells closest to the nucleus have the least energy; the electron shells farthest from the nucleus have the most energy. It is possible for electrons to move from one shell to another. A ball rolling from the top of a hill loses energy as it moves lower. So, too, does an electron as it moves from an outer, higher-energy level to an inner, lower-energy level. The movement of electrons between energy levels in atoms is what produces all changes or transformations of energy.

An organism is actually a very complex **system** for transforming energy. Through natural selection, evolution has produced species of organisms that are efficient energy transformers. Where does the energy come from that keeps organisms as different as grass, ants, and elephants alive? Sunlight is the main source of energy for most life on Earth. Plants capture this light energy, and then plants, animals, and other organisms use it to live, grow, and reproduce.

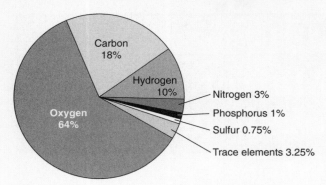

Figure 5-1 The graph shows the percentages of the main elements that make up organisms' bodies.

ATOMS BOND TOGETHER

There are only about 100 different types of atoms. Of these basic types, called *elements*, 92 are found naturally on Earth. **Carbon**, **hydrogen**, **nitrogen**, and **oxygen** are the four most important elements for organisms. Also important for living things are *phosphorus*, *sulfur*, and a few other elements. (See Figure 5-1.)

Atoms of most elements combine with other atoms to form larger structures called **molecules**. The atoms in a molecule are kept together by a kind of partnership called a *chemical bond*. One type of chemical bond is the covalent bond. In such a bond, atoms share electrons with each other, which makes each atom more stable.

Elements combine to form thousands of different *compounds*, but only a small number of compounds are important for living things. Many of the compounds found in living things are called **organic** compounds. All organic compounds contain the elements carbon and hydrogen. Carbon is of special importance; organic compounds usually have a skeleton of carbon atoms bonded to each other.

ORGANIC COMPOUNDS

Organic compounds accomplish many different, complex jobs that keep us and all other organisms alive. These include capturing and transforming energy, building new structures, storing materials, repairing structures, and keeping all chemical activities in the body

working properly. In organisms, chemical reactions are always putting things together or taking things apart. These chemical activities in an organism are called **metabolic** activities, or *metabolism*.

Organic molecules are often very large. A large molecule, such as the blood protein hemoglobin, is called a *polymer*. Polymers are formed by the linking together of smaller molecules, or *subunits*. In polymers, these subunits are held together by covalent bonds.

THE FAMILIES OF ORGANIC COMPOUNDS

Even though organisms contain many different molecules, there are relatively few different types of subunits that make up these molecules. The subunits located in the organic compounds found in living organisms, from bacteria to whales, are almost identical. However, the way the subunits are put together creates an enormous variety of polymers. Some polymers have more than 100 subunits; others have thousands of subunits. As a result, there are many different kinds of polymers.

Organic compounds are grouped into four major families: carbohydrates, lipids, proteins, and nucleic acids. These are the families of compounds found in all living things.

CARBOHYDRATES

Carbohydrates include the **simple sugars** as well as polymers, made up of sugar subunits. Carbohydrates are formed from the elements hydrogen, oxygen, and carbon. In organisms, the main functions of carbohydrates are energy storage and providing strong building materials for certain types of cells. (See Figure 5-2.)

Polysaccharides are made up of many sugar subunits

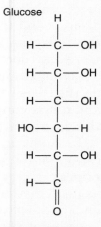

Figure 5-2 Glucose is a simple sugar. Like most other organic compounds, it has a skeleton of carbon atoms to which other atoms form bonds.

joined together. Energy is stored in these large molecules. Our muscles must contain stored energy to allow them to work at a moment's notice. In humans, energy is stored in the polysaccharide glycogen in muscles and in the liver. It is made up of many **glucose** subunits.

Plants store energy in the form of **starch**, a type of polysaccharide, which is contained in such foods as corn and rice. Cellulose, another important polysaccharide found in plants, helps build up tough structures such as wood.

Figure 5-3 Meats such as these contain high amounts of lipids and proteins.

LIPIDS

Oils and fats make up the second major family of organic compounds, the *lipids*. One type of lipid makes up the basic structure of the cell membrane. However, the main purpose of lipids is energy storage, which they do more efficiently than carbohydrates. Energy stored in lipids is for long-term use. During physical activity, the carbohydrate glycogen gets used up quickly. However, lipids, or fat deposits, do not disappear quickly.

PROTEINS

Hundreds of thousands of different **proteins** exist. No two people other than identical twins have exactly the same proteins. Proteins are responsible for a wide variety of functions in organisms. They are used to build materials, transport other substances, send signals, provide defense, and control chemical and metabolic activities. (See Figure 5-3.)

The building plan for proteins is the same as for other organic compounds. Proteins are large polymers that are made up of smaller subunits called **amino acids**. There are 20 different types of amino acid molecules. By combining these amino acids in a row, in different sequences, many types of proteins are made. Proteins can easily have more than 100 amino acid subunits.

The order in which its amino acids are linked determines the characteristics of a protein molecule. Every different sequence produces a different protein. However, the sequence does not behave like a string of let-ters; the protein chains twist, turn, and bend into specific three-dimensional shapes. The shape of a protein molecule is called its *conformation*. Every protein molecule has a very specific conformation, and that is what determines its function.

NUCLEIC ACIDS

Nucleic acids consist of *deoxyribonucleic acid (DNA)* and *ribonucleic acid (RNA)*. DNA and RNA are responsible for storing the genetic information that contains the directions for building every molecule that makes up an organism. (See Figure 5-4.)

The pattern for building nucleic acids is similar to that of the other organic molecules. Individual subunits, called *nucleotides*, are combined in a linear sequence to build the polymers DNA and RNA. DNA molecules usually contain thousands of nucleotides linked together in a specific sequence. In a ribosome,

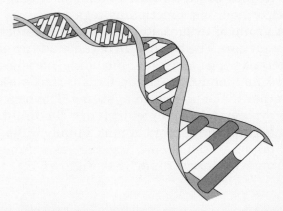

Figure 5-4 DNA contains the instructions for building the protein molecules that make up an organism.

the sequence in the DNA, which has been copied into RNA, is used as a building plan to construct a protein molecule. Specific portions of the DNA molecule are used to make specific proteins. Because proteins make us who we are—and because our DNA makes our proteins—it is really our DNA that makes us who we are.

CELLS ARE US

The idea that organisms are made up of **cells** is one of the central ideas of modern biology. Referred to as the *cell theory,* its main points are as follows:

♦ All organisms are made up of one or more cells.

♦ The cell is the basic unit of structure and function of all living things.

♦ All cells arise from previously existing cells.

LEVELS OF ORGANIZATION

The study of how living things are put together begins with atoms. The next **level of organization** above atoms is molecules, then the families of organic compounds, and then organelles and cells. Beyond that, the organization of living things continues. Plants and animals, which are composed of enormous numbers of cells, have groups of similar cells called **tissues** that work together to do specific functions. For example, the nervous tissue in your brain consists of billions of nerve cells, or neurons, functioning together.

A group of tissues, in turn, works together as an **organ**. The brain is an organ made up of different types of tissue, including nervous, blood, and connective tissue. Organs that work together make up **organ systems**. The nervous system, for example, includes the brain, spinal cord, and sensory organs. Finally, different organ systems work together as a functioning whole **organism**. (See Figure 5-5.)

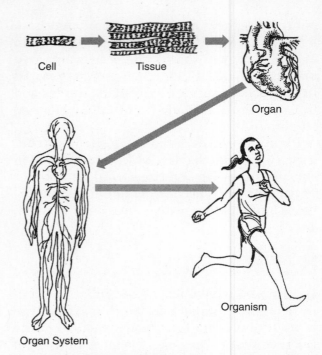

Figure 5-5 The levels of organization in a living organism start with the cell.

These levels of organization continue beyond ourselves. Individuals belong to specific **populations**. All the populations in one place make up a **community**. A community of living organisms, along with the nonliving environment of water, soil, and air, make up an **ecosystem**. And finally, all ecosystems on Earth together make up the *biosphere*.

To study life is to learn about all the levels of life's organization, sometimes one at a time, sometimes all together. At each level there are characteristics that were not present on the previous level. While we may not be able to define life, we can say what living things do. Properties of life include order, reproduction, growth and development, digestion, nutrition, excretion, respiration, **movement**, **coordination**, use of energy, response to the environment, regulation, immunity, and adaptation. We can see that the structures found in single-celled organisms also function like the tissues and systems found in multicellular organisms, thus allowing them to perform all of the life processes needed to maintain homeostasis.

Chapter 5 Review

Part A—Multiple Choice

1. As time passes, the universe becomes
    1. more organized
    2. less organized
    3. simpler
    4. smaller

2. To maintain organization, living things need
    1. time
    2. patience
    3. money
    4. energy

3. The densest region of an atom is the
    1. space around its nucleus
    2. nucleus at its center
    3. inner electron shells
    4. outer electron shells

4. Which event is most like an electron moving from an outer shell to an inner shell?
    1. a ball rolling down a hill
    2. a fish swimming upstream
    3. a soccer ball rolling across a field
    4. a pole-vaulter rising into the air

5. All transformations of energy involve
    1. a change in height
    2. the use of sunlight
    3. the organization of atoms
    4. the movement of electrons

6. How do molecules differ from atoms?
    1. Atoms occur naturally, whereas molecules do not.
    2. Atoms are individual particles, whereas molecules are combinations of atoms.
    3. There many different types of atoms but only a few different types of molecules.
    4. Atoms are found in living things, whereas molecules are found in nonliving things.

7. The two elements found in every organic compound are
    1. nitrogen and oxygen
    2. oxygen and hydrogen
    3. carbon and hydrogen
    4. carbon and oxygen

8. In a cell, all organelles work together to carry out
    1. diffusion
    2. information storage
    3. active transport
    4. metabolic processes

9. Which statement concerning simple sugars and amino acids is correct?
    1. They are both wastes resulting from protein synthesis.
    2. They are both needed for the synthesis of larger molecules.
    3. They are both building blocks of starch.
    4. They are both stored as fat molecules in the liver.

Refer to the diagram below to answer question 10.

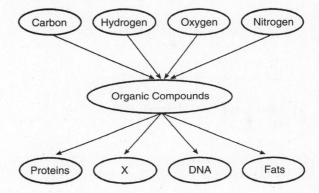

10. What substance could the letter X represent?
    1. carbohydrates
    2. carbon dioxide
    3. ozone
    4. water

11. Which family of organic compounds is used mainly to store energy and to build certain materials in cells?
    1. lipids
    2. carbohydrates
    3. proteins
    4. nucleic acids

12. An iodine test of a tomato plant leaf revealed that starch was present at 5:00 p.m. on a sunny afternoon in July. When a similar leaf from the same tomato plant was tested with iodine at 6:00 a.m. the next morning, the test indicated that less starch was present. This reduction in starch content most likely occurred because starch was

1   changed directly into proteins
2   transported downward toward the roots through tubes
3   transported out of the leaves through the guard cells
4   changed into simple sugars

13. The subunits that make up proteins are

1   amino acids
2   single atoms
3   fats and lipids
4   nucleic acids

Refer to the diagram below, which provides some information about proteins, to answer question 14.

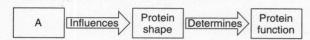

14. Which phrase does the letter A most likely represent?

1   sequence of amino acids
2   sequence of starch molecules
3   sequence of simple sugars
4   sequence of ATP molecules

15. The subunits of DNA are called

1   amino acids
2   nucleotides
3   polysaccharides
4   cell units

16. How is RNA related to proteins?

1   Proteins are made up of RNA molecules.
2   RNA determines which proteins are made.
3   RNA is copied into DNA to build a protein.
4   DNA is copied into RNA to build a protein.

17. DNA molecules are important because they store

1   fats for energy
2   genetic information
3   carbohydrates
4   polysaccharides

18. Which of the following is *not* an idea of the cell theory?

1   Organisms are made up of one or more cells.
2   Cells bond together much like atoms do.
3   The cell is the basic unit of structure in all living things.
4   All cells arise from previously existing cells.

19. Which sequence represents the correct order of levels of organization found in a complex organism?

1   cells→organelles→organs→organ systems→ tissues
2   organelles→cells→tissues→organs→ organ systems
3   tissues→organs→organ systems→ organelles→cells
4   organs→organ systems→cells→tissues→ organelles

20. In terms of levels of organization, a biosphere is most like

1   a cell
2   a tissue
3   an organ
4   an organism

Refer to the diagrams of the organisms shown below to answer question 21.

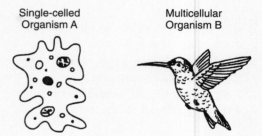

Single-celled Organism A   Multicellular Organism B

21. Which statement concerning organism A and organism B is correct?

1   Organism A contains tissues and organs, while organism B lacks these structures.
2   Organism A and organism B have structures that help them maintain homeostasis.
3   Organism A and organism B have the same organs to perform their life functions.
4   Organism A lacks structures that maintain homeostasis, while organism B has them.

22. Every single-celled organism is able to survive because it carries out

1   metabolic activities
2   heterotrophic nutrition
3   autotrophic nutrition
4   sexual reproduction

Part B—Analysis and Open Ended

23. How do living things maintain the high level of organization that they need to stay alive?

24. Why is carbon particularly important for the existence of life on Earth?

25. List four important functions of organic compounds in living things.

Refer to the diagram below to answer questions 26 and 27. (Molecule shown as a straight chain.)

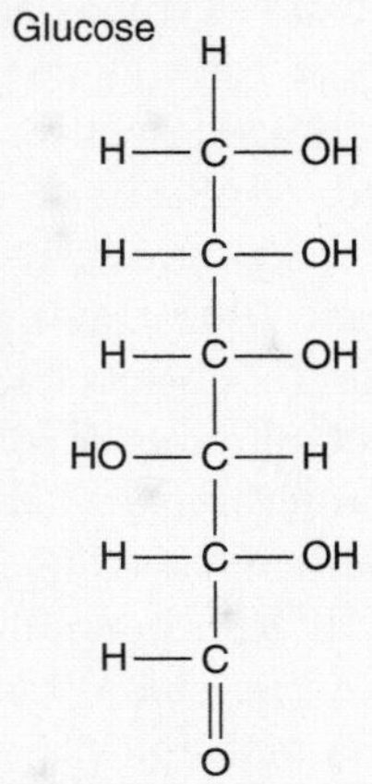

26. Based on the elements in glucose, and the way the atoms are attached, you could determine that glucose is an example of

1 a carbon molecule
2 a hydrogen molecule
3 an organic compound
4 an inorganic compound

Glucose

H
|
H—C—OH
|
H—C—OH
|
H—C—OH
|
HO—C—H
|
H—C—OH
|
H—C
‖
O

27. When many of these glucose subunits join together, they make up a

1 protein molecule
2 polysaccharide
3 lipid molecule
4 DNA molecule

28. Explain why athletes need to eat lots of complex carbohydrates during training.

29. The diagram below illustrates a reaction in which

1 several amino acids join to form a protein molecule
2 inorganic compounds form an organic compound
3 simple sugars join to form a larger sugar molecule
4 polysaccharides are broken down into simple sugars

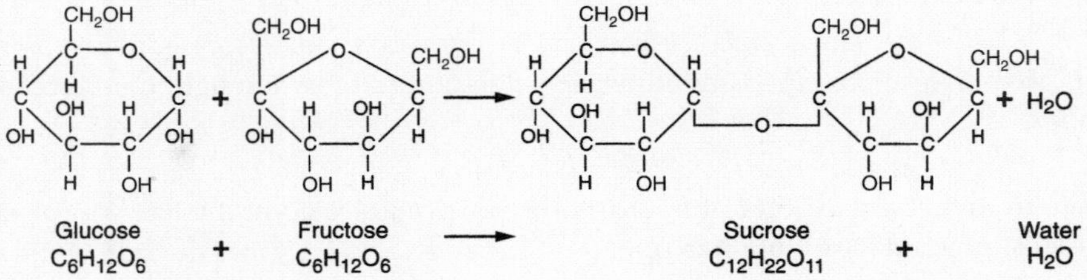

30. Identify three important characteristics of proteins. Your answer should include the following:

♦ what the subunits are that make up proteins
♦ four main functions of proteins in living things
♦ what determines structure and function of a protein

31. Hemoglobin and hair are both proteins, yet they have different structures. Explain.

32. In what way do the particular proteins in our bodies depend on our DNA?

33. In terms of levels of organization, what is the difference between a *tissue* and an *organ*; that is, how do their structures and functions differ? Give an example of each.

34. List the main levels of organization of living things, from atoms to organism.

Glucose $C_6H_{12}O_6$ + Fructose $C_6H_{12}O_6$ → Sucrose $C_{12}H_{22}O_{11}$ + Water H_2O

Part C—Reading Comprehension

Base your answers to questions 35 to 37 on the information below and on your knowledge of biology. Use one or more complete sentences to answer each question.

Light microscopes are used by scientists to view extremely small objects, such as cells. However, due to the physical properties of light, objects below a certain size cannot be seen in a sharp, focused image—no matter how well-made and powerful the light microscope is. Bacteria with a diameter of 0.5 micrometers are about the smallest living things that can be observed with a high-quality light microscope. (A micrometer is 1/1,000,000 of a meter.) During the twentieth century, scientists learned to use a beam of electrons (similar to what is used in a television set), instead of a beam of light, to produce images of very small objects. Electron microscopes have now been used to observe and produce photographs of the detailed, internal structures of plant and animal cells, as well as of the smallest living things, such as bacteria and viruses.

Modern electron microscopes are at least 1000 times more powerful than light microscopes, allowing clear observations of objects that are as small as 0.5 nanometer. (A nanometer is 1/1,000,000,000 of a meter.) Two types of electron microscopes are the transmission electron microscope (TEM) and the scanning electron microscope (SEM). In a TEM, a beam of electrons is passed *through* an object to show the details within it. In an SEM, a beam of electrons is passed *over* an object, producing a detailed, three-dimensional image of the surface of its cells. Electrons would bounce off the gas molecules that are present in the air. So there must be a vacuum where the specimen is placed inside an electron microscope. In addition, specimens are usually treated with stains that interact with the electron beams in order to produce the images. As a result, cells cannot be viewed with an electron microscope when they are still alive.

35. How have scientists been able to overcome the limitations of light microscopes?

36. Write a paragraph that describes the similarities and differences between light microscopes and electron microscopes.

37. If scientists were to invent a new kind of microscope, how might it combine the best of a light microscope with the best of an electron microscope?

6

Chemical Activity in the Cell

◆ INTRODUCTION TO THE CELL MEMBRANE

The inside of a single-celled organism is very much alive. However, the physical environment outside the cell is the opposite—a nonliving place where many changes occur. What stands between a cell and the potentially hostile environment that surrounds it? An ultrathin, extremely important layer separates the living world inside a cell from the nonliving world outside. This is the **cell membrane**, or plasma membrane.

The cell membrane performs two primary, yet very different, functions: it separates the cell from its environment and it enables communication and **movement** of materials between the cell and its environment. Without a cell membrane, there could be no cell. Protein molecules, which float within lipids in the membrane, enable much of the movement of materials across the cell membrane. (See Figure 6-1.) These protein molecules often extend from one side of the membrane to the other. In multicellular organisms, some cell membrane proteins function as *receptor molecules* to

which chemical signals may attach as one cell communicates with another cell.

◆ TRANSPORT ACROSS THE CELL MEMBRANE

For a cell to remain alive, it must have a very special collection of chemicals inside it. These chemicals may be quite different from the chemicals located in the outside environment. Some substances that are abundant outside the cell are not found inside the cell. Other substances that are scarce outside the cell are present in larger quantities inside the cell. The cell membrane creates and maintains this special environment inside the cell. How does it do this?

The cell membrane allows some substances —that is, molecules—to pass through but keeps other substances out. This ability to determine which molecules can pass through is called *selective permeability*. The cell membrane is selectively permeable—it determines which molecules move through it and whether the molecules go into or out of the cell. It also makes possible the rapid transport of some molecules across it, while other molecules pass through slowly.

◆ PASSIVE TRANSPORT

Typically, there is an overall or net movement of molecules from an area of high concentration—a place where molecules are crowded together—to an area of low concentration. This kind of movement is called **diffusion**. Molecules are constantly in motion and they naturally move from where they are more concentrated to where they are less concen-

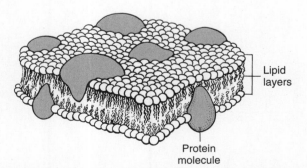

Figure 6-1 The cell membrane acts as a barrier, separating the inside of the cell from the outside. Protein molecules, floating within lipids, enable the movement of materials across the cell membrane.

Lipid layers

Protein molecule

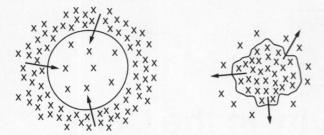

Figure 6-2 Diffusion is the movement of molecules from an area of higher concentration to an area of lower concentration.

trated. This movement happens automatically with a cell if its membrane is permeable to the molecules and if there is a difference in concentration of the molecules on either side of the membrane. This is called *passive transport*, because no energy is used by the cell and no work is done. For example, one of the basic needs of most cells is oxygen. There are few oxygen molecules inside a cell, but there is usually an abundance of oxygen molecules in the water or other liquid that surrounds the cell. Thus, oxygen molecules diffuse across the cell membrane into the cell by passive transport. (See Figure 6-2.)

The diffusion of water molecules across a cell membrane—so important for living cells—is given a special name, *osmosis*. When plant cells are put in a strong salt solution, the abundant fresh water inside the plant cells automatically moves out of the cells, to where there is more salt and relatively fewer water molecules. Plant cell membranes can be seen pulling away from the cell walls as the cells lose water.

The reverse happens when limp celery stems are put in fresh water. The celery stems are limp because their cells have too little water in them. When the celery is put in the water, osmosis occurs and water molecules move into the cells. The cells expand, the cell membranes push against the cell walls, and the cells—and thus the celery stems—become firm again.

◆ ACTIVE TRANSPORT

The movement of a substance against the concentration gradient is known as **active trans-**

port. When substances are moved from an area of low concentration to an area of high concentration, energy is used and work is done. This kind of transport of materials across the cell membrane is one of the most important activities of cells. Other than using energy from your food to keep you warm, the most important use of energy in your body is to help pump substances across the membranes of your cells by active transport—a process that goes on all the time. Cells get the energy for active transport from ATP molecules.

◆ ENERGY TRANSFORMATIONS INSIDE THE CELL

The energy used in the chemical reactions that take place inside cells is associated with the electrons of atoms. The greater an electron's distance from its nucleus, the more stored, or potential, energy it has. When some atoms join, the electrons shared between them form the covalent bonds of a new chemical compound. Each type of covalent bond has a specific amount of energy. Whenever covalent bonds are formed or broken, the amount of stored energy changes. Chemical reactions are mainly energy transformations in which the energy stored in chemical bonds is transferred to other newly formed chemical bonds or is released as heat or light. (See Figure 6-3.)

For example, the glucose in the food you eat is used by your cells after it is digested. In a chemical reaction that requires oxygen, the high-energy chemical bonds in the glucose molecules are broken. When glucose mole-

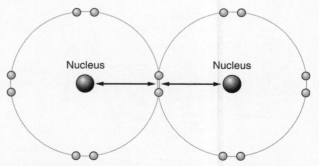

Figure 6-3 The energy in a covalent bond depends on how far the shared electrons are from the nuclei of the atoms. When covalent bonds are formed or broken, the amount of stored energy changes.

cules are broken apart, **carbon dioxide** and water are formed. The energy levels of the chemical bonds in the carbon dioxide and water are lower than the energy levels of the chemical bonds in the glucose. Thus, energy has been released. ATP is the substance in which cells store this released energy until it is needed.

ENZYMES: THE CELL'S MIRACLE WORKERS

For chemical reactions in the cell to take place, they must usually occur in a series of small steps rather than in a single large burst of activity, such as in a heated test tube. The steps must also be very precise. They must occur in the correct order, one after the other. One problem for the cell is that the reactants must get changed into exactly the right products and not into something else. Also, the reaction must occur at a relatively low temperature so that the cell is not harmed.

These problems are solved by substances called **enzymes**. Because a particular enzyme is needed for each type of chemical reaction, cells have thousands of different kinds of enzymes. The correct enzyme can enable a reaction to occur 10 times a second that otherwise might occur only once every 100 years. And the same enzyme molecule does its job over and over again without itself being changed.

Substances that are responsible for greatly changing the rate at which chemical reactions occur, without being changed themselves, are called **catalysts**. Enzymes are organic catalysts, because they are either proteins or nucleic acids. Enzymes are very accurate in their work because they are usually proteins that have a particular shape. The shape of an enzyme molecule includes a spot somewhere on its surface that is like a pocket. This pocket is exactly the right size and shape for a particular substance. If two different substances are involved in the chemical reaction, there will be a precise fit in the enzyme's pocket for each substance. This place on the enzyme molecule is called the *active site*. (See Figure 6-4.)

An enzyme does its work by joining with the substances in this close fit. This is only a

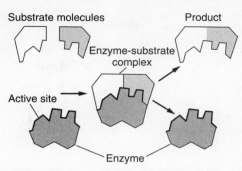

Figure 6-4 Enzymes speed up the rate of chemical reactions in the body by temporarily joining, at their active site, with other molecules.

temporary association. The chemical reaction occurs while the enzyme and substance are fitted together. The substance changes in a specific way, but the enzyme does not change. The product of the reaction is released from the active site and moves to the next step in the process. The unchanged enzyme, with its open active site, gets used again to catalyze the same reaction on another molecule of the same substance. In fact, the enzymes are recycled.

HELPING ENZYMES DO THEIR JOBS

For an enzyme to work properly, the protein molecule must maintain its correct shape. Two conditions in the cell, temperature and **pH**, are very important for maintaining the shape of an enzyme molecule. The main reason animals need to maintain a constant body temperature is to allow cellular enzymes to function properly. (See Figure 6-5.) The pH scale is used to measure how acidic or basic a

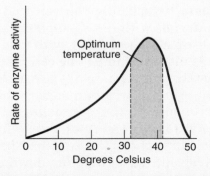

Figure 6-5 A typical enzyme in the body works best at temperatures between 35°C and 40°C.

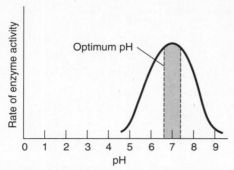

Figure 6-6 Most enzymes function best at about pH 7, which is neutral.

solution is. Slight changes in pH quickly change an enzyme's shape and its ability to affect a substance. Most enzymes function best at about pH 7, which is neutral. Our cells must maintain that pH in order to survive. (See Figure 6-6.)

✖ A TOUR OF THE CELL: ORGANELLES

Some living things are made up of a single cell, but most have many cells that work together on behalf of the organism. Yet, almost everything an organism does to stay alive is accomplished by each individual cell: getting food, using food for energy, transporting substances, growing, reproducing, and eliminating wastes. Each of these activities involves a large number of chemical reactions. Organization is needed for all of these reactions to take place under the precise control of so many enzymes. In many cells, these reactions take place in special internal structures called **organelles**.

Organelles are dispersed throughout the **cytoplasm**, which fills the cell and transports materials within it. Cytoplasm is a thin gel, made up mostly of water, with many other chemicals dissolved in it. The cell membrane encloses the cytoplasm. In a typical *eukaryotic* cell (that is, a cell with a nucleus), there are many **ribosomes**, the organelles at which proteins are built. The *lysosomes*, which are scattered throughout the cell, are the structures involved in the breaking down of food.

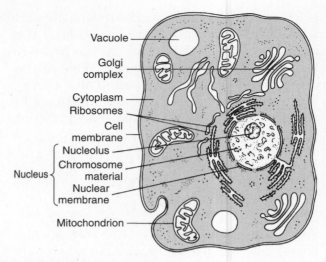

Figure 6-7 The organelles of a typical animal cell.

Complicated-looking **mitochondria**, shaped like tiny kidney beans, are the organelles in which the cell's energy is released. (See Figure 6-7.)

By far the largest structure in a cell is the **nucleus**, which is responsible for information storage. The nucleus often fills the entire central portion of the cell. In addition, other specialized structures may be present in a cell to handle other functions, such as: the *Golgi complex*, which packages many materials; **vacuoles**, which store materials such as wastes; and, in plants, *chloroplasts*, the organelles in which plants convert energy from the sun into food. (See Figure 6-8.)

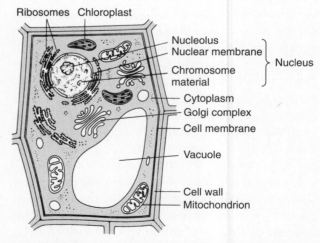

Figure 6-8 The organelles of a typical plant cell.

Chapter 6 Review

Part A—Multiple Choice

1. Which letter indicates the cell structure that directly controls the movement of molecules into and out of the cell?

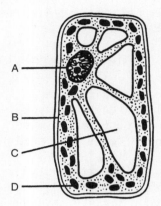

   1 A
   2 B
   3 C
   4 D

2. Which statement about the functioning of the cell membrane of all organisms is *not* correct?

   1 The cell membrane forms a boundary that separates the cell's contents from the outside environment.
   2 The cell membrane forms a barrier that keeps all substances that might harm the cell from entering it.
   3 The cell membrane is capable of receiving and recognizing chemical signals.
   4 The cell membrane controls the movement of molecules into and out of the cell.

3. What happens during diffusion?

   1 Molecules move automatically from an area of higher concentration to an area of lower concentration.
   2 Molecules are pumped from an area of lower concentration to an area of higher concentration.
   3 An enzyme joins with a particular molecule.
   4 A catalyst speeds up the rate of a chemical reaction.

Base your answer to question 4 on the following diagram, which represents a cell in water. Formulas of molecules that can move freely across the cell membrane are shown. Some molecules are located inside the cell and others are in the water outside the cell.

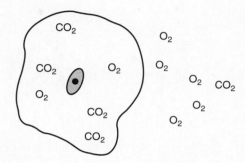

4. Based on the distribution of these molecules, what would most likely happen to them after a period of time has passed?

   1 The concentration of O_2 will increase inside the cell.
   2 The concentration of O_2 will remain the same outside the cell.
   3 The concentration of CO_2 will remain the same inside the cell.
   4 The concentration of CO_2 will decrease outside the cell.

5. A plant cell shrinks when placed in salt water due to the osmosis of

   1 water molecules out of the cell
   2 water molecules into the cell
   3 salt into the cell
   4 salt out of the cell

6. Placing limp celery in water will make the celery stalk firm again due to

   1 diffusion
   2 osmosis
   3 active transport
   4 a catalyst

7. A high concentration gradient means that the concentration of a substance is

   1 low on both sides of the cell membrane
   2 high on both sides of the cell membrane
   3 about the same on both sides of the cell membrane
   4 high on one side of the cell membrane and low on the other side

Base your answer to question 8 on the diagram below, in which the dark dots represent small molecules. These molecules are moving out of the cells, as indicated by the arrows. The number of dots represents the relative concentrations of the molecules inside and outside of the two cells.

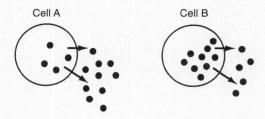

Cell A   Cell B

8. ATP is being used to move the molecules out of

1   cell A only
2   cell B only
3   both cell A and cell B
4   neither cell A nor cell B

Refer to the set of diagrams below, which shows the movement of a large molecule across a cell membrane, to answer question 9.

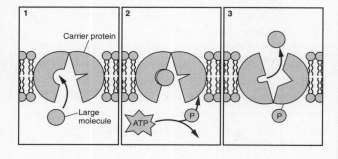

9. Which process is best represented by this set of diagrams?

1   active transport
2   protein building
3   diffusion
4   gene transfer

10. When covalent bonds are formed or broken,

1   the amount of stored energy changes
2   energy is always lost
3   energy is produced
4   new atoms are formed

11. Chemical reactions in the cell take place

1   over extremely long periods of time
2   in a series of small steps
3   all at once in a single burst
4   during short intervals of time

12. How do chemical reactions occur at the relatively low temperature found within cells?

1   Some energy is destroyed before it heats up the cell.
2   Some energy is stored temporarily in ATP molecules.
3   Enzymes are used to slow the rate of the reactions.
4   Enzymes are used to speed the rate of the reactions.

13. To carry out its chemical reactions, each cell contains

1   one specific type of enzyme
2   fewer than 10 different enzymes
3   thousands of different enzymes
4   thousands of copies of the same enzyme

14. The equation below represents a chemical reaction that occurs in humans. What data should be collected to support the hypothesis that enzyme C works best in an environment that is slightly basic?

ENZYME C

Substance X + Substance Y ⟶ Substance W

1   the amino acid sequence of enzyme C
2   the shapes of substances X and Y after the reaction occurs
3   the amount of substance W produced in 5 minutes at various pH levels
4   the temperature before and after the reaction occurs

Base your answer to question 15 on the diagrams below, which show an enzyme and four different molecules.

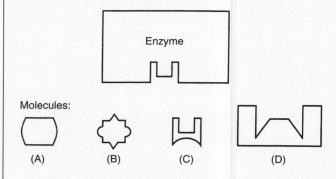

Enzyme

Molecules:

(A)   (B)   (C)   (D)

15. The enzyme would most likely affect reactions that involve

1   molecule A only
2   molecule C only
3   molecules B and D
4   molecules A and C

16. Which best describes the interaction between an enzyme and another substance?
    1 a temporary association in which the substance changes
    2 a temporary association in which the enzyme changes
    3 the final product of a series of chemical reactions
    4 the pocket into which the enzyme and the substance fit

17. Two conditions that must be maintained in a cell in order for enzymes to work properly are
    1 pH and oxygen content
    2 surface area and temperature
    3 temperature and pH
    4 volume and pressure

18. While viewing a slide of rapidly moving sperm cells, a student concludes that these cells require a large amount of energy to maintain their activity. The organelles that most directly provide this energy are known as
    1 vacuoles
    2 chloroplasts
    3 ribosomes
    4 mitochondria

19. The organelle that stores wastes for the cell is the
    1 vacuole
    2 chloroplast
    3 ribosome
    4 Golgi complex

20. Chloroplasts are important because they
    1 are necessary to release stored energy
    2 store wastes in both plants and animals
    3 use energy from the sun to make food
    4 are in the nuclei of both plant and animal cells

Part B—Analysis and Open Ended

Refer to the diagram in the next column to answer questions 21 and 22.

21. Which part of the diagram shows the cell membrane acting to allow materials into and out of the cell?
    1 A only
    2 B only
    3 both A and B
    4 neither A nor B

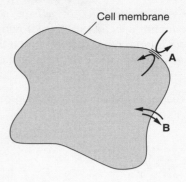

Cell membrane

22. Which part of the diagram shows the cell membrane acting to separate the inside of the cell from the outside environment?
    1 A only
    2 B only
    3 both A and B
    4 neither A nor B

23. Discuss the meaning of "selective permeability" for a cell membrane. Your answer should explain the following:
    ◆ why the cell membrane is said to be selectively permeable
    ◆ why this characteristic is important to the health of a cell

Base your answers to questions 24 and 25 on the diagram below, which represents a unicellular organism in a watery environment. The small triangles represent molecules of a specific substance.

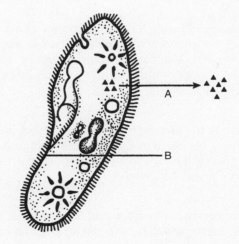

24. What kind of transport is represented by arrow A? State two ways in which active transport is different from diffusion (passive transport).

25. In the cells of multicellular organisms, structure B often contains special proteins known as *receptor molecules*. What specific function do these protein molecules carry out for the cell?

Refer to the diagrams below to answer questions 26 and 27.

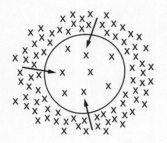 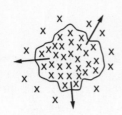

26. The diagrams represent the movement of molecules from an area of

1 low concentration to an area of high concentration

2 high concentration to an area of low concentration

3 low concentration to an area of equal concentration

4 high concentration to an area of equal concentration

27. The diagrams could be used to illustrate all of the following types of transport *except*

1 diffusion

2 osmosis

3 active

4 passive

28. Use the following terms to construct a concept map that shows how a cell regulates its internal environment: *active transport; diffusion; uses energy; does not use energy; cell membrane; osmosis.*

29. Identify the two main things that can happen (during chemical reactions) to the energy stored in chemical bonds. For what important activity do cells use this energy?

Base your answers to questions 30 and 31 on the diagram below, which represents an enzyme and

four types of molecules present within a solution in a flask.

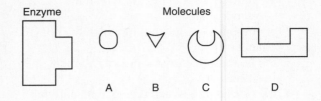

30. Which molecule would most likely react with the enzyme? Why?

31. What would most likely happen to the rate of reaction if the temperature of the solution in the flask were increased gradually from 10°C to 30°C?

32. Enzyme molecules are affected by changes in conditions within organisms. Explain how a prolonged, excessively high body temperature during an illness could be fatal to humans. Your answer must include:

♦ the role of enzymes in a human

♦ the effect of this high body temperature on enzyme activity

♦ the reason this high body temperature can result in death

33. Explain why it is important for the cells of our bodies to maintain a neutral pH.

34. What, specifically, would happen to a cell if its mitochondria were removed? Explain why.

35. Just like complex organisms, cells are able to survive by coordinating various activities. Complex organisms have a variety of systems, and cells have a variety of organelles that work together for survival. Describe the roles of two organelles. In your answer be sure to include:

♦ the names of two organelles and the function of each

♦ an explanation of how these two organelles work together

♦ the name of an organelle and the name of a system in the human body that have similar functions

Part C—Reading Comprehension

Base your answers to questions 36 to 38 on the information below and on your knowledge of biology. Use one or more complete sentences to answer each question.

The spleen is an organ that helps your body fight disease. It also helps break down old red blood cells. For John Moore, removal of his spleen cured the leukemia, a type of cancer of the blood, that he was found to have in 1976. Removing the spleen is a standard treatment for this disease. The leukemia did not return and Mr. Moore was very happy.

But his attitude was soon to change. For what Mr. Moore did not know was that the physicians at the University of California who removed his spleen kept some of the cells from the spleen alive. The physicians put the cells in a special nutrient-rich solution. As the cells continued to reproduce, the physicians studied them. They discovered that cells from Mr. Moore's spleen produced an interesting blood protein. The physicians received a patent on the cells, which made them the legal owners of the cells they removed from Mr. Moore. In turn, the cells and the blood protein they made were being sold to a company that planned to develop a new medicine from the cells—a medicine that would be sold for a large profit. Mr. Moore did not think this was right. He told the physicians, "Don't use my cells; I own them"—and then he went to court.

The lawsuit was heard by the Supreme Court of California, which ruled on July 9, 1990 in favor of the physicians. Later, the United States Supreme Court also ruled in the physicians' favor. The courts felt that scientific research would be threatened if researchers did not have the freedom to work with human cells. Besides, the courts said that John Moore never expected to get his cells returned to him when he gave permission for his spleen to be removed in the medical procedure. As of July 2001, Mr. Moore remains healthy. The drug company stock now owned by Mr. Moore's original doctor—in exchange for the rights to use the Moore cell line—is worth over five million dollars.

36. Explain why the removal of his spleen was a necessary procedure for the patient John Moore.

37. Why was the tissue from Mr. Moore's spleen of great interest to a company that produces medicines?

38. How was the legal system involved in the scientific research that arose from the removal of Mr. Moore's spleen?

The Flow of Energy: Photosynthesis and Respiration

FOOD: MATTER AND ENERGY

An apple is a type of food. It contains complex organic compounds; its atoms are held together as molecules by chemical bonds that are rich in stored energy. When you eat an apple, you get both the matter and the energy you need to build your body and to stay alive.

The apple tree that produced the fruit represents the group of organisms that are **autotrophic**, meaning "self-feeding." Like other plants, the apple tree makes its own food, taking in the **inorganic** substances carbon dioxide (CO_2) and water (H_2O) and changing them into **organic** compounds, such as sugars and starches. Humans represent the other group of organisms, which are **heterotrophic**, meaning "other-feeding." Since they cannot make their own food, humans and all other animals must get their complex organic compounds by eating other organisms. (See Figure 7-1.)

For the apple tree to combine the inorganic

Figure 7-1 The grass, like all other plants, is autotrophic because it makes its own food. The cows, like all other animals, are heterotrophic because they have to eat other organisms in order to survive.

raw materials of CO_2 and H_2O into organic compounds such as sugar and starch, it needs a source of energy. The rays of sunlight, as they fall on the leaves of the apple tree, provide that energy. The process of making this food, by using light as the source of energy, is called **photosynthesis**. All green plants are photosynthetic autotrophs. Without plants to capture the energy of sunlight and convert it into the chemical forms that are edible, most animals would have no constant source of food and could not exist.

PHOTOSYNTHESIS

Plants are able to make their own energy-rich carbon compounds. In particular, they make the simple sugar glucose, whose chemical formula is $C_6H_{12}O_6$. Plants get the carbon for these glucose molecules from inorganic CO_2 in the air. In addition, plants release oxygen (O_2) to the air. (See Figure 7-2.) Scientists discovered that photosynthesis requires the green pigment *chlorophyll*. The chemical reactions of photosynthesis occur within the chlorophyll-containing organelles called **chloroplasts**, found within plant leaves and stems. Some scientists consider this process of photosynthesis the single most important chemical reaction that occurs on Earth. This all-important reaction can be summarized by the following chemical equation:

$$CO_2 + H_2O \xrightarrow[\text{CHLOROPHYLL}]{\text{LIGHT ENERGY}} C_6H_{12}O_6 + O_2$$

CARBON DIOXIDE + WATER   GLUCOSE + OXYGEN

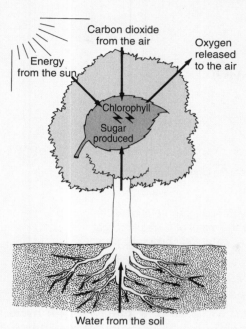

Figure 7-2 The diagram illustrates the basic process of photosynthesis—plants take in inorganic substances from the environment and produce organic substances such as glucose (a sugar).

LEAVES: PHOTOSYNTHETIC FACTORIES

The structures present inside the leaf are well organized. Such organization allows cells that contain chlorophyll to get maximum exposure to light. At the same time, the leaf controls the amount of water lost to the air. It also makes possible the movement of CO_2 and O_2 into and out of the leaf. (See Figure 7-3.)

THE RATE OF PHOTOSYNTHESIS

As with any chemical reaction, the reactions of photosynthesis can occur at different rates. The factors that affect the rate at which photosynthesis occurs are temperature, light intensity, CO_2 concentration, availability of water, and the presence of certain minerals. (See Figure 7-4 on page 50.)

CELLULAR RESPIRATION: RELEASING THE STORED ENERGY IN FOOD

Consider the relationship between the sunlight that falls on the leaves of an apple tree and the chemical process of photosynthesis. During photosynthesis, the light energy of the sun is converted into the stored chemical energy of glucose in the apple. After you eat the apple, your cells are ready to use that stored chemical energy. How does this happen?

The release of energy cannot occur all at once. Too much heat would be released inside your cells. Instead, the release of energy occurs in a series of enzyme-controlled small steps. The energy stored in glucose is converted into a usable form, the energy source of all cells, *adenosine triphosphate*, or **ATP**. This process is known as **cellular respiration**. Cellular respiration is basically the opposite process of photosynthesis. Instead of being produced in the cells, the energy-rich glucose

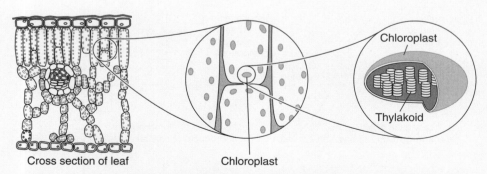

Figure 7-3 Cells inside a leaf contain chloroplasts, which capture the sunlight that is used for photosynthesis. The structure of the leaf allows these cells to get maximum exposure to the light.

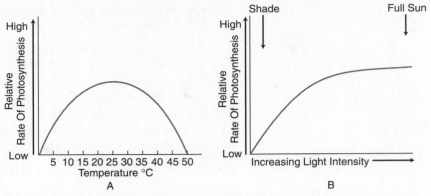

Figure 7-4 Factors that influence the rate of photosynthesis: graph A shows the effect of different temperatures, while graph B shows the effect of increasing light intensity. Other factors, such as the availability of CO_2, water, and minerals, also affect the rate of photosynthesis.

molecules are taken apart to release their stored energy. Oxygen is used, and CO_2 and water are released as wastes. (Because oxygen is used to produce ATP, this is referred to as an *aerobic* process.) Cells use the energy from ATP to perform many functions, such as obtaining materials and eliminating wastes. (See Figure 7-5.)

■ A FINAL VISIT BACK TO PLANTS

A plant does not specifically go through the process of photosynthesis in order to make food for people and other animals. The apple tree, for example, is simply making food for itself to live long enough to reproduce successfully. The apples contain seeds, which may get

carried away to new places by animals that eat the apples. This makes it possible for the apple tree to produce more apple trees in other places. However, most of the glucose made in the leaves of the tree does not go into storage in the form of apples. Rather, it gets taken to different parts of the tree and used by the tree to stay alive. In fact, the tree uses the same process as you do to get the energy it needs from the glucose it has made—cellular respiration.

To summarize: All plants are autotrophs and are able to make their own food by photosynthesis and use it for energy through cellular respiration. All animals are heterotrophs and must obtain food energy from other organisms. They use cellular respiration, just as plants do, to obtain energy from the food they eat.

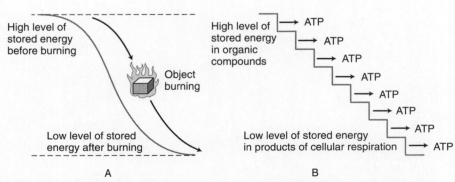

Figure 7-5 The burning of an object (diagram A) is due to the sudden release in one step of the energy stored in that object. By contrast, in cellular respiration (diagram B), energy is released from organic compounds as ATP in a series of small, enzyme-controlled steps.

Chapter 7 Review

Part A—Multiple Choice

1. Which of the following is an autotroph?
   1 lizard
   2 cactus
   3 shark
   4 antelope

2. In heterotrophs, energy for the life processes comes from the chemical energy stored in the bonds of
   1 water molecules
   2 organic compounds
   3 oxygen molecules
   4 inorganic compounds

3. During photosynthesis,
   1 animals use sunlight to convert the starch in plants into food
   2 animals use the oxygen released by plants to make carbon dioxide
   3 plants use the energy of sunlight to convert carbon dioxide and water into glucose and oxygen
   4 plants use the energy of sunlight to convert glucose and oxygen into carbon dioxide and water

4. The equation below represents an important biological process. This process is carried out within a cell's

   carbon dioxide + water $\longrightarrow$

   glucose + water + oxygen

   1 mitochondria
   2 cell membranes
   3 ribosomes
   4 chloroplasts

5. The source of energy for photosynthesis is
   1 oxygen
   2 sunlight
   3 carbon dioxide
   4 glucose

6. To occur, photosynthesis requires the presence of the green substance
   1 tree sap
   2 glucose
   3 chlorophyll
   4 copper

7. The approximate mass of a field of corn plants at the end of its growth period was 3 tons per hectare. Most of this mass was produced from
   1 water and organic compounds absorbed from the soil
   2 minerals and organic materials absorbed from the soil
   3 minerals from the soil and oxygen from the air
   4 water from the soil and carbon dioxide from the air

8. The diagram below represents part of the life process that occurs inside a leaf chloroplast. If the process were to be interrupted by a chemical at point X, there would be an immediate effect on the release of which substance?

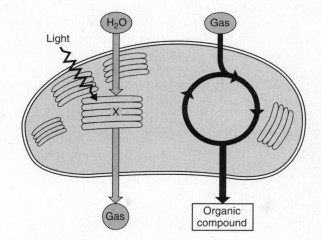

   1 chlorophyll
   2 carbon dioxide
   3 nitrogen
   4 oxygen

9. The food produced by plants during photosynthesis is used
   1 by the plants themselves only
   2 by animals that eat them only
   3 by both the plants and the animals that eat them
   4 up at the end of the reaction

10. If stored energy were to be released too quickly, a cell would
    1 release too much heat
    2 produce ATP molecules
    3 become an autotroph
    4 become a heterotroph

Answer question 11 based on the following information and diagram.

The flow of energy through an ecosystem involves many energy transfers. The diagram below summarizes the transfer of energy that eventually powers muscle activity.

Sun ⟶ Food ⟶ ATP ⟶ Muscle Activity
       A   B   C

11. The process of cellular respiration is represented by
    1 arrow A only
    2 arrow B only
    3 arrow C only
    4 arrows A, B, and C

12. How do humans and plants interact in terms of the two gases involved in photosynthesis?
    1 Humans take in the CO_2 released by plants and release O_2 to the plants.
    2 Humans take in the O_2 released by plants and release CO_2 to the plants.
    3 Plants and humans usually compete for the same O_2 available in the air.
    4 Plants and humans usually compete for the same CO_2 available in the air.

13. Cellular respiration occurs in
    1 autotrophs only
    2 heterotrophs only
    3 autotrophs and heterotrophs
    4 humans only

14. Eating a sweet potato provides energy for human metabolic processes. The original source of this energy is the energy
    1 in protein molecules stored within the potato
    2 that is made available by photosynthesis
    3 from starch molecules absorbed by the potato plant
    4 in vitamins and minerals found in the soil

15. In nature, during a 24-hour period, green plants *continuously* use
    1 carbon dioxide only
    2 oxygen only
    3 both carbon dioxide and oxygen
    4 neither carbon dioxide nor oxygen

16. Plant leaves contain openings known as stomates, which are opened and closed by specialized cells, allowing for gas exchange between the leaf and the outside environment. Which phrase best describes the net flow of gases in-

volved in photosynthesis into and out of the stomates on a sunny day?
    1 carbon dioxide moves in, oxygen moves out
    2 oxygen moves in, nitrogen moves out
    3 carbon dioxide and oxygen move in, ozone moves out
    4 water and ozone move in, carbon dioxide moves out

Part B—Analysis and Open Ended

Answer question 17 based on the following information and graph.

As the depth of the ocean increases, the amount of light that penetrates to that depth decreases. At about 200 meters, there is almost no light present. The graph below illustrates the population size of four different species at different water depths.

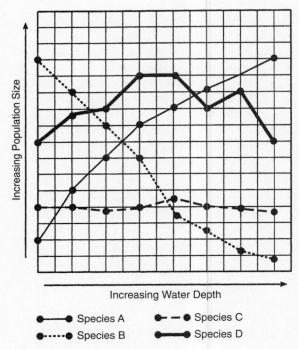

17. Which species most likely performs photosynthesis?
    1 species A
    2 species B
    3 species C
    4 species D

18. Explain why plants are defined as autotrophs and why animals are defined as heterotrophs.

19. Why might the process of photosynthesis be considered a "bridge" between the living and nonliving parts of the world?

20. Briefly describe three ways in which the structures of a leaf enable the process of photosynthesis to occur. Your answer should include the following factors:

- ◆ light
- ◆ water
- ◆ gases

Base your answers to questions 21 and 22 on the summary equations of two processes shown below and on your knowledge of biology.

Photosynthesis

water + carbon dioxide $\xrightarrow{\text{ENZYMES}}$

glucose + oxygen + water

Respiration

glucose + oxygen $\xrightarrow{\text{ENZYMES}}$

water + carbon dioxide

21. Choose *one* of the processes shown above and identify the following:

a the source of the energy in the process you chose; and

b where the energy ends up at the end of that process.

22. State *one* reason why *each* of the following processes is important for living things:

a respiration; and

b photosynthesis.

Base your answers to questions 23 to 25 on the information and diagram below and on your knowledge of biology.

The diagram represents a system in a space station that includes a tank containing algae. An astronaut from a spaceship boards the space station.

23. Identify *one* process that is being controlled in the setup shown in the diagram.

24. State *two* changes in the chemical composition of the space station atmosphere as a result of the astronaut coming on board the station.

25. State *two* changes in the chemical composition of the space station atmosphere that would result from turning on more lights.

Base your answers to questions 26 and 27 on the information and diagram below.

The diagram represents a single-celled organism known as *Euglena*. This organism is able to carry out both photosynthesis and cellular respiration.

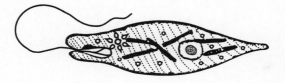

26. Choose *one* of the two processes that *Euglena* carries out. Write down the word for it; then use words or chemical symbols to summarize the reaction for the process you chose.

27. State *one* reason why the process you chose is essential for the survival of the *Euglena*.

28. Look at the chart on page 54 to answer this question. Which phrase would you choose to fill in the missing title?

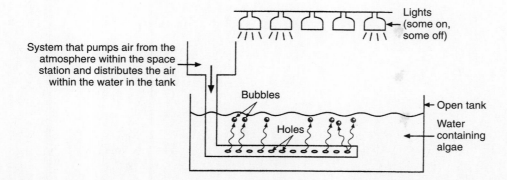

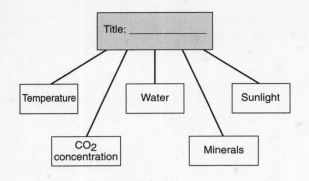

1 Some Living Factors in the Environment
2 The Chemical Process of Photosynthesis
3 Factors That Affect the Rate of Photosynthesis
4 The Nonliving Things That Make Up a Plant

29. Look at the following diagrams (A and B) to answer this question. The energy change in diagram B is different from the energy change in diagram A because, in diagram B,

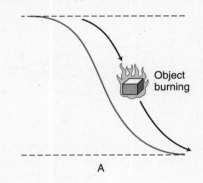

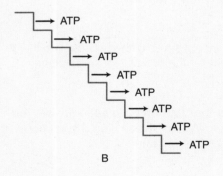

1 the energy is released suddenly in one step
2 the energy is released in a series of steps
3 there is less stored energy at the beginning
4 there is less stored energy remaining at the end

Refer to the chemical equation below to answer questions 30 and 31.

$$CO_2 \;+\; H_2O \;\xrightarrow[\text{CHLOROPHYLL}]{\text{LIGHT ENERGY}}\; C_6H_{12}O_6 \;+\; O_2$$

30. What important life process is described by this equation? What are the two vital products of this reaction?

31. Explain why "cellular respiration is basically the opposite" of the process shown in the equation. What are the two waste products of cellular respiration?

Part C—Reading Comprehension

Base your answers to questions 32 to 34 on the information below and on your knowledge of biology. Use one or more complete sentences to answer each question.

We walk on land. Even the very name Earth is used to mean land. But look at a world map and you will see a lot of blue space. In fact, more than 70 percent of Earth's surface is covered by water, mostly oceans. Unseen in these waters— drifting along with waves and currents—are countless numbers of tiny organisms. Photosynthetic bacteria, protists, and plants are included in these drifters. Some of these unicellular species are so small that if 12 million cells were lined up in a row, the line would be only about 1 centimeter long. In some places in the oceans, these microscopic organisms are so numerous that a cup of seawater may hold 24 million individuals of a single species, and that cup would contain other species as well!

These species are very small, but their importance to the overall life on the planet is huge. Tiny sea-dwelling organisms are the beginning food source for almost all living things in the oceans. It is easy for us land dwellers to understand that many animals eat plants to get food. We have seen cattle and sheep grazing on grasses in a pasture. The drifting cells in the ocean could be called the *grass fields* or *pastures* of the sea. Just like grass on land, the sea drifters capture energy from the sun and convert inorganic CO_2 and water into organic molecules, which become important foods for other organisms. On land, plants bloom with wild displays of colorful flowers in spring. The photosynthetic drifters in the pastures of the seas are said to "bloom" in the spring, too, as the water warms and nutrients from ocean depths are brought to the surface by currents. A great deal has been learned recently about the seasonal explosive growth of these photosynthetic cells in the ocean from photographs taken by orbiting satellites.

32. Explain why the drifting cells in the ocean can be called the grass fields or pastures of the sea.

33. Describe three ways in which microscopic drifting cells in the ocean are similar to plants on land.

34. How has modern technology improved our ability to study life in the ocean?

8

Getting Food to Cells: Nutrition

✛ THE NEED FOR DIGESTION

The food you eat provides the matter your body needs to build cells, tissues, and organs. In addition to being used to build cells and tissues, food provides the energy an organism needs to remain alive. The process of taking in and utilizing food is called **nutrition**. (See Figure 8-1.)

For all organisms, food must be digested. Why? Every organism is either a single cell or a collection of cells. The cell theory states that cells are the basic units of structure and function of all living things. Therefore, what an organism needs is what its cells need. In order to nourish you, the food you eat must get into each of your cells.

That is why digestion is necessary. No matter what the food is—an earthworm for a bird or a hamburger for a person—the food is too large, and its molecules too complex, to get inside a single cell. To get inside the cells, food must be broken down into relatively simple molecules. **Digestion** is the process of break-

ing down food particles into molecules small enough to be absorbed by cells.

✛ THE HUMAN DIGESTIVE SYSTEM

The digestive system in humans has the same purpose as the digestive system in other organisms: to get **nutrients** from food into cells. Food does not really enter your body when you put it into your mouth; not even when it goes into your stomach or intestine. These locations are actually connected to the outside world. Only when the food is absorbed across the membranes of cells that line your digestive system is it truly inside you. (See Figure 8-2.)

Follow the path of a hamburger as it travels through the human digestive system.

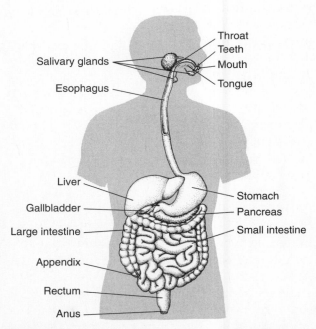

Figure 8-2 The human digestive system—it functions to get the nutrients from your food into your cells.

Figure 8-1 Food provides you with matter and energy to keep you alive.

First, the hamburger must enter the mouth. This is the process of *ingestion*. Then teeth in the mouth provide mechanical digestion; they grind up the meat and bun into smaller and smaller pieces. Smaller pieces of food provide more surface area for the body's enzymes to work on. Chemical digestion begins as an enzyme in saliva starts the digestion of starch in the hamburger bun.

The food is swallowed and pushed by the tongue to the back of the mouth where it enters the *esophagus*, a muscular tube that carries mouthfuls of hamburger down to the stomach. As this begins, the pathways up to the nose and down to the lungs are closed off. The *epiglottis*, a small flap of tissue in the throat, is pushed down by the food to close off the trachea. The *trachea* is the pathway for air to the lungs; if food enters it by mistake, coughing and difficult breathing occur. (See Figure 8-3.)

When the food arrives at the stomach, a muscular valve quickly opens and closes. The valve allows food to enter the stomach without letting out the acidic contents of the stomach. Occasionally, acid backs up through the valve, causing a painful feeling in the chest, commonly called "heartburn." The *stomach* is a large muscular organ located in the abdomen. The muscular movement of the stomach churns the food and continues the process of mechanical digestion. As muscles in the stomach wall contract and relax, food gets mixed with *gastric juice*, which contains hydrochloric acid, pepsinogen, and water. The hydrochloric acid kills many bacteria that may be present in the food. The acid also turns the pepsinogen into the protein-digesting enzyme pepsin. Therefore, a second step in the chemical digestion of the hamburger begins in the stomach when pepsin breaks down proteins in the meat.

Muscular contractions of the stomach walls push the food contents into the *small intestine*, which is a muscular tube about seven meters in length. Most of the chemical digestion of food occurs in the small intestine, not in the stomach. Large quantities of different enzymes accomplish this task. These enzymes come from two main sources: the lining of the small intestine itself and the pancreas. The *liver* (the largest organ in the abdomen) helps

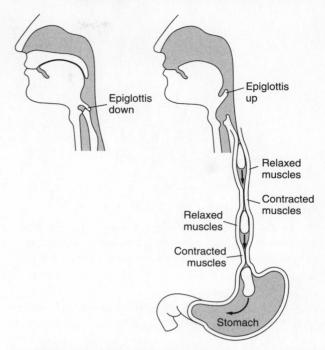

Figure 8-3 The process of swallowing food—muscular contractions move food down the esophagus to the stomach.

the body digest fat by producing bile, which is stored in the gallbladder until it is needed. Then it is emptied into the small intestine through a small duct. When it comes in contact with fat, bile acts like a detergent to break down the fat into smaller droplets. This process, called *emulsification*, allows the fat droplets to be more effectively digested by enzymes.

Now the hamburger is very much changed. The starch that was in the bun has been changed into simple sugar molecules, the meat protein exists as amino acids, and the fat is fatty acid molecules and glycerol. Yet the food is still outside of the body. Only now are its molecules tiny enough to pass through the wall of the small intestine and into the blood vessels. This process is known as *absorption*. Once they are in the blood, the food molecules are carried to all the cells and then are, finally, really inside the body. Once inside the cells, the nutrients are used in the **synthesis** of compounds needed for life. (See Figure 8-4.)

Almost all of the useful nutrients get absorbed in the small intestine. Anything that passes on from there into the large intestine is primarily indigestible material, such as the

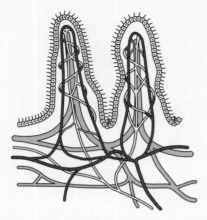

Figure 8-4 Tiny projections on the lining of the small intestine increase its surface area. This makes absorption of food molecules into the blood vessels more efficient.

cellulose in lettuce. Compared to the small intestine, the *large intestine* is large in diameter but short in length. In the large intestine, water from the remaining material is reabsorbed into the body. Then, the muscles of the rectum force feces (the solids that remain) out through the anus. If too much water is reabsorbed, the feces cannot move easily through the large intestine, and constipation occurs. If too little water is reabsorbed, the feces are too liquid, and diarrhea occurs.

Throughout the intestines, huge numbers of **bacteria** help break down the food. In addition to helping the process of digestion,

these intestinal bacteria make several important vitamins. They also help rid the body of harmful bacteria.

❖ WHEN THINGS GO WRONG: DISEASES OF THE DIGESTIVE SYSTEM

To prevent the stomach from digesting itself, cells in the wall of the stomach produce a thick layer of mucus. Sometimes, the layer of mucus protection fails. Gastric juice reaches the wall of the stomach and begins to break it down. The eating away of tissues produces an *ulcer*, a painful and serious condition. When this happens in the stomach, it is called a peptic ulcer. Infection by acid-resistant bacteria is the main cause of ulcers. Treatment with antibiotics is most effective for eliminating ulcers.

One of the most common types of cancer in North America is *colon cancer*. The large intestine is made up of the colon and the very end of the intestine, the rectum. In North America, the typical diet contains low levels of fiber. As a result, the feces move too slowly through the colon. This digestive problem may be related to the high incidence of colon cancer. Physicians also suggest that there may be a strong hereditary predisposition to colon cancer.

Chapter 8 Review

Part A—Multiple Choice

1. The main function of the human digestive system is to
   1. rid the body of cellular waste materials
   2. break down glucose in order to release energy
   3. process organic molecules so they can enter cells
   4. change amino acids into proteins and carbohydrates

2. At what point does food really enter your body?
   1. when you bite into the food
   2. when you chew the food
   3. when it enters your stomach

   4. when it is absorbed into your cells

3. The teeth play an important role in digestion because they
   1. begin chemical digestion by releasing an enzyme in the mouth
   2. begin mechanical digestion by grinding food into smaller pieces
   3. break the food down so that it can be absorbed by the cells
   4. force the epiglottis to close off the trachea

4. The epiglottis is important because it
   1. pushes food to the back of the mouth
   2. prevents food from entering the esophagus

3 lets food into the stomach without letting acid out

4 prevents food from entering the trachea

5. What causes heartburn?

1 stomach acid backing up to the esophagus

2 food blocking the trachea

3 undigested food reaching the heart

4 stomach acid reaching the heart

6. Which statement is true of the stomach's role in digestion?

1 The process of digestion is completed in the stomach.

2 Aside from the mouth, all chemical digestion takes place in the stomach.

3 Both mechanical and chemical digestion occur in the stomach.

4 Nutrients from food are absorbed by the cells of the stomach.

7. What happens when the muscles in the stomach wall relax and contract?

1 Food is pulled through the esophagus.

2 Food is mixed with gastric juice.

3 Nutrients in food are moved into body cells.

4 Wastes are removed from the body.

8. The stomach kills bacteria that may be present in food by

1 churning the food repeatedly

2 releasing hydrochloric acid

3 producing pepsinogen

4 pushing the food into the small intestine

Base your answer to question 9 on the information below and on your knowledge of biology.

A student completed a series of experiments and found that a protein-digesting enzyme (intestinal protease) functions best when the pH is 8.0 and the temperature is 37°C. During one experiment, the student used some of the procedures listed below:

a adding more protease

b adding more protein

c decreasing the pH to 6.0

d increasing the temperature to 45ºC

e decreasing the amount of light

9. Which procedure would have the *least* effect on the rate of protein digestion?

1 procedure a

2 procedure b

3 procedure d

4 procedure e

10. Most of the chemical digestion of food takes place in the

1 stomach

2 small intestine

3 pancreas

4 large intestine

11. A sample of food containing one type of a large molecule was treated with a specific digestive enzyme. Nutrient tests performed on the resulting products showed the presence of simple sugars only. Based on these test results, you could determine that the original large molecules contained in the sample were molecules of

1 protein

2 starch

3 glucose

4 DNA

12. The pancreas is an organ connected to the digestive tract of humans by a duct (tube) through which digestive enzymes flow. These enzymes are important to the digestive system because they

1 form proteins needed in the stomach

2 change the food into molecules that can pass into the bloodstream

3 form the acids that break down the food

4 change food materials into wastes that can be passed out of the body

13. During the process of emulsification,

1 gastric juices break down proteins

2 bile breaks down fat into tiny droplets

3 enzymes break down fat into droplets

4 bacteria present in food are killed

14. The liver and the gallbladder are involved in digestion because the

1 food from the small intestine is passed to the liver and then to the gallbladder

2 food from the small intestine is passed to the gallbladder and then to the liver

3 gallbladder produces bile that is stored in the liver until needed

4 liver produces bile that is stored in the gallbladder until needed

15. During digestion, fat is changed into

1 fatty acid molecules and glycerol

2 amino acids

3 monosaccharides

4 bile

Base your answer to question 16 on the diagram and chart below. The diagram represents one metabolic activity of a human.

Metabolic Activity A

| Protein | → | B | B | B | B |

Row	Metabolic Activity A	B
(1)	Respiration	Oxygen molecules
(2)	Reproduction	Hormone molecules
(3)	Excretion	Simple sugar molecules
(4)	Digestion	Amino acid molecules

16. Letters A and B in the diagram are best represented by which row in the chart?

1 row 1

2 row 2

3 row 3

4 row 4

17. Food that reaches the large intestine is

1 basically indigestible

2 small enough to be absorbed by cells

3 carried by the blood to the cells

4 broken down by bile

18. Solids that remain in the large intestine are

1 returned to the small intestine for further digestion

2 forced out of the body through the anus

3 diluted with water until they can be digested

4 stored until they break down further

19. Bacteria that live in the intestines

1 help break down food and get rid of harmful bacteria

2 are all harmful and are killed by gastric juices

3 cause illness unless they are released with feces

4 exist in small enough numbers to be ignored

20. A peptic ulcer may result when

1 helpful bacteria are removed from the digestive system

2 too little fiber passes through the digestive system

3 too much water is removed from the large intestine

4 gastric juice eats away at the tissue of the stomach lining

Part B—Analysis and Open Ended

21. Distinguish between the processes of ingestion and digestion.

22. According to the cell theory, the cell is the basic unit of structure and function in the body. In two or more sentences, explain how this fact relates to the need for digestion.

23. Identify and describe the two types of digestion that take place in the mouth. Your answer should include:

♦ the role of the teeth

♦ the role of saliva

24. Use the figure below and your knowledge of digestion to answer this question. How is food moved from the esophagus to the stomach?

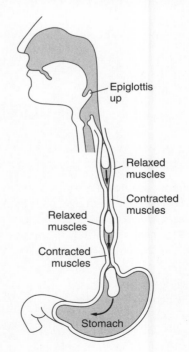

1 The tongue pushes the food down.

2 The epiglottis pushes the food down.

3 The esophagus muscles move the food down.

4 The stomach muscles pull the food down.

25. Describe the ways muscles are important to the functioning of our digestive system. Your answer should include at least:

♦ *two* ways muscles function in the stomach

♦ *two* ways muscles function in the intestines

26. Complete the following chart by indicating where digestion begins for each of the following nutrients: proteins; fats; and starches.

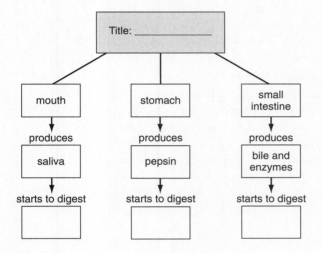

27. The best title for this concept map probably would be
    1 The Basic Food Groups
    2 The Types of Enzymes
    3 Chemical Digestion
    4 Mechanical Digestion

Base your answers to questions 28 to 30 on the information and diagram below and on your knowledge of biology.

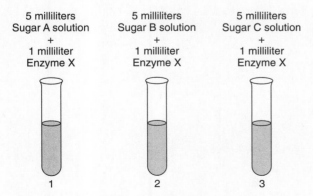

An investigation was performed to determine the effects of enzyme X on three different disaccharides at 37°C. Three test tubes were set up as shown in the diagram. At the end of five minutes, the solution in each test tube was tested for the presence of disaccharides (double sugars) and monosaccharides (simple sugars).

	Test Tube 1	Test Tube 2	Test Tube 3
Monosaccharide	not present	not present	present
Disaccharide	present	present	not present

The results of these tests are shown in the data table above.

28. What can be concluded about the activity of enzyme X from the data table?

29. With only the materials listed below and common laboratory equipment, design an investigation that could show how a change in pH would affect the activity of enzyme X. Your design need only include a detailed procedure and a data table.
    Materials: enzyme X; sugar C solution; indicators; substances of various pH values (vinegar [acidic], water [neutral], baking soda [basic]); data table.

30. State *one* safety precaution that should be used during this investigation.

31. How is the human digestive system similar to a factory assembly line? In what important way is it *not* similar (aside from it being a living system)?

32. Explain why the liver, pancreas, and gallbladder are all considered to be part of the digestive system, even though food does not pass through them.

33. Briefly describe the roles of the following three liquids in the digestive system: saliva; gastric juice; and bile.

34. Explain why absorption is considered to be the final step in digestion.

35. Why are some bacteria considered to be "friendly" to our digestive system?

Part C—Reading Comprehension

Base your answers to questions 36 to 38 on the information below and on your knowledge of biology. Use one or more complete sentences to answer each question.

People try to lose weight by dieting. Researchers now know that dieting may reduce the amount of fat in the body, but it also changes how the body functions. When people diet, they usually limit the amount of food calories they take in. A dieter's body reacts to protect itself. Because of our evolutionary history, the body thinks that there is an actual shortage of food—that starvation is imminent. The body doesn't know that the dieting person is intentionally limiting the amount of food taken in, and it reacts by slowing down to survive the food shortage. The result? Fewer calories are burned during normal activities, and not much weight loss occurs. The fat cells that normally store fat are being emptied, but they still remain in the body. The person feels hungry and may end the diet. The "starved" fat cells quickly refill their reserves and a type of on again, off again dieting may result. The fluctuating weight loss and gain that occurs can be dangerous.

Most researchers now realize that the best way to avoid becoming overweight is to reduce the amount you eat somewhat and to increase physical activity. Exercise increases the amount of energy used by the body. It also increases the amount of muscle tissue, which even when resting burns more calories than other types of body tissues.

There are other serious health risks involved in severe weight loss that is caused by a refusal to eat. The disorder called *anorexia nervosa* is most common in young women. Abnormal fears of being overweight, as well as other fears, may lead to anorexia nervosa. An anorexic person appears unhealthy. This disorder can be fatal.

Bulimia is another eating disorder. Unlike most anorexics, a person with bulimia might appear healthy. However, this person swings between overeating and getting rid of the food, often by taking laxatives or inducing vomiting. Some studies show that as many as 20 percent of college-age women suffer from some form of bulimia. This disorder can be dangerous. It can damage the heart, kidneys, or digestive system. Counseling to help a person understand the reasons behind these eating disorders is important. It is also important to learn how to make wise choices about what one eats. In some severe cases, hospital treatment may be necessary.

36. In a paragraph, describe how evolution explains the unintended effects of dieting.

37. Explain why exercising is a healthier way to lose weight than dieting is.

38. Compare and contrast the eating disorders *anorexia nervosa* and *bulimia.*

Matter on the Move: Gas Exchange and Transport

THE RESPIRATORY SYSTEM, BREATHING, AND CELLULAR RESPIRATION

Animals move oxygen and carbon dioxide into and out of their bodies with a respiratory system. Although cellular respiration refers to the energy-releasing chemical reactions that occur in cells, the word **respiration** also means the process of exchanging gases. In many animals, air is physically pumped into and out of the body by the process of breathing. In multicellular animals, breathing or respiration that involves a respiratory system is necessary to allow the life-sustaining activities of cellular respiration to occur.

GAS EXCHANGE SURFACES

All aerobic organisms, both plants and animals, exchange oxygen and carbon dioxide with their environment. (The oxygen is used to produce ATP during cellular respiration.) Respiration in all organisms involves the diffusion of gases across cell membranes. This occurs for plants in the spongy layer of cells within the plants' leaves. Although gas exchange in animals may involve a complex respiratory system, the actual process of taking in and getting rid of gases is identical to that of the ameba. Gases must cross a barrier to be moved in or out of the animal. This barrier, a part of the animal's body, is known as the *respiratory surface*. Respiratory surfaces in different animal species vary in shape and size. However, they all share certain requirements:

♦ The respiratory surface has to remain moist at all times so that gases can diffuse across the cell membranes.

♦ The respiratory surface must be very thin so that gases are able to pass through it.

♦ There must be a source of oxygen, either in the air or dissolved in the water.

♦ The respiratory surface must be closely connected to the transport system that delivers gases to and from cells.

THE HUMAN RESPIRATORY SYSTEM

The human respiratory system is similar to the respiratory system of other mammals. (See Figure 9-1.) Air moves through the nostrils into the nasal cavity, where dirt and

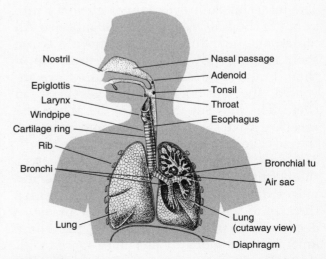

Nostril — Nasal passage
Epiglottis — Adenoid
Larynx — Tonsil
Windpipe — Throat
Cartilage ring — Esophagus
Rib
Bronchi — Bronchial tu
— Air sac
Lung — Lung (cutaway view)
— Diaphragm

Figure 9-1 The human respiratory system (showing cutaway view of a lung).

other particles in the air are trapped by tiny hairs and mucus. The air is also warmed, humidified, and tested for odors. The nasal cavity leads to the *pharynx*, where it meets air and food arriving from the mouth. Air continues flowing down, passing by the vocal cords in the larynx. Farther down the tube is the *trachea*, which is surrounded by rings of stiff cartilage that help maintain its tubelike shape, keeping it open for airflow.

The trachea branches into two *bronchi*. Each bronchus leads to a lung. In the two lungs, the bronchi continue branching into smaller and smaller tubes called *bronchioles*. Most of these tubes are covered on the inside by mucus and by tiny hairlike extensions called *cilia*. The cilia help keep clean the delicate tissues that line the lungs. When cilia are paralyzed, such as by the harmful effects of cigarette smoking, the lungs lose much of their ability to keep themselves clean.

The tiniest bronchioles end in bunches of microscopic air sacs, the *alveoli*. It is the lining of the alveoli that acts as the respiratory surface. Blood vessels surround the alveoli. It is only when oxygen molecules diffuse across this lining—from the alveoli into the blood vessels—that they really enter the body. (See Figure 9-2.)

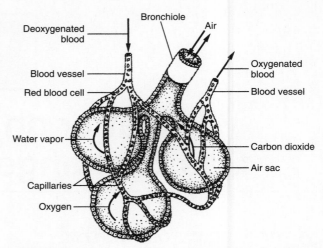

Figure 9-2 The lining of the alveoli acts as our respiratory surface—it is where gas exchange occurs in our bodies.

To move air out, the diaphragm and the muscles between the ribs relax. When they return to their original positions, the volume of the chest cavity decreases. With less space to fill, the air pressure in the lungs increases and air moves out. Exhalation occurs. Breathing is the physical process of *inhalation* (an active process that occurs when the muscles contract) and *exhalation* (a passive process that occurs when the muscles relax). Humans inhale and exhale at a rate of about 12 times a minute, automatically, 24 hours a day.

◼️ BREATHING: MAKING THE AIR MOVE

The lungs on their own cannot move air into or out of the body. Lungs have no muscles. All mammals, including humans, move air into their lungs by lowering the air pressure in their lungs. Two sets of structures are involved: the ribs and the muscles between them. The muscles move the ribs. When the muscles move the ribs upward and outward, the rib cage expands. At the same time, the *diaphragm*, a large flat muscle that lies across the bottom of the chest cavity, contracts and moves down. This movement also increases the size of the chest cavity. The air pressure decreases in the lungs because the same amount of air suddenly has more space to fill. Inhalation occurs when air rushes into the lungs through the respiratory tubes, and the lungs fill with air. (See Figure 9-3.)

◼️ WHEN THINGS GO WRONG: DISEASES OF THE RESPIRATORY SYSTEM

Cigarette smoking causes many deaths from respiratory diseases. Evidence has shown that the chemicals in tobacco smoke can cause can-

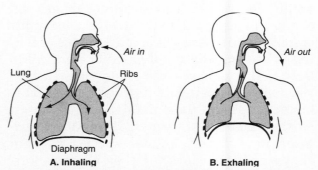

Figure 9-3 The process of breathing is controlled by muscles of the ribs and the diaphragm.

cer in the lungs, esophagus, larynx, and mouth. Other diseases of the respiratory system, which smokers are more likely to get, include:

♦ *Bronchitis.* This is an inflammation and swelling of the inside of the bronchial tubes.

♦ *Asthma.* The walls of the bronchi contract, restricting the flow of air. As a result, the lungs do not fill or empty normally.

♦ *Emphysema.* This is a very serious chronic disease in which the alveoli break down, greatly reducing the total area of the respiratory surface.

♦ *Pneumonia.* This disease results from a bacterial or viral infection that causes the alveoli to fill with fluid.

WANTED: A TRANSPORT SYSTEM

For an aerobic organism to live and grow, matter in the form of food molecules must be delivered to every cell. To release the energy stored in those molecules, oxygen must make that same trip to all cells. Finally, to prevent harmful buildups, carbon dioxide gas and other wastes must be taken away from all cells. An organism needs a transport system to accomplish these tasks. Transport involves two processes: absorption and circulation. Absorption occurs when materials cross cell membranes from the outside environment into the body. Then, during the process of **circulation**, materials are circulated throughout the body to where they are needed.

All organisms transport materials inside of them within a liquid, and the transport fluid must reach every cell in the body. Thus, there have to be vessels that are able to deliver ma-

terials around the body—similar to the water pipes in a house. Finally, there must be something that forces the fluid through the vessels—the transport system must have a pump.

THE HUMAN CIRCULATORY SYSTEM

The heart is the pump that moves blood through the body. In mammals, the heart has four separate chambers. Our circulatory system has the same layout as that of all other mammals and is highly efficient at the job it does.

The circulatory system accomplishes the vital task of transport through the many thousands of kilometers of *blood vessels* in the body. (See Figure 9-4.) Blood from the lungs enters the left atrium of the heart, then passes to the left ventricle. This oxygen-rich blood then begins its journey throughout the body by passing through the aorta (the body's largest artery) to the other arteries.

The *arteries* are vessels that have thick, muscular walls. They are very elastic, expanding and contracting as blood from the heart is pumped through them in pulses—one pulse for every heartbeat. The elasticity of the arteries' walls exerts pressure against the blood inside. This pressure is measured as your *blood pressure*.

The arteries become smaller and smaller as their distance away from the heart increases. Eventually the blood enters *capillaries*, the smallest vessels. Capillaries are close to every body cell. It is at the capillaries that the exchange of nutrients and gases between the blood and the cells takes place. After moving through capillaries, blood returns through the thin-walled *veins*, which get larger and larger

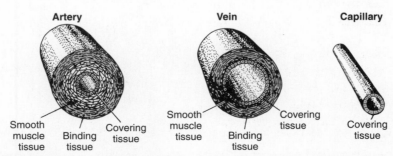

Figure 9-4 Cross sections of the three kinds of blood vessels in humans.

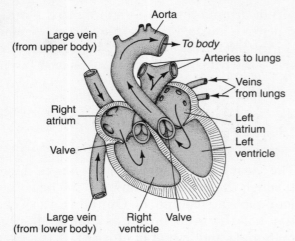

Figure 9-5 All mammals, including humans, have an efficient four-chambered heart to pump blood through the body.

closer to the heart. Unlike the walls of arteries, the walls of veins have little elasticity; blood is under low pressure in them. It would be very easy for blood to flow backward in the veins of the legs. To prevent the blood from moving backward, one-way valves work to trap it. The blood remains stationary in the veins until the beating of the heart and other muscular activities, such as leg muscle contractions, force the blood back up toward the heart.

The oxygen-poor blood from the body enters the right atrium of the heart, passes through the right ventricle, and then enters the arteries that take it to the lungs. After the exchange of gases that occurs in the lungs, oxygen-rich blood once again returns by veins to the left atrium of the heart. (See Figure 9-5.)

▩ BLOOD TISSUE AND BLOOD FLOW

The entire purpose of the transport system is to move materials to and from cells. A capillary is so small that red blood cells must travel through it in single file. Molecules, including water, diffuse through the capillary walls and enter the spaces around body cells. These intercellular spaces are filled with a fluid that surrounds all cells. Molecules diffuse between this *intercellular fluid* (ICF) and the body cells. (See Figure 9-6.)

Blood itself is a tissue. But unlike any other tissue in the body, blood is a liquid. Blood is made up of cells, cell parts, and a clear, light-yellow-colored liquid called *plasma*. The plasma is 90 percent water, plus many important proteins, salts, vitamins, hormones, gases, sugars, and other nutrients. One of the proteins, *fibrinogen*, helps in the clotting process that stops bleeding caused by an injury. (See Figure 9-7.)

The cells in the blood include *red blood cells*, which contain the oxygen-carrying protein hemoglobin, and *white blood cells*. There are five types of white blood cells, all of which are involved in protecting the body from disease-causing foreign substances. Blood also contains *platelets*—fragments of cells that plug "leaks" wherever an injury occurs. Platelets also begin the complex chemical process that results in production of a clot. A blood clot stops the flow of blood out of a damaged blood vessel. (See Figure 9-8.)

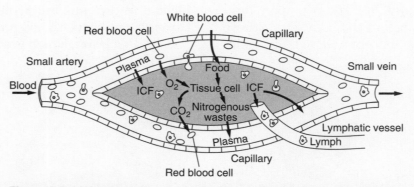

Figure 9-6 Molecules diffuse between the capillaries, ICF, and body cells.

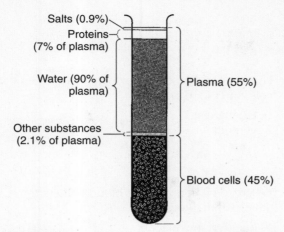

Salts (0.9%)
Proteins (7% of plasma)
Water (90% of plasma)
Plasma (55%)
Other substances (2.1% of plasma)
Blood cells (45%)

Figure 9-7 Blood is a tissue that is made up of plasma and blood cells.

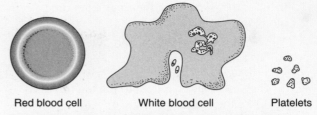

Red blood cell   White blood cell   Platelets

Figure 9-8 Red blood cells carry oxygen, white blood cells protect us from disease, and platelets begin the clotting process.

▚ WHEN THINGS GO WRONG: CARDIOVASCULAR DISEASE

The human transport system is called the *cardio-* (heart) *vascular* (vessels) *system*. Cardiovascular disease includes several important, and potentially fatal, conditions. A *heart attack* occurs when the vessels that bring blood to the heart get blocked and the heart tissue beyond that point is not supplied with blood. The muscle tissue in that area of the heart then dies.

Unlike a heart attack, which often occurs suddenly, some forms of cardiovascular disease develop slowly over a long period of time. Clogged arteries result from the gradual buildup of layers of fatty deposits inside the arteries. Cardiovascular disease also includes strokes that may block an artery in the brain. Depending on what area of the brain is affected, strokes can damage a person's ability to feel things or to speak and move. Scientists think that both diet and heredity play a part in the development of cardiovascular disease.

Chapter 9 Review

Part A—Multiple Choice

1. What happens during respiration in all animals?
   1 Air must be physically pumped into and out of the body.
   2 Gases (O_2 and CO_2) must move into and out of the body.
   3 Blood must move gases throughout the body.
   4 The diaphragm must relax and contract.

2. Which is *not* a requirement of a respiratory surface?
   1 It must be thin.
   2 It must be moist.
   3 It must be thick and solid.
   4 It must be closely connected to the system that delivers gases.

3. How are dirt and other small particles removed from the air humans breathe?
   1 They are trapped by hairs and mucus in the nasal cavity.
   2 They are filtered by the pharynx at the back of the mouth.
   3 They become stuck on the rings of the trachea.
   4 They are caught by the lining of the alveoli.

4. Moving from the nasal cavity to the lungs, which sequence is correct?
   1 larynx, bronchi, bronchioles, trachea, alveoli
   2 trachea, larynx, alveoli, bronchioles, bronchi
   3 larynx, trachea, bronchi, bronchioles, alveoli
   4 pharynx, trachea, alveoli, bronchi, bronchioles

5. What is the purpose of the rings extending along the trachea?

   1 They allow the trachea to rotate sideways.
   2 They maintain its tubelike shape to allow for airflow.
   3 They relax and contract to force air in and out of the lungs.
   4 They allow both food and air to pass through the same tube.

6. Cilia are important to the lungs because the tiny hairs

   1 help clean the tissues that line the lungs
   2 allow air to pass from the trachea into the lungs
   3 act as our respiratory surface
   4 hold the end of the trachea open

7. Oxygen in the air we breathe enters our blood through the lining of the

   1 bronchi
   2 bronchioles
   3 alveoli
   4 cilia

8. You physically inhale when the

   1 muscles of the lungs relax, increasing the size of the lungs
   2 diaphragm relaxes, increasing the size of the chest cavity
   3 diaphragm contracts, increasing the size of the chest cavity
   4 diaphragm contracts, decreasing the size of the chest cavity

9. During inhalation, the air pressure in the chest cavity is

   1 greater than the pressure outside the body
   2 lower than the pressure outside the body
   3 the same as the pressure outside the body
   4 exactly balanced by the pressure inside the lungs

10. The process of circulation serves to

    1 break down gases absorbed in the lungs
    2 move air into and out of the body
    3 absorb gases inhaled through the mouth or nose
    4 transport absorbed materials throughout the body

11. The pump that pushes blood through the body is the

    1 heart

    2 artery
    3 brain
    4 plasma

12. Nutrients and gases are exchanged between the cells and blood across the

    1 capillaries
    2 ventricles
    3 aorta
    4 veins

13. When compared with the blood in the arteries, the blood in the veins

    1 does not move at all
    2 is under lower pressure
    3 contains more gases
    4 contains more water

14. Which shows the path of blood after it leaves the lungs?

    1 aorta → left atrium → left ventricle
    2 left atrium → right atrium → aorta
    3 aorta → right atrium → right ventricle
    4 left atrium → left ventricle → aorta

15. Molecules diffuse from capillaries to body cells by moving through the

    1 veins
    2 arteries
    3 intercellular fluid
    4 plasma

16. The main component of plasma is

    1 oxygen
    2 water
    3 protein
    4 gases

17. Red blood cells are important because they

    1 repair damaged blood vessels
    2 contain hemoglobin
    3 protect the body from disease
    4 produce blood clots

18. In the blood, both fibrinogen and platelets work to

    1 pump blood through the veins
    2 carry blood to the heart
    3 help in the clotting process
    4 transport nutrients into the cells

19. The immediate cause of a heart attack is

    1 blood pressure drops in the veins
    2 layers of fatty deposits build up slowly

3 an artery in the brain becomes blocked by a clot

4 vessels that bring blood to the heart get blocked

20. In an investigation to determine the change in heart rate with increased activity, a biology teacher asked students to take their pulses immediately before and immediately after exercising for 2 minutes. The data showed an average heart rate of 72 beats per minute before exercising and 90 beats per minute after exercising. If a valid conclusion is to be made from the results of this investigation, which assumption must be made?

1 In most students, the average heart rate is not affected by exercise.

2 Each student exercised with the same intensity.

3 Exercise causes the heart rate to slow down.

4 The heart rate of each student goes up 18 beats after jogging for 2 minutes.

Part B—Analysis and Open Ended

21. Briefly explain the difference between "breathing" and "respiration."

22. Describe the four main characteristics of an animal's respiratory surface.

23. Trace the pathway of air through the human respiratory system, from the nostrils to the lungs. At what point within the lungs do the oxygen molecules really enter the body?

24. Refer to the diagram below to explain the process of breathing. Your answer should include the following:

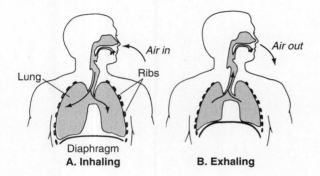

Lung

Ribs

Air in

Air out

Diaphragm
A. Inhaling   **B. Exhaling**

♦ what happens to the rib cage during inhalation and exhalation

♦ what happens to the diaphragm during inhalation and exhalation

♦ how these movements affect the volume of the chest cavity

♦ how these movements affect the air pressure in the lungs

25. Why is inhalation an active process whereas exhalation is a passive process?

26. Transport involves absorption and circulation. Define these two processes.

27. In general, what are three main components of a transport system?

28. Identify and describe the three types of blood vessels in the human body. Your answer should explain the following: (You may want to refer to Figure 9-4 on page 65.)

♦ the differences in their size (width) and structure

♦ the differences in their main functions and direction of flow

♦ if they usually carry oxygen-rich and/or oxygen-poor blood

29. In what way do the lungs depend upon the circulatory system?

30. In Texas, researchers gave a cholesterol-reducing drug to 2335 people and an inactive substitute (placebo) to 2081. Most of the volunteers were men who had normal cholesterol levels and no history of heart disease. After 5 years, 97 people getting the placebo had suffered heart attacks compared to only 57 people who had received the actual drug. The researchers are recommending that to help prevent heart attacks, all people (even those without high cholesterol) take these cholesterol-reducing drugs. In addition to the information above, which piece of information should the researchers have *before* support for the recommendation can be justified?

1 Were the eating habits of the two groups similar?

2 Did the heart attacks result in deaths?

3 How does a heart attack affect cholesterol levels?

4 What chemical is in the placebo?

31. According to the diagram, how many chambers does the human heart have?

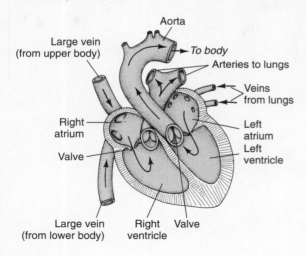

1 one
2 two
3 three
4 four

32. Why is intercellular fluid so important to the functioning of the transport system? (Explain its role in relation to capillaries and body cells.)

Base your answers to questions 33 and 34 on the diagram of blood cells shown below.

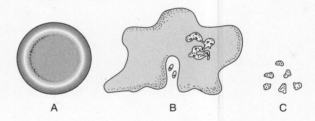

A   B   C

33. Identify the three types of blood cells shown in the diagram. What is the main function of each type of blood cell?

34. How does each type of blood cell help to maintain homeostasis?

35. Imagine that you are a red blood cell. Describe your trip through the human body, going from one foot, through the circulatory system, to the lungs, through the circulatory system, then back to the foot. Identify which parts of the heart you would pass through and how often you would pass through the heart on one round-trip.

Part C—Reading Comprehension

Base your answers to questions 36 to 38 on the information below and on your knowledge of biology. Use one or more complete sentences to answer each question.

While it may seem out of place, certain health-related posters are required by law to be posted in New York City restaurants. People eating in restaurants need to know a certain procedure—called the *Heimlich maneuver*—because a piece of food may cause a person to choke. However, food is not the only cause of choking. It is just as important for people at home to know how to perform the Heimlich maneuver. Chewing gum or, in the case of a child, small toys or a piece of balloon can block the air passage.

The Heimlich maneuver uses the air that is already present in the lungs to push an object up and out of the throat. The Heimlich maneuver is actually quite simple. First, you must determine if a person has an object blocking the throat or is, in fact, experiencing another condition, such as a heart attack. Ask the person, "Are you choking?" If so, he or she will not be able to speak to answer you. For a person who is standing, you administer the Heimlich maneuver by standing behind the choking person and grasping him or her with both arms around the waist. Place one hand, now closed as a fist, just below the bottom of the rib cage and above the navel. Grasp the closed hand with your other hand and make an upward, not inward, thrust. This thrusting motion puts pressure on the victim's diaphragm. The increased pressure forces air up and out of the lungs in order to force the blockage out of the throat.

36. Describe how the Heimlich maneuver works.

37. Make a list that shows each of the steps in the Heimlich maneuver.

38. Why is it necessary in this procedure to make an upward, rather than an inward, thrust?

THEME III
Maintaining a Dynamic Equilibrium

The Need for Homeostasis

:: HOMEOSTASIS

Organisms live in a world of changing conditions. But, to remain alive, every organism needs to keep the conditions inside of itself fairly constant. An organism must have ways to keep its internal conditions from changing as its external environment changes. This ability of all living things to detect **deviations** and to maintain a constant internal environment is known as **homeostasis**.

An obvious change that has occurred in the course of evolution is the development of larger **multicellular** organisms from microscopic, single-celled ones. Is there an advantage to being multicellular? Being microscopic and single-celled makes it difficult for an organism to maintain homeostasis. Having a multicellular body makes possible many types of protection against changes in the environment. In other words, an organism with many cells is able to have structures and systems that protect its individual cells from external changes, thus helping it to stay alive. (See Figure 10-1.)

To maintain homeostasis, organisms actually must make constant changes. That is why

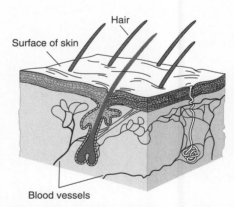

Figure 10-1 Multicellular organisms have systems and structures that help them maintain homeostasis. For example, our skin has features that detect and respond to changes in external temperature.

homeostasis is often referred to as maintaining a **dynamic equilibrium**. *Dynamic* means "active," and *equilibrium* means "balanced." Homeostasis requires active balancing.

:: THE CELL AND ITS ENVIRONMENT

One of the most fascinating facts about our bodies is that each of our many, many cells is

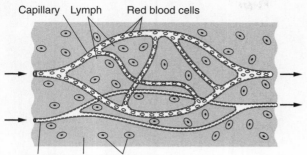

Figure 10-2 All the cells in our body are surrounded by intercellular fluid (ICF). Materials are exchanged between the cells and the fluid, which helps to maintain stable conditions inside each of the cells.

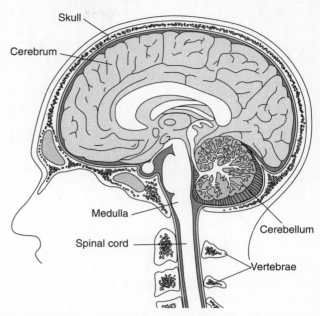

Figure 10-3 A structure in the brain (the medulla) monitors the amount of CO_2 in the body, adjusting the breathing rate to maintain proper levels.

surrounded by liquid. The smallest blood vessels in our bodies, the capillaries, are close to every cell. There is a small amount of space between the capillaries and the body cells. This space is filled with fluid. The fluid that surrounds cells is made up mostly of water, with many substances dissolved in it. This intercellular fluid is important in helping to maintain stable conditions inside each of our cells. Many materials are exchanged between the cells and the fluid. In turn, materials may be exchanged between the fluid and the blood in the capillaries. All of this is done to make sure that each and every body cell is able to maintain homeostasis and remain healthy. (See Figure 10-2.)

❖ MAINTAINING HOMEOSTASIS WHEN WE EXERCISE

Exercise involves increased muscle activity. This activity creates changes within the body. To maintain homeostasis, the body needs to be able to respond to these changes.

An example of a change that occurs when we exercise is the increase in the body of carbon dioxide (CO_2), produced by muscle cells as a result of cellular respiration. The level of CO_2 increases in both the intercellular fluid and the blood. To maintain homeostasis, the body first must be able to detect this change and then respond to the change.

A structure in the brain detects the increased CO_2 level in the blood passing through the brain and in the fluid around the brain cells. As a result, this part of the brain

sends signals to the chest to increase the rate of breathing and the amount of air taken in on each breath. These changes in breathing increase the exchange of gases in the lungs, lowering the CO_2 levels in the body. These lower levels are then detected in the brain, which in turn sends a signal to reduce the breathing rate. This process is an example of a **feedback mechanism**. Feedback mechanisms are important in maintaining homeostasis. (See Figure 10-3.)

❖ FEEDBACK MECHANISMS

Carbon dioxide levels in your body are regulated somewhat as a thermostat regulates the temperature of your house. A thermostat measures the temperature of the air in a room. When the air temperature in the house falls below a preset figure, the thermostat turns a furnace on. The furnace produces heat, and the temperature of the air in the house increases. When the temperature of the air rises above the preset temperature, the thermostat tells the furnace to shut down. The temperature in the house stops rising, the air begins to cool, and the thermostat continues the cycle of telling the furnace to produce heat or to shut down. (See Figure 10-4.)

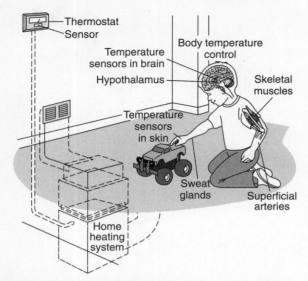

Figure 10-4 Both CO_2 levels and body temperature are regulated by feedback mechanisms, much as a thermostat controls the temperature in a room.

In this type of feedback mechanism, a change occurs that produces another change, which in turn reverses the first change. This is an important process in maintaining homeostasis.

The following are parts of a feedback mechanism used in maintaining homeostasis:

♦ *Sensor.* Something must be able to detect a change. A thermometer attached to a thermostat is a sensor. In the body, structures in the brain detect changes in CO_2 levels.

♦ *Control unit.* Something must know what the correct level should be. A thermostat in a house is set to a particular comfort level. Information in the brain is preset at the correct CO_2 level.

♦ *Effector.* Something must take instructions from the control unit and make the necessary changes. In a house, the effector would be a furnace or an air conditioner. In the body, the effector for CO_2 levels would be the muscles in the chest that are used for breathing.

❖ MAINTAINING HOMEOSTASIS: WATER BALANCE IN PLANTS

Maintaining water balance is a major concern for all living things. Plants as well as animals

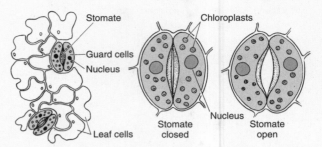

Figure 10-5 Special openings in the surface of a leaf function to maintain water balance in plants.

must maintain water balance. Openings in the surface of a leaf are adapted to control the loss of water. Each opening is surrounded by two *guard cells*. These guard cells, like any cells, allow water to diffuse through their cell membrane. When water is abundant, it moves into the guard cells. The increased quantity of water increases the pressure within the cells. Guard cells are somewhat curved in shape; when they are filled with water, they become even more curved. The space between them expands, the opening widens, and excess water is allowed to evaporate out of the air spaces inside the leaf to the air that surrounds the plant. (See Figure 10-5.)

When water becomes scarce, the guard cells become less curved in shape and the opening closes. Water loss is reduced and the plant is able to maintain its water balance.

❖ SYSTEMS FOR MAINTAINING HOMEOSTASIS

Multicellular animals have evolved highly organized, complex organ systems especially suited to maintaining a relatively constant internal environment. These organ systems include the excretory system, which regulates the chemistry of the body's fluids while removing harmful wastes; the nervous system, which uses electrochemical impulses to regulate body functions; the endocrine system, which produces hormones—chemical messengers essential in regulating the functions and behavior of the body; and, finally, the immune system, which uses a set of defenses to protect the body from dangerous substances and microorganisms that could upset the internal balance on which life itself depends.

Chapter 10 Review

Part A—Multiple Choice

1. Organisms undergo constant chemical changes as they maintain an internal balance known as

   1 interdependence
   2 synthesis
   3 homeostasis
   4 recombination

2. What characteristic has evolved that helps to maintain homeostasis?

   1 taller bodies with larger cells
   2 shorter bodies with fewer cells
   3 multicellular bodies with many cells
   4 multicellular bodies with fewer cells

3. A system in dynamic equilibrium

   1 makes constant changes
   2 changes in intervals or steps
   3 changes very infrequently
   4 never changes at all

4. Intercellular fluid is made up mostly of

   1 water
   2 blood
   3 mineral salts
   4 cytoplasm

5. Intercellular fluid is important for the exchange of materials between

   1 body cells and arteries
   2 body cells and veins
   3 veins and capillaries
   4 body cells and capillaries

6. As a result of exercise, CO_2 levels increase in the

   1 blood only
   2 intercellular fluid only
   3 blood and intercellular fluid
   4 muscles only

7. The brain sends a signal to increase the breathing rate when the CO_2 level has

   1 not changed for a while
   2 decreased
   3 increased
   4 increased, then decreased

8. The increased breathing rate signaled by the brain serves

   1 to increase the CO_2 level in the body
   2 to decrease the CO_2 level in the body
   3 to decrease the O_2 level in the body
   4 no function in changing O_2 and CO_2 levels

9. In adjusting the CO_2 level, the part of the body that acts like a thermostat in the home is the

   1 brain
   2 chest
   3 lungs
   4 muscle tissue

10. If an organism fails to maintain homeostasis, the result may be

    1 disease only
    2 death only
    3 disease or death
    4 none of the above

11. A change in the body results in another change. This second change reverses the first change in order to maintain homeostasis. This describes a type of

    1 control mechanism
    2 feedback controller
    3 feedback mechanism
    4 effector mechanism

12. The effector for adjusting the CO_2 level in the body would be the

    1 blood tissue
    2 brain
    3 lungs
    4 chest muscles

13. Why might a blood clot be important to maintaining homeostasis?

    1 It slows the flow of blood through the body.
    2 It prevents the loss of blood from the body.
    3 It increases the amount of water in the blood.
    4 It adds more cells to the blood tissue.

14. The changing shape of a plant's guard cells helps to

    1 allow the plant to grow stronger
    2 prevent the plant from losing food
    3 regulate the temperature of the plant
    4 maintain the plant's water balance

Base your answer to question 15 on the table below, which shows the rate of water loss in three different plants.

Plant	Liters of Water Lost Per Day
Cactus	0.02
Potato plant	1.00
Apple tree	19.00

15. One reason each plant loses a different amount of water from the other plants is that each has

1 different guard cells that are adapted to maintain homeostasis
2 the same number of chloroplasts but different rates of photosynthesis
3 different types of insulin-secreting cells that regulate water levels
4 the same rate of photosynthesis but different numbers of chloroplasts

16. The nervous system helps to maintain homeostasis by

1 using electrochemical impulses to regulate functions
2 regulating the chemistry of the body's fluids
3 releasing hormones directly into the bloodstream
4 protecting the body from harmful bacteria

17. Which homeostatic adjustment does the human body make in response to an increase in environmental temperatures?

1 a decrease in glucose levels
2 an increase in perspiration
3 a decrease in fat storage
4 an increase in urine production

18. Which situation is *not* an example of the maintenance of a dynamic equilibrium in an organism?

1 Guard cells contribute to the regulation of water content in a geranium plant.
2 The release of insulin lowers the blood sugar level in a human after eating a big meal.
3 Water passes into an animal cell, causing it to swell.
4 A runner perspires while running a race on a hot summer day.

Part B—Analysis and Open Ended

Base your answer to question 19 on the photograph below, which shows a microscopic view of the underside (lower surface) of a leaf.

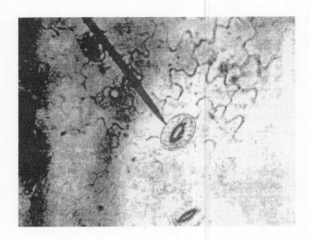

19. What is the main function of the cells indicated by the black pointer?

1 to regulate the rate of gas exchange
2 to store food for winter dormancy
3 to undergo mitotic cell division
4 to give support to the leaf's veins

20. How does being multicellular increase an organism's ability to maintain homeostasis and survive?

21. Write a brief essay comparing the life of a cell in your body with that of an ameba in the soil. Why is it more likely that the body cell will survive for a long time, but the ameba will not?

Refer to the diagram below to answer questions 22 and 23.

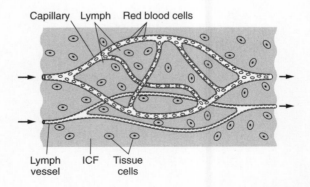

Capillary   Lymph   Red blood cells

Lymph vessel   ICF   Tissue cells

22. Which analogy most accurately describes the location of the body's tissue cells?

1 cities within states
2 islands within oceans
3 chains of mountains
4 clouds in the air

23. Use your knowledge of biology and the diagram to explain the purpose of intercellular fluid (ICF). Why is it so important for homeostasis?

Base your answer to question 24 on the information and diagrams below.

To survive, an organism must maintain the health of its cells. The normal internal environment of a human's cells would include a temperature of 37°C, a pH of 7, and a water/salt balance of 0.1 percent.

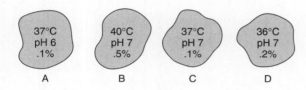

24. Which of the cells shown above would belong to someone who is *not* maintaining homeostasis?

25. List, and describe the roles of, the three components of a homeostatic process.

26. Use the diagram below to explain how feedback mechanisms maintain homeostasis.

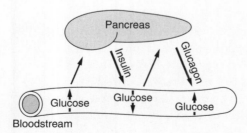

27. Use the following terms to replace the definitions given within the boxes in the following chart: *Higher CO$_2$ levels; Lower CO$_2$ levels; Muscles in the chest; Structures in the brain (with preset information).*

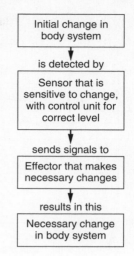

28. The best title for this concept map probably would be:

1 The Respiratory System and CO$_2$ Levels
2 The Circulatory System and CO$_2$ Levels
3 Feedback Mechanisms and CO$_2$ Levels
4 The Bloodstream and Its CO$_2$ Levels

29. Briefly explain the way our bodies adjust breathing rates in order to maintain homeostasis.

Base your answer to question 30 on the data in the graph below.

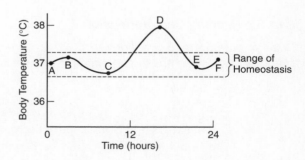

30. The graph shows evidence of disease in the human body. A disruption in the dynamic equilibrium is indicated by the temperature change that occurs between points

1 A and B
2 B and C
3 C and D
4 E and F

Study the following graph to answer questions 31 and 32.

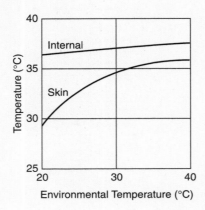

Environmental Temperature (°C)

31. The graph shows the effect of external (environmental) temperatures on a student's skin and internal temperatures. Which statement best describes what happens as the environmental temperature increases?

 1 The skin temperature increases, then decreases to 20°C.
 2 The internal temperature increases abruptly to about 30°C.
 3 The skin temperature decreases, due to sweating, to 30°C.
 4 The skin temperature increases, then levels off at about 36°C.

32. What is the difference between the effects of rising external temperatures on the student's internal temperature and skin temperature? Explain how homeostatic processes are responsible for the effects seen in the graph.

33. In desert environments, organisms that cannot maintain a constant internal body temperature, such as snakes and lizards, rarely go out during the hottest daylight hours. Instead, they stay in the shade, under rocks, or in burrows. Explain how this behavior helps these organisms to maintain homeostasis.

34. Describe how plants maintain their water balance. Your answer should include the following:

 ◆ *one* reason why water balance is important to plants
 ◆ the structure that plants have to perform this function
 ◆ how this structure works to maintain water balance

35. In what way are the functions of the contractile vacuoles of an ameba and the guard cells of a plant similar?

36. Identify the four main organ systems that are involved in maintaining homeostasis. Briefly describe each of their roles in this process.

Part C – Reading Comprehension

Base your answers to questions 37 to 40 on the information below and on your knowledge of biology. Use one or more complete sentences to answer each question.

In 2002, flight engineers Carl Walz and Dan Bursch set the record for the longest United States space flight, with 196 days in space as members of Expedition 4 on the International Space Station (ISS). Typically, ISS crews have six or seven members who live on the station for 3 to 6 months. The crews live in a world of weightlessness—the station has no up or down, so there are no real ceilings or floors. While the total inside space of the station is about equal to that of a jumbo jet, the individual spaces in which the astronauts actually live and work are relatively small, each about the size of a school bus's interior. Crews sleep standing up or camping out where they feel comfortable by attaching their sleep restraints to the wall with Velcro.

Biomedical researchers are interested in studying the effects of weightlessness on humans. Being "weightless" is a brand-new challenge never experienced before in the millions of years humans have lived on Earth. And yet, time and again, space travel has demonstrated the marvelous, and often subtle, abilities of the human body to adapt. The body's reactions to weightlessness are teaching us a great deal about its normal responses to gravity. Astronauts report that when they grab the wall of a spacecraft and move their bodies back and forth, they feel as if they are staying in one place and that the spacecraft is moving. Being free of gravity's effects makes us aware of new things. Humans have evolved many automatic reactions to deal with the constant pressure of living in a downward-pulling world. Until we leave that world, we are usually not aware of such reactions.

These reactions include the use of signals from our eyes, from the fluid-filled tubes in our ears, from pressure receptors on the bottom of our feet, and from the distribution of liquids in our blood vessels. A sophisticated control system has evolved to keep gravity from pulling all the liquid in our body to our legs. Within minutes of being in a weightless environment, the veins in an astronaut's neck begin to bulge. The astronaut's face begins to fill out and become puffy. In this situation, the fluids in an astronaut's body are not being pulled down by gravity. The fluids spread throughout the body. Because the body seeks to maintain homeostasis, this new distribution of fluid causes other changes in the body in order to control fluid movement. Included in these are changes in hormone levels, kidney function, and red blood cell production. Keeping things stable even when conditions change—that is, dynamic equilibrium—is as necessary for life in space as it is on Earth. The unexpected result of "living" in space is a better understanding of how the body works here on Earth.

37. Describe three ways in which life on the ISS is very different from everyday life on Earth.

38. Why are the effects of weightlessness on humans of interest to researchers?

39. How do the body's responses to weightlessness help explain homeostasis?

40. Describe some adaptations of the body related to living in a world with gravity.

Integration and Control: Nervous and Hormonal Regulation

COMMUNICATION BETWEEN CELLS

All living things interact with their environment in many ways. Conditions outside and inside an organism are constantly being checked and, when needed, adjustments are made to maintain homeostasis. Whatever the interaction is—finding food, maintaining body temperature, or protecting oneself from disease—communication is required. Information must be received from the environment, processed, and responded to. Organisms, particularly complex, multicellular ones, must organize the information they receive and respond to it. This makes it necessary for all parts of an organism to work in a coordinated fashion. Therefore, to maintain homeostasis, an organism must have a means for *integration*—making all of its body parts work together—and a means for *control*—acting in an organized and appropriate fashion.

Every function of an organism must involve cells. This includes the communication of an organism with its environment and the communication within an organism among all of its parts. Most important, the only way that cells communicate is chemically. Communication for a cell means having chemicals moving into and out of it. The work of the two organ systems responsible for integration and control—the nervous system and the endocrine system—is based on the chemical communication between cells.

THE NERVE CELL: A CELL FOR RAPID COMMUNICATION

How does a message travel through the nervous system? The cell theory tells us that the messages must travel along pathways composed of cells. The specialized cells that make up these pathways are the **nerve cells**. The message itself is a *nerve impulse*. Every nerve cell does three things: it receives, conducts, and sends impulses. The most important part of a nerve cell involved in conducting an impulse is the cell membrane. Through the rapid movement of positive ions across the cell membrane, an electrical voltage is created. Electrical voltage is a form of energy. In a nerve cell, the voltage changes that occur at one place on the membrane trigger the same kind of changes at the next spot on the membrane. This movement of voltage changes along the length of the nerve cell, or *neuron,* is the nerve impulse. (See Figure 11-1.)

CROSSING THE GAP BETWEEN NERVE CELLS

If you accidentally touch a hot pan on a stove, you immediately pull your hand away. The nerve pathway from your finger to your spinal cord and brain and then back to your finger consists of many nerve cells. Close examination shows that nerve cells do not touch each other. They are separated by a gap. How does the impulse get from one nerve cell to another? Extremely important chemicals are released by the ends of one nerve cell. The

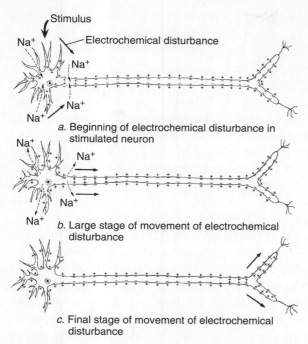

a. Beginning of electrochemical disturbance in stimulated neuron

b. Large stage of movement of electrochemical disturbance

c. Final stage of movement of electrochemical disturbance

Figure 11-1 A nerve impulse is the movement of cell membrane voltage changes along the length of a nerve cell (neuron).

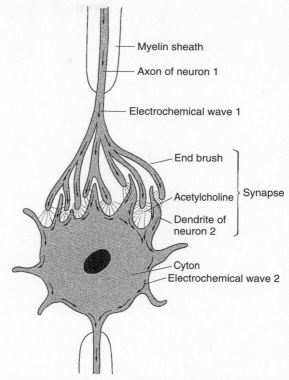

Figure 11-2 Nerve impulses travel from one nerve cell (neuron) to the next by way of chemicals that diffuse across the gap (at the synapse).

chemicals are released as the impulse arrives at the nerve cell endings. These chemicals diffuse across the gap to the next nerve cell. Once received by the next nerve cell, the chemicals make a new nerve impulse possible. In this way, the message continues along the entire nerve pathway, moving from one nerve cell to another. (See Figure 11-2.)

✖ THE HUMAN NERVOUS SYSTEM

Any event, change, or condition in the environment that causes an organism to react is called a **stimulus** (plural, *stimuli*). The resulting reaction of the organism is a *response*. A hot pan on the stove is a stimulus; pulling your hand away is a response. The nervous system of an organism makes possible both detection of a stimulus and response to a stimulus. In vertebrates, such as humans, the nervous system is a complex organization of cells and organs and includes both a central and a peripheral nervous system.

The *central nervous system* consists of the brain and spinal cord. The spinal cord runs from the base of the brain to the lower portion of the back. In most vertebrates, the spinal cord is surrounded by hollow, bony vertebrae that make up the backbone. The *peripheral nervous system* consists of nerve cells that travel out of the central nervous system to all parts of the body. Signals are carried by these nerve cells to muscles or glands. Other nerve cells carry signals to the central nervous system from *sensory receptors*, such as those in the ears and in the eyes. (See Figure 11-3 on page 82.)

✖ DISEASES THAT AFFECT THE NERVOUS SYSTEM

Cerebral palsy results from brain damage and is the collective name for a group of disorders that affect a person's ability to control body movements.

Multiple sclerosis occurs when cells in the brain and spinal cord do not function normally. A wide variety of symptoms, including shaking of the hands, blurred vision, and

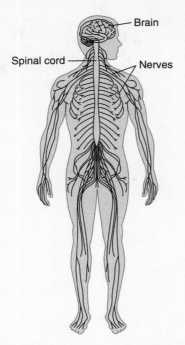

Figure 11-3 The human nervous system: the central nervous system consists of the brain and spinal cord; the peripheral nervous system consists of the nerve cells that connect the spinal cord to all parts of the body.

slurred speech, occur in people with multiple sclerosis.

Alzheimer's disease is a progressive degenerative disease. Eventually, memory loss and the inability to think, speak, or care for oneself occur.

Parkinson's disease involves a particular group of nerve cells in the brain. Loss of function in these nerve cells produces the typical shaking motion, poor balance, lack of **coordination**, and stiffening of the muscles that occur with this disease.

■ THE ENDOCRINE SYSTEM: ANOTHER COMMUNICATION NETWORK

A system of glands and hormones makes up the endocrine system. The key function of this important system is maintaining homeostasis. The *endocrine glands* produce **hormones**, chemical messengers that are released into the blood and carried throughout the body by the circulatory system. At some place in the body—often far from the gland that made it—a hormone arrives at its special *target*

cells and puts into effect whatever changes it has been designed to produce.

How do hormones do their work? Some hormones bind to specific **receptor molecules** (proteins) found in the cell membrane. The binding of a hormone with a receptor protein then causes a change inside the cell, usually involving the cell's enzymes. Other hormones pass right through the cell membrane and bind to receptor proteins in the cytoplasm. The hormone-receptor complex may then move to the nucleus and interact with the cell's DNA, affecting gene activity.

Both the nervous system and the endocrine system are communication networks. However, there are important differences in the two systems. Impulses sent by the nervous system usually produce rapid responses, frequently produced by the actions of muscles. Hormones generally produce slower, more long-lasting changes, which often involve metabolic activity within the target cells.

An important characteristic of hormones is that only small amounts are usually needed to produce the required effect. The group of target cells for a hormone is usually very sensitive to its particular hormone. Feedback mechanisms work to control the amounts of many hormones that are released. In this way, the release of the hormone has the effect of stopping any further release of it until the hormone is needed once again.

■ THE HUMAN ENDOCRINE SYSTEM

There is a close link between the human nervous and endocrine systems. The *hypothalamus* is a part of the brain. The hypothalamus receives information about conditions in the body as blood passes through it. It also receives information from nerve impulses that are carried to it by nerve cells. In turn, the hypothalamus uses the information it receives to control hormones that are released from its neighbor in the brain, the pituitary gland. (See Figure 11-4.)

The *pituitary gland*, only about the size of a pea, is sometimes called the "master gland" because it controls the activities of so many other glands of the endocrine system. When the hypothalamus detects a need for one of the

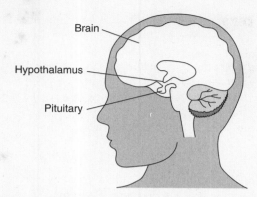

Figure 11-4 The actions of the hypothalamus connect the nervous and endocrine systems—it receives information from the body and uses it to control hormones that are released from the pituitary gland.

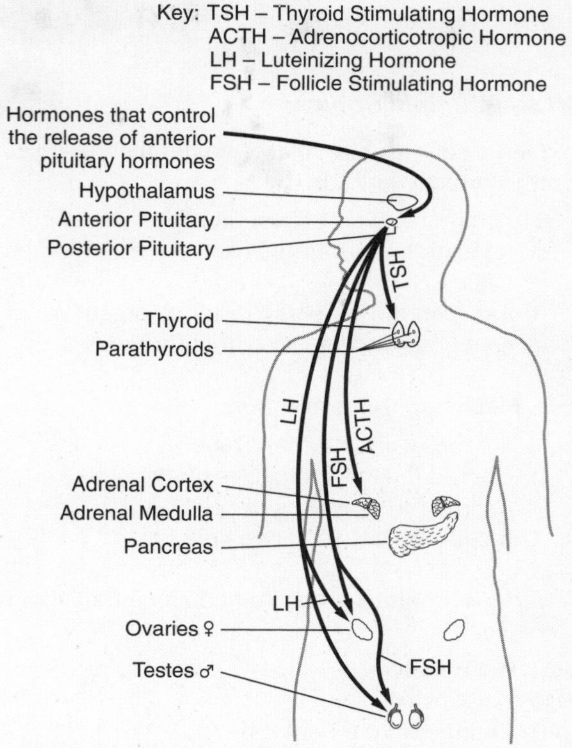

pituitary hormones in the body, it sends a tiny amount of a releasing factor to the pituitary to secrete the correct hormone into the blood.

The *adrenal gland* is attached to the top of each kidney. The most important hormone from this gland, cortisol, is released only after another hormone, ACTH, from the pituitary triggers it to do so. This process of the pituitary hormone triggering another gland to release its hormone occurs throughout the body.

Other major glands of the endocrine system are the thyroid gland in the neck, the four small parathyroids connected to the thyroid, the pancreas, and the ovaries and testes. (See Figure 11-5.)

WHEN THINGS GO WRONG: DIABETES—A DISEASE OF THE ENDOCRINE SYSTEM

Diabetes is a disease that occurs for a variety of reasons. In all cases, something goes wrong

Key: TSH – Thyroid Stimulating Hormone
ACTH – Adrenocorticotropic Hormone
LH – Luteinizing Hormone
FSH – Follicle Stimulating Hormone

Hormones that control the release of anterior pituitary hormones
Hypothalamus
Anterior Pituitary
Posterior Pituitary
TSH
Thyroid
Parathyroids
LH
FSH
ACTH
Adrenal Cortex
Adrenal Medulla
Pancreas
LH
Ovaries ♀
Testes ♂
FSH

Figure 11-5 The human endocrine system—the pituitary gland controls the release of hormones by many other glands in the body.

with the metabolism of carbohydrates. Carbohydrate metabolism involves the hormone **insulin**, which is released from the **pancreas**. Thus, diabetes is a disease of the endocrine system. Normally, when an increase in blood sugar level is detected, insulin is released from the pancreas. This results in a drop in the blood sugar level, and then the pancreas stops releasing more insulin. This is, therefore, a feedback mechanism. Interruption of this process for any number of reasons may result in diabetes.

Chapter 11 Review

Part A—Multiple Choice

1. Nerve cells are essential to an animal because they directly provide

   1 communication between cells
   2 regulation of reproductive rates within other cells
   3 transport of nutrients to various organs
   4 an exchange of gases within the body

2. A nerve impulse results from

   1 the removal of fluid from between cells
   2 electrical voltage changes on the cells
   3 the motion of groups of cells
   4 a collision between two cells

3. A nerve impulse is transmitted along the length of a cell's

   1 membrane
   2 nucleus
   3 endoplasmic reticulum
   4 Golgi complex

4. How do nerve impulses cross the gap between nerve cells?

   1 The impulse pulls the two nerve cells together to close the gap.
   2 The impulse forms a bridge across the gap.
   3 One nerve cell releases chemicals that diffuse across the gap.
   4 One nerve cell releases chemicals that build cells to fill the gap.

5. When you pull your finger away after touching a sharp pin, the stimulus is the

   1 sharp pin itself
   2 message your finger sends to your brain
   3 motion of your finger away from the pin
   4 motion of your finger toward the pin

6. The central nervous system consists of the

   1 stomach and chest cavity
   2 brain and spinal cord
   3 eyes, ears, and mouth
   4 legs and arms

7. The endocrine system maintains homeostasis by

   1 maintaining physical coordination
   2 controlling the size of blood vessels
   3 sending nerve impulses to the brain
   4 releasing hormones into the blood

8. Hormones begin their work by

   1 replacing the nucleus of a cell
   2 breaking down the membrane of a cell
   3 binding to receptor proteins in a cell's membrane or cytoplasm
   4 changing the shape of a cell and then becoming part of the cell

9. The endocrine system differs from the nervous system in that the endocrine system

   1 produces faster, short-term changes
   2 produces slower, long-term changes
   3 operates in isolated regions of the body
   4 produces changes in the brain only

10. Hormones and secretions of the nervous system are chemical messengers that

    1 store genetic information
    2 extract energy from nutrients
    3 carry out the circulation of materials
    4 coordinate system interactions

11. Which of the following statements is true?

    1 Hormones are carried by the respiratory system.
    2 Hormones are produced by the nervous system.
    3 Hormones are generally needed in small amounts.
    4 Hormones are generally needed in large amounts.

12. The body does not produce too much of most hormones because

    1 the excretory system controls the amount released
    2 feedback mechanisms control the amount released
    3 cells can store any excess hormones they receive
    4 the body can produce only fixed amounts of each hormone

13. The hypothalamus controls the release of hormones directly from the

    1 pituitary gland
    2 adrenal gland
    3 thyroid gland
    4 ovaries

14. Which statement describes a feedback mechanism involving the human pancreas?

1 The production of estrogen stimulates the formation of gametes for sexual reproduction.
2 The level of sugar in the blood is affected by the amount of insulin in the blood.
3 The level of oxygen in the blood is related to heart rate.
4 The production of urine allows for excretion of cell waste.

15. The pancreas produces one hormone that lowers blood sugar level and another that increases blood sugar level. The interaction of these two hormones most directly helps humans to

1 maintain a balanced internal environment
2 dispose of wastes formed in other body organs
3 digest needed substances for other body organs
4 increase the rate of cellular communication

16. What process is represented by the boxed sequence below?

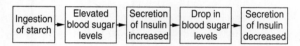

| Ingestion of starch | Elevated blood sugar levels | Secretion of Insulin increased | Drop in blood sugar levels | Secretion of Insulin decreased |

1 a feedback mechanism in multicellular organisms
2 the differentiation of organic molecules
3 an immune response by cells of the pancreas
4 the disruption of cellular communication

Part B—Analysis and Open Ended

17. Explain the importance of both "integration" and "control" for maintaining homeostasis in an organism. What two systems accomplish this in humans?

18. Briefly describe the means by which cells communicate with each other.

19. What are the three main functions of every nerve cell?

20. Describe how a nerve impulse travels along the length of a nerve cell. Your answer should explain the following:

♦ what the nerve impulse actually is

♦ what the role of the cell membrane is

♦ what happens to the impulse at the end of the cell

21. How does a nerve impulse travel from one nerve cell to the next?

Base your answers to questions 22 through 24 on the diagram below, which illustrates one type of cellular communication, and on your knowledge of biology.

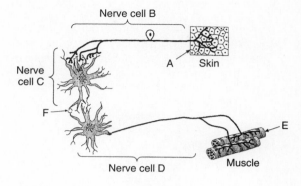

22. In region F, there is a space between nerve cells C and D. Nerve cell D is usually stimulated to respond by

1 a chemical produced by cell C moving to cell D
2 the movement of a virus from cell C to cell D
3 the flow of blood out of cell C to cell D
4 the movement of material through a blood vessel that forms between cell C and cell D

23. If a stimulus is received by the skin cells in area A, the muscle cells in area E will most likely use energy obtained from a reaction between

1 fats and enzymes
2 glucose and oxygen
3 ATP and pathogens
4 water and carbon dioxide

24. State *one* possible cause for a failure of muscle E to respond to a stimulus at area A (skin).

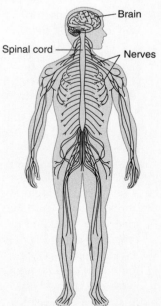

25. The vertebrate nervous system consists of the central nervous system and peripheral nervous system. The main components of the human nervous system are shown in the diagram at right. To which system, or systems, do the labeled parts belong?

26. Compare and contrast the central nervous system and the peripheral nervous system. Where do they carry their signals (to and from) in the body?

27. An important method of communication between cells in an organism is shown in the diagram below. What type of chemical is referred to in the diagram?

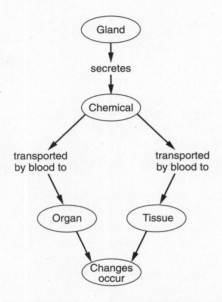

1 a hormone important in maintaining homeostasis
2 DNA necessary for regulating cell functions
3 an enzyme detected by a cell membrane receptor
4 a food molecule taken in by an organism

28. Explain what hormones are and tell the following facts about them:

♦ what body system releases them into the blood

♦ what body system carries them throughout the body

♦ what body system controls the release of hormones

29. Use the following terms to fill in the boxes of the following chart, which compares the endocrine system and nervous system: *electrical impulse; chemical messenger; nerve cells; bloodstream; to muscles and glands; from glands to target cells; rapid (muscle) responses; slow (metabolic) changes.*

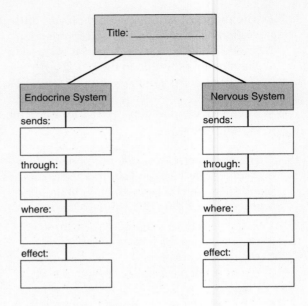

30. The best title for this concept map probably would be
1 Why the Endocrine System Is Better Than the Nervous System
2 Why the Nervous System Is Better Than the Endocrine System
3 Contrasting Features of the Body's Communication Networks
4 The Role of Chemicals in the Body's Feedback Mechanisms

31. Which graph of blood sugar level over a 12-hour period best illustrates the concept of dynamic equilibrium in the body?

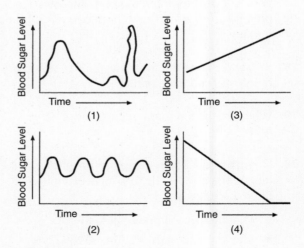

1 graph 1
2 graph 2
3 graph 3
4 graph 4

32. Why are only small amounts of a hormone usually needed to produce an effect? By what means does the body stop the release of too much of a hormone?

Base your answers to questions 33 and 34 on the diagram below, which illustrates a function of hormones.

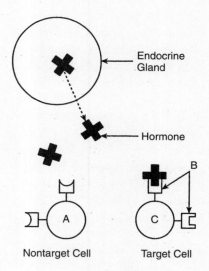

Endocrine Gland

Hormone

B

A

Nontarget Cell

C

Target Cell

33. The structures identified by letter B, which are attached to cell C, represent

1 ribosomes
2 receptor molecules
3 specialized tissues
4 inorganic substances

34. How can you tell that cell A is *not* a target cell for the hormone shown in the diagram?

35. Why is the pituitary called the "master gland" of the endocrine system?

36. The following diagram represents a function of the thyroid gland. State *one* effect of an increasing level of TSH-releasing factor.

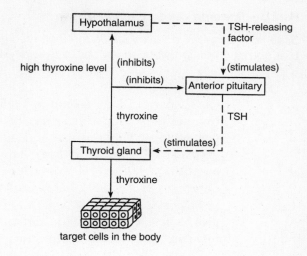

Hypothalamus

TSH-releasing factor

high thyroxine level   (inhibits)   (stimulates)

(inhibits)   Anterior pituitary

thyroxine   TSH

(stimulates)

Thyroid gland

thyroxine

target cells in the body

37. Parts of the brain and the endocrine system interact to release hormones. Use the following terms to fill in the flowchart below so that it shows which two body parts control the release of cortisol by another body part: *adrenal gland; pituitary gland; hypothalamus.*

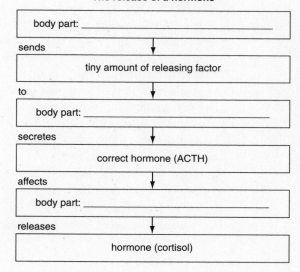

The release of a hormone

body part: _____

sends

tiny amount of releasing factor

to

body part: _____

secretes

correct hormone (ACTH)

affects

body part: _____

releases

hormone (cortisol)

Part C—Reading Comprehension

Base your answers to questions 38 to 40 on the information below and on your knowledge of biology. Use one or more complete sentences to answer each question.

On Memorial Day weekend in 1995, the actor Christopher Reeve was thrown from his horse Buck during a riding competition. The actor who had played the part of Superman in movies during the 1980s had suddenly become paralyzed with a spinal-cord injury. Years later, Reeve regained some very limited movement through intensive exercise, but his paralysis largely continued. Unfortunately, in October 2004, Reeve died of an infection due to complications from his paralysis. However, just as he had fought hard for his own recovery, Reeve had also become an important leader in the fight to support research that could lead to a cure for others, if not for him.

This research had taken a large step forward in 1998 when scientists at the University of Wisconsin isolated the first human embryonic stem cells. Stem cells are very unique. Found in tiny quantities, they are the cells in an organism that have not yet developed to do their specific jobs. In fact, stem cells have the ability to become almost any kind of tissue in the body. Researchers hope that they can use stem cells to produce specific tissues such as heart, lung, kidney, or nerve tissue. There is some evidence now that tissues grown from stem cells may offer cures to millions of people who suffer from conditions such as diabetes, Alzheimer's disease, Parkinson's disease, and spinal-cord injuries.

However, there is much controversy about stem-cell research. Since the 1998 discovery, scientists have had the means to collect stem cells that are very easy to grow from human embryonic tissue. But is it right to use cells from human embryos? In 2001, a committee of scientists was formed by the National Academy of Sciences and the National Research Council to study this problem. It concluded that public policy should keep as many methods of research open as possible, including the use of adult and embryonic human stem cells, to speed the way toward finding cures. In the same year, because of the ethical questions, the United States government placed strict limits on stem-cell research. It was because of this controversy that, by the year 2002, Christopher Reeve began to be very vocal in support of stem-cell research. Though not in time to help Reeve, such research may someday provide a cure for the paralysis of others with spinal-cord injuries. Regardless of one's opinion on stem-cell research, you can be sure that much more will be heard about it in the years ahead, from scientists and politicians, here and abroad.

38. Explain the unique nature of embryonic stem cells.

39. Why are stem cells of great interest to medical researchers?

40. Why is there a great deal of controversy about stem-cell research?

Animal Behavior

▓ THE PURPOSE OF ANIMAL BEHAVIOR

Everything an organism does is its **behavior**. This includes finding a place to live, avoiding predators, finding food, mating and reproducing, and even dying. (See Figure 12-1.)

Figure 12-1 Everything an animal does is its behavior. This frog, shown leaping with its tongue extended to catch its insect prey, is carrying out feeding behavior.

The purpose of animal behavior is to allow the organism to maintain homeostasis, survive, and reproduce. For example, birds build nests, keep their eggs warm until they hatch, then feed and protect their hatchlings. In Africa, huge herds of wildebeests travel northward and then southward again each year, following the seasonal rains, in a behavior pattern known as *migration*. When they are very young, discus fish swim around their parents' bodies and feed off a nutritious slime produced on their body surfaces. All of these behaviors aid survival. (See Figure 12-2.)

A variety of animal species—from insects to mammals—live in organized social groups, or *societies*, in which different members of the group have different tasks or levels of impor-

Figure 12-2 Animal behaviors, such as nest building, allow an organism to maintain homeostasis, survive, and successfully reproduce and raise offspring.

tance. Many animals have another social behavior that involves defending the area in which they live, known as a *territory*. These kinds of behaviors also aid survival, either of the individuals or of the group as a whole. (See Figure 12-3.)

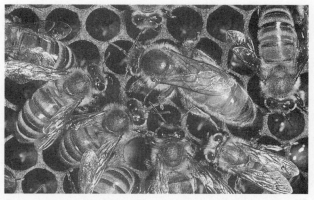

Figure 12-3 The social behavior of the members of a honeybee colony serves to maintain stable conditions within the hive—conditions that enhance the health and survival of the bees.

▓ BEHAVIOR AND HOMEOSTASIS

Maintaining a constant temperature, water balance, and level of nutrition are all behaviors that control aspects of homeostasis. Homeostasis requires that the conditions within an organism remain relatively constant. Animals have developed a variety of behaviors that enable them to maintain these life-sustaining conditions.

Honeybees lay eggs in individual wax cells that make up the structure of the beehive. The eggs hatch and develop through larval and pupal stages in the cells before they develop into adults. Sometimes the developing bees die in their individual cells. Some types of honeybees remove young pupae that die in their cells. Other types of honeybees do not. How has this particular behavior developed? It has evolved over time, like any other adaptation that has survival value. Although requiring more effort, removing dead pupae may have the survival value of keeping the beehive free from disease—a condition that helps to maintain homeostasis.

◆ EVOLUTION OF ANIMAL BEHAVIOR

How have animals evolved all the behaviors they now have? Within a population of organisms, there are variations among the individuals—in behavior as well as in appearance. The behaviors may actually be determined by genetic traits already present at birth. Those individuals whose behavior makes them more

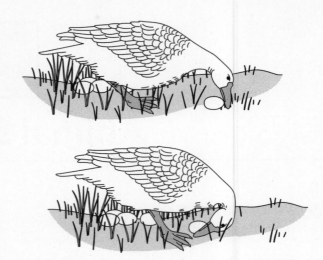

Figure 12-4 Animal behaviors evolve over time because they have survival value. Here, a goose sees that an egg has rolled from the nest, so she rolls it back—an inborn behavior that increases her reproductive success, and which will be passed along to her offspring.

likely to survive will pass those genetic traits on to their offspring. Other individuals, whose behavior is not as adaptive—particularly in a changing environment—will be less likely to survive and reproduce. In this way, the environment naturally selects certain behaviors (just as it selects physical traits). Thus, the evolution of behavior occurs over time. More and more individuals with adaptive behaviors live to reproduce; and these behaviors appear with increasing frequency in the population. (See Figure 12-4.)

Chapter 12 Review

Part A—Multiple Choice

1. An organism's behavior can best be described as
   1 how the organism acts when it is angry
   2 its response to a stimulus only
   3 everything the organism does to survive
   4 actions that are not related to its survival

2. Wildebeests of Africa migrate northward and southward each year because they
   1 are forced to relocate by humans
   2 need to follow the seasonal rains
   3 try to add members to their herds
   4 enjoy changes in body temperature

3. To aid their survival, many animals live in organized groups called
   1 societies
   2 territories
   3 tribes
   4 clubs

4. Some animals defend the area they live in, known as a
   1 hometown
   2 core area
   3 territory
   4 neighborhood

5. Behaviors that control homeostasis include maintaining all of the following *except*

   1 a constant temperature
   2 water balance
   3 level of nutrition
   4 vocalizations

6. Removing dead pupae from a beehive may have the survival value of

   1 making the beehive look cleaner
   2 keeping the beehive free from disease
   3 regulating the beehive's temperature
   4 preventing wax buildup in the beehive

7. A behavior probably will be passed on to off-spring if it

   1 separates the organism from others in the population
   2 makes the organism more likely to survive and reproduce
   3 makes the organism change its external environment
   4 is a characteristic of the organism for a long period of time

8. Which of the following statements is true?

   1 Physical traits only—*not* behaviors—can be naturally selected.
   2 Behaviors only—*not* physical traits—can be naturally selected.
   3 Behaviors as well as physical traits can be naturally selected.
   4 Behaviors are *never* determined by genetic traits present at birth.

9. Some behaviors such as mating and caring for young are genetically determined in certain species of birds. The presence of these behaviors is most likely due to the fact that

   1 birds do not have the ability to learn
   2 these behaviors helped birds to survive in the past
   3 individual birds need to learn to survive and reproduce
   4 within their lifetimes, birds developed these behaviors

Part B—Analysis and Open Ended

10. Give three examples of the general types of behavior in animals.

11. How does animal behavior help to maintain homeostasis?

Refer to the following map to answer questions 12 and 13.

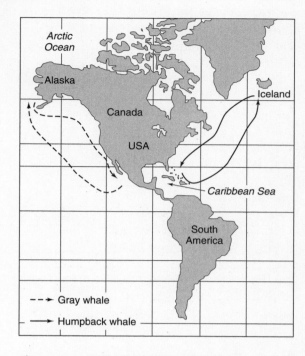

12. The arrows on the map represent the movements of groups of whales over the course of one year. This kind of behavior is known as

    1 societal
    2 territorial
    3 migration
    4 defensive

13. In what way might such long-range movements be adaptive for the whales?

14. Explain why defending the area in which they live might aid the survival of a group of animals.

Refer to Figure 12-3 on page 89 to answer questions 15 and 16.

15. The honeybees live together in a group, also called a

    1 unit
    2 society
    3 nest
    4 culture

16. How is this way of living similar to the way in which humans live?

17. What are some conditions in an animal's body that may affect its behavior (in terms of its need to maintain homeostasis)? Explain how.

18. Use the following terms to fill in the boxes of the concept map below: *low survival rate; high reproductive rate; natural selection; a changing environment.*

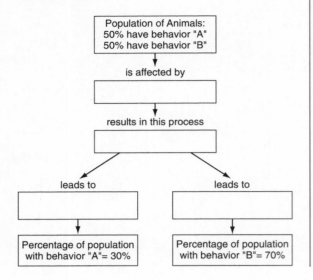

Population of Animals:
50% have behavior "A"
50% have behavior "B"

is affected by

results in this process

leads to   leads to

Percentage of population with behavior "A"= 30%

Percentage of population with behavior "B"= 70%

19. An appropriate title for this concept map probably would be:

    1 Competition for Resources
    2 The Defense of a Territory
    3 Evolution of Animal Behavior
    4 Behavior in Different Species

20. Describe how an animal's behavior may develop as a result of natural selection.

21. Suppose that a monkey starts to beg for food scraps from tourists in India. Later, after observing this, her offspring start to beg for food from tourists, too. The offspring grow well nourished and survive to reproduce. Their offspring, in turn, learn to survive by begging for food. Is this an example of natural selection acting on inherited behavioral traits? Explain why or why not.

Part C—Reading Comprehension

Base your answers to questions 22 to 24 on the information below and on your knowledge of biology. Use one or more complete sentences to answer each question.

What could be more fascinating than trying to learn about the mind of a baby? During their first year, infants gain physical skills, such as crawling and walking. They also take their first steps in being able to think and to make sense of the world around them. This field of study is called infant cognition. It includes the study of the increasing abilities of infants to perceive, recognize, categorize, and remember things in their environment.

In one research project, several patterns, which were made up of either straight or curved lines, were shown to infants. By measuring the amount of time the infants looked at each pattern, scientists were able to conclude that infants under one year of age look longer at concentric than at non-concentric forms and longer at curved than at straight lines.

Why? What do these results mean? What later developments occur in the mind of an infant? These and many other questions are being asked and examined by researchers that study infant cognition.

22. Explain what is meant by the term "infant cognition."

23. Describe an experiment that has been used to learn about infant cognition.

24. Suggest questions in addition to those mentioned that might be asked about infant cognition.

13

Excretion and Water Balance

■ REGULATION OF BODY CHEMISTRY

The human body is about 70 percent water. This water contains many types of solutes, including table salt. Yet we cannot drink salt water, only fresh water. Drinking salt water disrupts the careful balance that our bodies need internally, and the results could be very dangerous. That is because the amounts of these solutes in our bodies must remain constant day to day and hour to hour. In other words, one of the most important functions of homeostasis is to maintain a constant internal chemical environment.

The real meaning of the cell theory is that we are only as healthy as our cells are. Therefore, the chemical contents of our cells are regulated carefully. All body cells are surrounded by intercellular fluid, which is mostly water. (Refer to Figure 10-2 on page 73.) The level of chemicals in cells must remain within very narrow limits, so it is essential that the composition of intercellular fluid remains within narrow limits, too. How is its chemical composition regulated? Capillaries deliver blood to every cell in the body and exchange materials with the intercellular fluid that surrounds each cell. The blood controls the levels of substances in the fluid that surround each cell. So one of the most important jobs of homeostasis is to regulate the chemical composition of blood. (See Figure 13-1.)

In vertebrates that live on land, the organ primarily responsible for regulating the chemical composition of blood is the *kidney*. When the kidneys do not work properly, the result is blood that does not have the proper balance of chemicals. In time, this can cause death, unless the kidneys' functions are taken over by mechanical means.

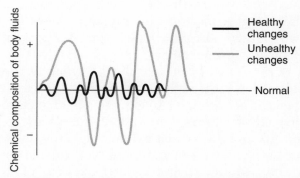

Figure 13-1 If our cells are to remain healthy, the chemical composition of our body fluids (blood and intercellular) must remain within very narrow limits.

■ THE KIDNEYS

The job of regulating both salt levels and water levels is often called maintaining a *water balance* in the body. This is one of the functions of the kidneys.

The kidneys are also part of the sanitation system of the body. All metabolic activities in the body produce wastes. These wastes are toxic and must not be allowed to accumulate in the body. So another important function of the kidneys is to carefully select the chemical wastes for removal while keeping useful nutrients. This process of ridding the body of metabolic wastes is called **excretion**.

By far the most important and potentially dangerous metabolic wastes produced in the body contain the element nitrogen. These *nitrogen wastes* result when amino acids, the building blocks of proteins, are broken down. In the case of one-celled aquatic organisms, such as the ameba, the nitrogen wastes diffuse into the water through the cell membrane. But most animals, which are multicellular, have some type of internal organ that contains tubelike structures to carry out chemical regulation. As body fluids pass around and inside these tubes, materials are exchanged between the tubes and the blood in capillaries that surround the tubes. Useful materials stay in the blood and wastes are collected in the tubes and passed to the external environment.

THE HUMAN EXCRETORY SYSTEM

The kidneys are the most important part of the excretory system in humans. There are two kidneys, located in the lower rear portion of the abdominal cavity.

A large artery brings blood to each of the kidneys. A large vein leaves each kidney with the newly cleansed blood. As blood passes through the kidneys, metabolic wastes are removed, and the correct balance of salt and water in the blood is maintained. As a result, the kidneys produce urine. Urine leaves each kidney through a tube, which drips the urine into the *bladder*. Here the urine is stored until it is passed from the body. (See Figure 13-2.)

There are about one million tiny, tubular structures in each kidney. Around each structure, a complex network of capillaries is wrapped. Blood is cleansed, water balance is maintained, and urine is produced through the exchange of materials between the capillaries and the tubular structures.

The result of all this activity in the kidneys is that the blood returns through veins to the body, cleansed of wastes and properly balanced

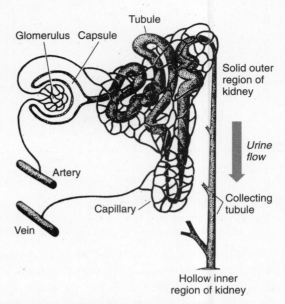

Figure 13-3 Within each kidney, materials are exchanged between about one million tiny tubular structures and the network of capillaries that surrounds them.

anced with salts and water. In addition, urine, which contains wastes as well as any excess salts and water, is collected and removed from the body. (See Figure 13-3.)

REGULATION OF THE EXCRETORY SYSTEM

The purpose of the excretory system is to maintain homeostasis by regulating the chemical composition of the blood and, in turn, of all the body's cells. At times, to carry out this extremely important job, both the endocrine system and the nervous system have to help.

Regulation of the level of water in the body is very important. For example, when you are exercising a great deal and sweating a lot, your body loses water. The result is that the volume of your blood decreases, since it consists mostly of water. It is the blood, and not your cells, that loses the water. Our bodies "know" that it is absolutely necessary to keep the right amount of water inside our cells.

The lower volume of the blood acts as a stimulus to the endocrine and nervous systems. Glands in the brain detect the lower blood volume and send a hormone to the kidneys. The hormone signals the kidneys to re-

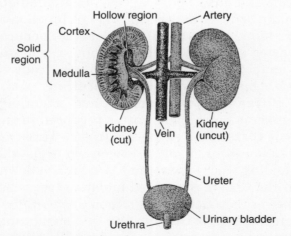

Figure 13-2 The human excretory system—the kidneys, which are the most important part of this system, cleanse the blood of wastes and maintain water balance.

turn more water from the urine back into the blood. The blood volume returns to normal. Meanwhile, wastes become more concentrated in the urine. The urine becomes darker, which shows that the body is automatically readjusting itself by not releasing as much water. This is another example of a feedback mechanism.

✖ HOMEOSTASIS AND THE SKIN

The *skin* is the largest organ of the body. It is made up of a variety of different types of cells and tissues. This organ, which surrounds and protects the body, also has an important role in keeping the conditions that exist within the body fairly constant.

There are two separate layers in the skin: an outer layer and an inner layer. The outer layer is a very thin layer where new skin cells push their way toward the surface, replacing the dead skin cells there. This process occurs all the time. These skin cells are very tough and form a protective layer that prevents bacteria and harmful chemicals from entering, and water from leaving, the body. Beneath this layer are cells that produce the pigments that give skin its color.

The inner layer is a thick, complex layer that contains a variety of structures that help maintain homeostasis: nerve endings that can detect temperature and touch, hairs, capillar-

ies that help regulate body temperature, glands that keep skin and hair soft by releasing oils, and *sweat glands*. Perspiration from sweat glands contains some nitrogen wastes, so these glands can be considered excretory structures. However, the main role of perspiration is to assist in regulating temperature by cooling the body through evaporation. (See Figure 13-4.)

✖ HOMEOSTASIS AND THE LIVER

No theme on homeostasis and excretion would be complete without talking about the *liver*. This organ is involved in homeostasis by assisting most of the important systems of the body. The liver helps in excretion by removing nitrogen from waste amino acids. It breaks down all red blood cells. Chemical poisons are also made harmless by the liver. For example, an excessive intake of alcohol, which is considered a poison by the body, makes the liver work very hard and can, in time, cause liver disease.

An example of homeostasis is the need to keep blood sugar levels constant. The liver helps the body do this because it stores excess glucose in the form of a starch and then releases the sugar into the bloodstream when it is needed. In addition, the liver helps the digestive process by making bile, the substance used

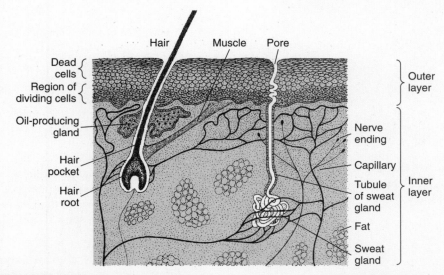

Figure 13-4 Skin, the largest organ of the body, helps us maintain homeostasis through a variety of structures. The sweat glands help carry out excretion, since perspiration contains some nitrogen wastes.

to digest fats; packages fats for transport in the blood; controls the level of cholesterol in the blood; stores vitamins A, D, and E; and helps maintain water balance by making important proteins for the blood. (See Figure 13-5.)

✸ WHEN THINGS GO WRONG: DISEASES OF THE EXCRETORY SYSTEM

One of the most common diseases of the kidney, a condition called *nephritis,* can be detected by a microscopic examination of the urine. Cells and chemicals that normally should not be in the urine are indicators of the disease. It is caused by a bacterial infection and can usually be treated with antibiotics. Kidney stones occur when a chemical compound that contains calcium builds up in the kidney. Gout, a condition caused by high levels of nitrogen wastes in the blood, produces severe pain in the joints.

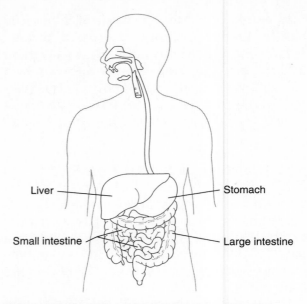

Figure 13-5 The liver performs many functions that maintain homeostasis: it removes nitrogen from waste amino acids; breaks down old red blood cells; detoxifies chemical poisons; stores excess glucose; and helps maintain the body's water balance.

Chapter 13 Review

Part A—Multiple Choice

1. Humans cannot drink salt water because

    1 it tastes so bad
    2 the human body does not contain any salt
    3 even small amounts of salt cause cells to die
    4 it disrupts the balance of salt in the body

2. The levels of substances in the intercellular fluid are controlled by the

    1 blood
    2 lungs
    3 heart
    4 skin

3. In land vertebrates, the organ primarily responsible for regulating the chemical composition of the blood is the

    1 liver
    2 kidney
    3 pancreas
    4 heart

4. During excretion, the kidneys function to

    1 remove solid wastes from the body
    2 cause the skin to perspire, thereby removing excess water
    3 select wastes for removal while keeping useful nutrients
    4 remove water from the blood that passes through them

5. Wastes that contain nitrogen are produced when

    1 polluted air is breathed in
    2 proteins are formed
    3 fats are digested
    4 amino acids are broken down

6. In the ameba, nitrogen wastes

    1 are released by the kidneys
    2 are pumped out through specialized tubes
    3 diffuse out through the cell membrane
    4 are stored in a special saclike structure

7. In humans, wastes and useful materials are *directly* exchanged between the kidneys'

    1 tiny tubes and the skin surface
    2 tiny tubes and the bladder
    3 veins and arteries and the bladder

4 tiny tubes and the blood in surrounding capillaries

8. Blood that returns to the body through veins from the kidneys has

   1 a higher salt content than when it entered the kidneys
   2 been cleansed of wastes and has a proper salt balance
   3 lower levels of oxygen and carbon dioxide in it
   4 more waste products, including a high level of urine

9. In humans, urine that is darker than normal indicates

   1 excess water and a high volume of blood in the body
   2 loss of water and a low volume of blood in the body
   3 a high amount of nitrogen wastes in the body
   4 a low amount of nitrogen wastes in the body

10. Perspiration from sweat glands in the skin helps maintain homeostasis by

    1 excreting nitrogen wastes and salts only
    2 cooling the body and regulating its temperature only
    3 adjusting the salt and water balance of the body
    4 excreting nitrogen wastes and regulating body temperature

11. When a person does strenuous exercise, small blood vessels (capillaries) near the surface of the skin increase in diameter. This change allows the body to be cooled. These statements best illustrate

    1 synthesis
    2 excretion
    3 homeostasis
    4 locomotion

12. The liver helps the body maintain homeostasis by doing all of the following *except*

    1 removing nitrogen from waste amino acids
    2 producing red blood cells and vitamins for the body

3 making bile, packaging fats, and controlling cholesterol levels
4 storing excess sugar and then releasing it when it is needed

13. The diagram below represents one metabolic activity of a human. Which row in the table under it correctly represents the letters A and B in the diagram?

Metabolic Activity A

| Protein ⟶ | B | B | B | B |

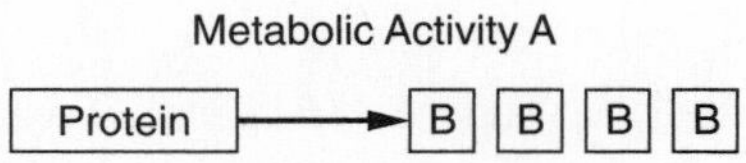

Row	Metabolic Activity A	Product B
1	Respiration	Oxygen molecules
2	Reproduction	Hormone molecules
3	Excretion	Simple sugar molecules
4	Digestion	Amino acid molecules

1 row 1
2 row 2
3 row 3
4 row 4

14. The diagram at the bottom of this page represents events involved as energy is ultimately released from food. Which row in the table below correctly represents the X's and letters A and B in the diagram?

Row	X-X-X-X-X-X-X-X	Substances A and B
1	Nutrient	Antibodies
2	Nutrient	Enzymes
3	Hemoglobin	Wastes
4	Hemoglobin	Hormones

1 row 1
2 row 2
3 row 3
4 row 4

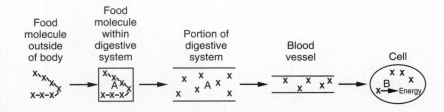

Part B—Analysis and Open Ended

15. Explain why it is harmful for a person to drink salt water.

16. Why is it necessary for the body to regulate the chemical composition of the blood?

17. Describe the two main functions of the kidneys.

18. What is the source of nitrogen wastes in the body?

19. Briefly compare and contrast the removal of nitrogen wastes by one-celled and multicellular organisms. Your answer should include the following:

♦ the basic structures involved for each type of organism

♦ the process by which each removes the nitrogen wastes

♦ the term that describes this removal of metabolic wastes

20. List the four main structures of the human excretory system. In what sequence are the parts connected to carry out waste removal?

Refer to the following figure to answer questions 21 and 22.

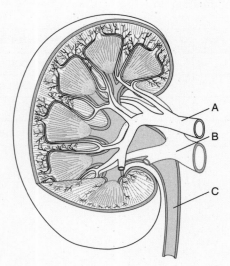

21. The diagram represents a cross section of what excretory organ? Identify the three main tubes—A, B, and C—that are connected to it.

22. Briefly describe the function of each of these tubes. Your answer should explain the following:

♦ what kind of body fluid is transported in each of these tubes

♦ which way the fluid is flowing in relation to the organ

♦ what process occurs as each fluid passes through the organ

23. What kind of blood vessel surrounds the millions of tiny tubules in the kidneys? Describe what occurs between these blood vessels and the tubules.

24. Use the following terms to replace the phrases in the chart below so that they show how the endocrine, nervous, and excretory systems work together to maintain a healthy water balance in the body: *hormone to the kidneys; normal blood volume; lower blood volume; concentrated wastes in urine; glands in the brain; return water from urine to blood.*

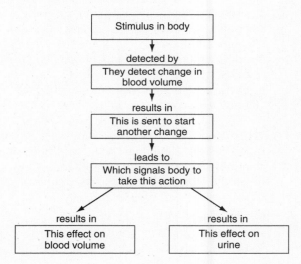

25. Refer to Figure 13-4 on page 95 to answer the following questions.

a What structure in the skin can be considered as part of the excretory system?

b In what part of the skin is this structure found—the inner layer or outer layer?

c What job does this structure perform for the excretory system? Is this the structure's *main* function? (If not, what is?)

26. The liver carries out many functions that help to maintain homeostasis. Briefly describe the two ways that it helps in (a) waste excretion and (b) water balance.

27. Why is excessive alcohol consumption harmful to a person's health? Explain.

Base your answers to questions 28 to 30 on the diagram below, which shows some of the specialized organelles in a single-celled organism, and on your knowledge of biology.

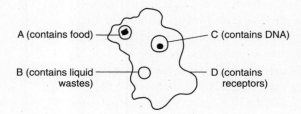

28. Write the letter of *one* of the labeled organelles and state the name of that organelle.

29. Explain how the function of the organelle you selected in question 28 assists in the maintenance of homeostasis.

30. Identify a system in the human body that performs a function similar to that of the organelle you selected in question 28.

31. The graph below shows the relationship between kidney function and arterial pressure in humans. State how a steady decrease in arterial pressure will affect homeostasis in the body.

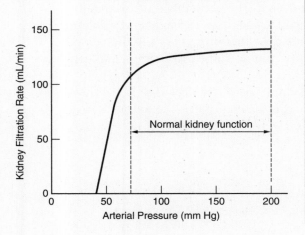

32. The skeletal system of an animal is shown in the photograph below. List *three* systems (other than the skeletal system) that the animal had when alive, which helped it to survive. Describe how each of these three systems contributed to maintaining homeostasis.

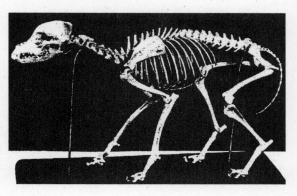

33. Several systems interact in the human body to maintain homeostasis. Four of these systems are the circulatory, the digestive, the respiratory, and the excretory. Based on these, answer the following:

a Select *two* of the systems listed. Identify each system selected and state its function in helping to maintain homeostasis in the body.

b Explain how a malfunction of *one* of the four systems listed disrupts homeostasis and how that malfunction could be prevented or treated. In your answer be sure to:

◆ name the system and state *one* possible malfunction of that system

◆ explain how the malfunction disrupts homeostasis

◆ describe *one* way the malfunction could be prevented or treated

Part C—Reading Comprehension

Base your answers to questions 34 to 36 on the information below and on your knowledge of biology. Use one or more complete sentences to answer each question.

Several years ago, a young lady named Jeanette fell and ruptured her patella, the knee bone. Surgery was planned to repair it, but this was abruptly cancelled when blood tests showed that Jeanette was in kidney failure. A few decades ago, this would have meant certain death. Now, however, Jeanette had options. The one she chose was at-home hemodialysis.

Kidney failure, now called *ESRD* (*e*nd-*s*tage *r*enal *d*isease), leads to a buildup of wastes in the blood, higher concentrations of salt and water, and unbalanced pH levels. Homeostasis is seriously disrupted. Without kidney function, life cannot continue. Fortunately, however, treatments have been developed to artificially cleanse the blood. The treatments involve the exchange of substances across membranes in a process called *dialysis*. In the 1960s, when the first artificial kidney was developed, a patient had to be connected to a dialysis machine for 8 to 10 hours at a time, several times a week. There were also serious side effects from the treatment, such as severe infections and inflammation around the heart.

The many options that Jeanette and other ESRD patients today can choose from are remarkable. The two basic choices are hemodialysis or peritoneal dialysis. With hemodialysis, the patient's blood passes through tubes into a machine next to them, where materials are exchanged between the blood and fluids in the machine. The patient has the option to undergo hemodialysis at home, like Jeanette did, or in the hospital, and at different times during the week. Some people have dialysis treatment every day, others two or three times a week.

The other kind of treatment, peritoneal dialysis, amazingly uses the lining of the patient's own abdomen to filter the blood. A cleansing solution is transported into the abdomen, where it stays for awhile. As the person's blood passes in capillaries through the solution, the blood gets cleansed. Then the solution is drained out, and a fresh solution is put into the patient's abdomen to start the process again. Peritoneal dialysis can be done in several ways. One method uses no machine; it just goes on continuously, even as the patient walks around. Another way uses a machine to move the cleansing solution through a tube, in and out of the person's abdomen. The machine does this at night while the person sleeps. Another method also uses a machine, but only at certain times, and usually in the hospital.

So there are choices for people with kidney failure, choices that allow those who no longer have the use of an essential internal organ—the kidney—to go on living.

34. In what ways does kidney failure disrupt homeostasis?

35. Describe the options available to a person undergoing hemodialysis.

36. How does peritoneal dialysis use the patient's body to cleanse the blood?

14

Disease and Immunity

⚏ DISEASE: A LACK OF HOMEOSTASIS

Homeostasis, the theme you have been studying, emphasizes the need for organisms to maintain a carefully controlled internal set of conditions, a dynamic equilibrium. Maintaining these conditions—including pH, temperature, water and salt balance, and levels of carbon dioxide and oxygen—allows an organism's cells to function normally. Organisms can tolerate changes that occur within very definite limits. But changes outside normal limits disrupt homeostasis, producing illness, disease, and even death.

There are many reasons why the body may be pushed beyond its normal limits. These reasons, or factors, are often the causes of disease. An inherited defect in a genetic trait might be the cause of a disease. The disruption of homeostasis in such a disease would be caused, in a sense, by a factor inside the body. Many other diseases result from some influence outside the body, that is, from the environment.

⚏ FACTORS THAT CAUSE DISEASES

Diseases may be caused by one of the following factors, or by a combination of several of these.

♦ *Inheritance.* Defective genetic traits can be passed from parents to offspring. Often, neither parent has the disease, but both may carry a single form of a gene for the disease. It is the combination of these two defective genes in the child that gives him or her the disease. A well-known example of an inherited disease is sickle-cell

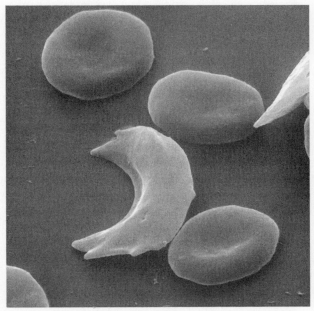

Figure 14-1 Sickle-cell anemia is an inherited disease that disrupts homeostasis because the protein that carries oxygen in red blood cells is flawed.

anemia, in which the protein that carries oxygen in red blood cells is flawed. (See Figure 14-1.)

♦ *Microorganisms.* Microscopic organisms that cause diseases are called **pathogens**; they include certain **fungi**, **bacteria**, **viruses**, and protozoa. Some diseases caused by pathogenic microorganisms may be passed in a variety of ways from one person to another. These are called *infectious diseases*. Microorganisms most often enter the body through respiratory pathways, the digestive system, or pathways of the excretory system. Infections may also occur through breaks in the skin. Tuberculosis is an infectious disease caused by certain bacteria. (See Figure 14-2 on page 102.)

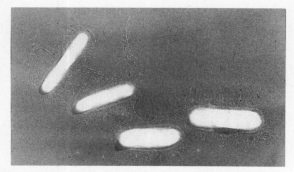

Figure 14-2 These bacteria are an example of micro-organisms that can cause infectious diseases, such as tuberculosis.

♦ *Pollutants* and *poisons*. Chemical agents present in the environment may upset the body's normal functioning and produce disease. These pollutants and poisons include coal dust, asbestos, lead, phosphorus, mercury, and many others. For example, when asbestos fibers enter the respiratory system, they cause asbestosis, a disease of the lungs; years later, this may result in cancer in the lungs and chest.

♦ *Organ malfunction*. A disease may develop when one or more of the body's organs **malfunction**. When an organ such as the liver, lung, heart, stomach, or kidney does not function properly, serious effects on the body result.

♦ *Harmful lifestyles*. The way one lives can also be an important factor in causing dis-

ease. Specifically, tobacco, alcohol, and drugs in the body can disrupt homeostasis, producing illness. In addition, overeating, poor nutrition, not exercising, having unsafe sexual experiences, and living with stress can lead to certain diseases. Hypertension, or high blood pressure, is one such disease.

❖ THE BODY'S DEFENSES AGAINST DISEASE

Our bodies are surrounded by microorganisms trying to get into us. Some of them succeed, through the nose, through cuts in our skin, or along with the food we eat. Many of these microorganisms cause serious problems if they survive and reproduce inside us without challenge. Controlling these microscopic invaders is as important to homeostasis as is regulating body temperature and chemistry.

The first line of defense against infection consists of *physical barriers* that block the entry of microorganisms. The skin is the main physical barrier in our body. A second line of defense, called *inflammation*, is present when microorganisms get through our physical barriers. For example, when we get a cut or scrape on the skin, the injured area may become warm, reddened, and perhaps swollen with pus. (See Figure 14-3.) Chemicals released by the damaged tissues are acting like

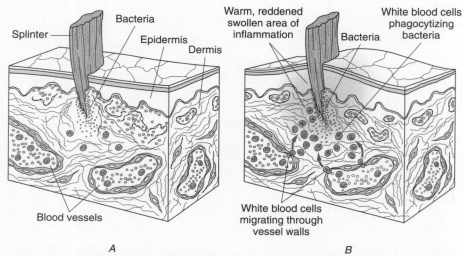

Figure 14-3 Inflammation is the body's second line of defense against infectious bacteria that get through the skin's first defense.

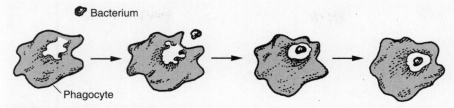

Figure 14-4 During inflammation, the white blood cells engulf and destroy the invading bacteria, which prevents the development of a more serious infection.

an alarm, causing an increase in blood flow to the site of the injury. Special white blood cells that arrive engulf microorganisms, destroying them by ingesting them. All of this activity helps prevent a more serious infection from developing. (See Figure 14-4.)

Vertebrates have evolved a very important system that attacks specific invaders. This is the **immune system**. The immune system recognizes who the "bad guys" are and goes after these invaders to try to keep them from disrupting normal body functions.

THE HUMAN IMMUNE SYSTEM

The immune system defends our bodies against very specific microscopic invaders. Each invader—usually a bacterium or virus—has specific protein molecules attached to its surface. These protein molecules are called **antigens**. It is these molecules that are detected by the body's immune system.

When the immune system detects an antigen, it produces **antibodies**, molecules that bind to that antigen. Once the antibodies bind to the antigen, the invader can be destroyed by the body. (See Figure 14-5.) **Vaccinations** use weakened **microbes** (microorganisms), or parts of them, to stimulate the immune sys-

tem to react by recognizing specific antigens. This reaction provides the body with **immunity**—the ability to resist an infection—by preparing it to fight subsequent invasions by the same microbes (by producing the appropriate antibodies). Vaccines offer protection against a number of diseases. People are now given harmless antigens in a vaccine, which cause the body to produce antibodies.

B CELLS AND T CELLS

The immune system also includes B cells and T cells, special kinds of *white blood cells* that are produced in bone marrow, the thymus gland, the spleen, the lymph nodes, and the tonsils. (See Figure 14-6.) *B cells* are the ones

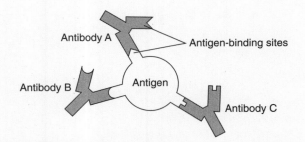

Figure 14-5 Antibodies, produced by the immune system, bind to antigens, which are specific protein molecules on an invading microbe's surface.

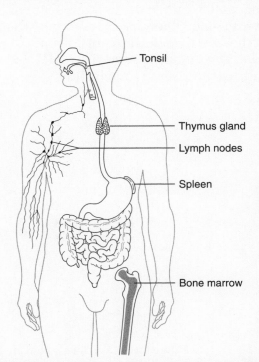

Figure 14-6 Special white blood cells, called B cells and T cells, are produced in the tonsils, thymus gland, lymph nodes, spleen, and bone marrow.

that respond to specific antigens by beginning to produce antibody proteins that will bind only with that antigen.

As time goes on, the body comes to have many different types of B cells, each producing antibodies for one specific antigen. After having been invaded once by an antigen, some special B cells that recognize that antigen remain in the body for the rest of one's life. These are called memory B cells. Because they are already present in the body, you instantly start making antibodies the moment you encounter the same invading microorganisms again. That is why individuals usually do not get measles or chicken pox a second time. The immune system remembers the first exposure to the disease and is ready to defend the body. (See Figure 14-7.)

One type of T cell is called *killer T cells*. Through protein receptors on their surface, they can recognize cells in the body that have been infected with invading microorganisms. The killer T cells punch holes in the membranes of the infected cells, sometimes injecting poison into them. (See Figure 14-8.) Another important type of T cell, called *helper T cells*, assists both B cells and killer T cells. Without helper T cells, the other members of the immune system cannot do their job. Just how important helper T cells are, is shown by the fact that they are the cells that are destroyed by the *h*uman *i*mmunodeficiency *v*irus (HIV), which results in the disease called AIDS.

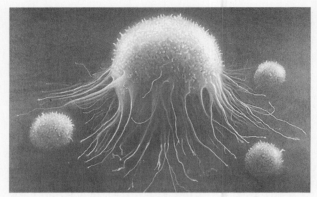

Figure 14-8 Killer T cells can recognize cells in the body that have been infected by invading microorganisms. Here, some killer T cells are shown attacking an infected body cell.

▚ WHEN THINGS GO WRONG: DISEASES OF THE IMMUNE SYSTEM

The immune system helps maintain the internal dynamic equilibrium necessary for life. However, the immune system can become out of balance. It can be overactive or underactive, and in either case the body's equilibrium is upset.

Allergic reactions result from overactivity of the immune system. The body responds inappropriately to common substances such as dust, mold, pollen, or certain foods. The immune system begins making a special type of antibody to these substances, which normally

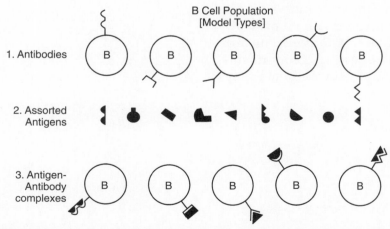

Figure 14-7 Over time, a person's body comes to have many different B cells. The memory B cells, which remain in the body after their first exposure to an antigen, can instantly make antibodies when they encounter the same antigen again.

would not stimulate it. These antibodies cause cells in the body to release substances, including *histamines*, which cause many allergic symptoms, such as extra fluid in the nasal pathways, difficulty breathing, or hives. The allergies are often treated with *antihistamines*, drugs that stop the release of histamines.

Sometimes an overactive immune system begins to attack its own normal body tissues. These are called *autoimmune diseases* and they are very serious. They include rheumatoid arthritis and lupus erythematosus. In all autoimmune diseases, the body is literally rejecting its own tissues. The immune system may also attack transplanted organs; medications are then taken to try to prevent organ rejection.

Also inflammation, which protects us when we are young, may actually be contributing to crippling diseases when we get older. For example, researchers now suspect that many heart attacks occur when a rupture develops in the wall of an artery, brought on by overactive immune system cells causing inflammation.

The condition known as **AIDS** (*a*cquired *immuno*deficiency *s*yndrome) is a type of *immunodeficiency disease*, which means that the body's immune system is underactive because it is weakened, in this case by HIV, the human immunodeficiency virus. As a result, the body cannot protect itself from other diseases (such as pneumonia, tuberculosis, and cancer) that may attack it—a condition that is usually fatal.

Chapter 14 Review

Part A—Multiple Choice

1. A disruption of homeostasis can result in all of the following *except*
   1. illness
   2. death
   3. disease
   4. stability

2. Infectious diseases result from
   1. genetic defects
   2. microorganisms
   3. pollutants
   4. organ malfunctions

3. The inhalation of particles such as asbestos fibers and coal dust can result in respiratory diseases. In such a case, the main cause of the disease would be
   1. microorganisms
   2. pollutants
   3. genetic defects
   4. organ malfunction

4. The body's main physical barrier against infection is
   1. the skin
   2. white blood cells
   3. red blood cells
   4. inflammation

5. Scientific studies have indicated that there is a higher percentage of allergies in babies fed formula containing cow's milk than in breastfed babies. Which statement represents a valid inference made from these studies?
   1. Milk from cows causes allergic reactions in all human infants.
   2. There is no relationship between drinking cow's milk and having allergies.
   3. Breastfeeding by humans prevents all allergies from occurring.
   4. Breast milk most likely contains fewer substances that trigger allergies.

6. Allergic reactions are most closely associated with
   1. the action of circulating hormones
   2. immune responses to usually harmless substances
   3. a low blood sugar level
   4. the shape of red blood cells

7. White blood cells can prevent a serious infection by
   1. filling the damaged tissues with pus
   2. repairing the skin after it has been cut
   3. ingesting the harmful microorganisms
   4. constructing physical barriers against microorganisms

8. Certain microbes, foreign tissues, and some cancerous cells can cause immune responses in the human body because all three contain

   1 antigens
   2 lipids
   3 enzymes
   4 cytoplasm

9. Which activity would stimulate the human immune system to provide protection against an invasion by a microbe?

   1 receiving antibiotic injections after surgery
   2 being vaccinated against chicken pox
   3 choosing a well-balanced diet and following it throughout life
   4 receiving hormones contained in mother's milk while nursing

10. When the immune system detects an antigen, it

    1 pushes it out of the body immediately
    2 produces antibodies that bind to the antigen
    3 produces antigens that cancel the effects of it
    4 destroys the antigen by cutting it in half

11. Many vaccinations stimulate the immune system by exposing it to

    1 antibodies
    2 mutated genes
    3 enzymes
    4 weakened microbes

12. A part of the hepatitis B virus can be synthesized in the laboratory. This viral particle can be identified by the immune system as a foreign substance, but the viral particle is not capable of causing the disease. Immediately after this viral particle is injected into a human, it

    1 stimulates the production of enzymes that are able to digest the hepatitis B virus
    2 synthesizes specific hormones that provide immunity against the hepatitis B virus
    3 triggers the formation of antibodies that protect against the hepatitis B virus
    4 breaks down key receptor molecules so that the hepatitis B virus can enter body cells

13. The diagram at the top of the next column represents one possible immune response that can occur in the human body. The structures that are part of the immune system are represented by

    1 A, only
    2 A and C only
    3 B and C only
    4 A, B, and C

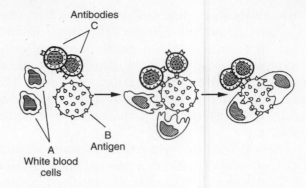

14. Which of the following is *not* a characteristic of white blood cells?

    1 They destroy some microbes by engulfing them.
    2 They carry oxygen atoms throughout the body.
    3 They make antibodies that bind with antigens.
    4 They punch holes in membranes of infected cells.

15. Which statement does *not* identify a characteristic of antibodies?

    1 They are produced in response to the presence of foreign substances.
    2 They are nonspecific, acting against any foreign substance in the body.
    3 They may be produced in response to an antigen.
    4 They may be produced by the white blood cells.

16. Which is the best procedure for determining if a vaccine for a disease in a particular bird species is effective?

    1 Vaccinate 100 birds; then expose all 100 birds to the disease.
    2 Vaccinate 50 birds, do not vaccinate 50 other birds; expose all 100 birds to the disease.
    3 Vaccinate 100 birds and expose only 50 of them to the disease.
    4 Vaccinate 50 birds, do not vaccinate 50 other birds; expose only the vaccinated birds to the disease.

17. Produced in several parts of the body, B cells and T cells are special kinds of

    1 blood platelets
    2 red blood cells
    3 white blood cells
    4 microorganisms

18. The killer T cells function to

1 produce antibodies that kill invading microorganisms

2 destroy the cells that are infected by microorganisms

3 bind with the infected cells and repair their membranes

4 destroy invading microorganisms before they infect any cells

19. Which condition would most likely result in a human body being unable to defend itself against pathogens and cancerous cells?

1 a genetic tendency toward a disorder such as diabetes

2 the production of antibodies in response to an infection in the body

3 a parasitic infestation of ringworm on the body

4 the presence in the body of the virus that causes AIDS

20. The human immunodeficiency virus (HIV) is particularly devastating to the immune system because it destroys

1 all the white blood cells in the body

2 all the red blood cells in the body

3 the special B cells and killer T cells, only

4 helper T cells, which assist B cells and killer T cells

21. Overactivity of the immune system due to a common substance can lead to an

1 equilibrium

2 allergic reaction

3 antihistamine

4 immunodeficiency

22. When an overactive immune system starts to attack its own body tissues, it causes serious conditions known as

1 antihistamine diseases

2 allergic reaction diseases

3 autoimmune diseases

4 immunodeficiency diseases

23. A characteristic shared by all enzymes, hormones, and antibodies is that their function is determined by the

1 shape of their protein molecules

2 inorganic molecules they contain

3 age of the organism that makes them

4 organelles present in their structure

24. A researcher needs information about antigen–antibody reactions. He could best find information on this topic by searching for which phrase on the Internet?

1 Protein Synthesis

2 Energy Sources in Nature

3 White Blood Cell Activity

4 DNA Replication

Part B—Analysis and Open Ended

25. How can a child inherit a disease if neither parent appears to have it?

26. Explain what a pathogen is. Your answer should include the following information:

♦ what the *four* types of pathogens are that can cause diseases

♦ the term for such diseases when they are passed from one person to another

♦ the *three* common ways that pathogens can enter the body

27. What is the difference between an inherited disease and an infectious disease?

28. Briefly describe the relationship between organ malfunction and disease.

29. Explain why a harmful lifestyle can lead to disease. Give *one* example. In what way are the factors that cause such a disease preventable or controllable?

30. Describe the first line of defense against infection in the body. Include the following:

♦ what *kind* of defense it consists of

♦ what *organ* carries out this function

♦ what it is defending *against* and how

31. Refer to the diagram below to answer this question. When bacteria enter a cut, this process occurs as part of the body's second line of defense, known as

1 infection

2 inflammation

3 invasion

4 immunity

32. Use the following terms to fill in the missing words in the flowchart below, which describes the immune system's reaction to microscopic invaders: *body develops immunity*; *microbes destroyed by body*; *antibodies*; *antigens.*

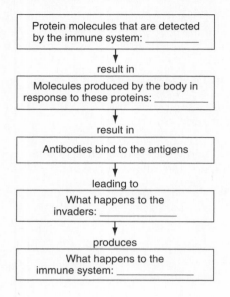

Protein molecules that are detected by the immune system: _____

↓ result in

Molecules produced by the body in response to these proteins: _____

↓ result in

Antibodies bind to the antigens

↓ leading to

What happens to the invaders: _____

↓ produces

What happens to the immune system: _____

33. How does the immune system help to maintain homeostasis?

34. Briefly define the term *immunity*. How do vaccinations provide people with immunity?

Base your answers to questions 35 and 36 on the table below and on your knowledge of biology.

Volunteer	Injected with Dead Chicken Pox Virus	Injected with Dead Mumps Virus	Injected with Distilled Water
A	✔		
B		✔	
C			✔
D	✔	✔	

35. None of these volunteers ever had chicken pox. After the injection, there most likely would be antibodies to chicken pox in the bloodstream of

1 volunteers A and D only
2 volunteers A, B, and D
3 volunteer C only
4 volunteer D only

36. Volunteers A, B, and D underwent a medical procedure known as

1 cloning
2 vaccination
3 electrophoresis
4 chromatography

Refer to the following list, which describes three ways of controlling viral diseases in humans, to answer question 37:

♦ administer a vaccine containing dead or weakened viruses, which stimulates the body to form antibodies against the virus;

♦ use chemotherapy (chemical agents) to kill viruses, similar to the way in which sulfa drugs or antibiotics act against bacteria; and

♦ rely on the action of interferon, which is produced by cells in the body and protects it against pathogenic viruses.

37. Based on this information, which activity would provide the greatest protection against viruses?

1 producing a vaccine that is effective against interferon
2 using interferon to treat a number of diseases caused by bacteria
3 developing a method to stimulate the production of interferon in cells
4 synthesizing a drug that prevents the destruction of bacteria by viruses

38. Compare and contrast the functions of B cells and killer T cells.

39. Explain how HIV affects the immune system. Your answer should include the following:

♦ what "HIV" stands for and what the term *immunodeficiency* means

♦ which cells in the immune system are affected and in what way

♦ how this affects the rest of the immune system's functioning

Base your answers to questions 40 and 41 on the following image, which shows a slide of normal human blood cells (magnified several times), and on your knowledge of biology.

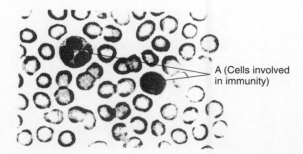

A (Cells involved in immunity)

40. An increase in the production of the cells labeled A occurs in response to an internal environmental change. State *one* possible change that could cause this response.

41. Describe *one* possible immune response, other than an increase in number, that one of the cells labeled A could carry out.

42. In what way are allergic reactions and autoimmune diseases similar to one another (in terms of the immune system)? In what important way are they different?

Base your answers to questions 43 and 44 on the information below and on your knowledge of biology.

Immunization protects the human body from disease. The United States is now committed to the goal of immunizing all children against common childhood diseases. In fact, children must be vaccinated against certain diseases before they can enter school. However, some par-

ents feel that vaccinations may be dangerous and do not want their children to receive them. Many parents are choosing not to immunize their children against childhood diseases such as diphtheria, whooping cough, and polio. For example, the mother of a newborn baby is concerned about having her child receive the DPT (diphtheria, whooping cough, and tetanus) vaccine. Since bacteria cause these diseases, she believes antibiotic therapy is a safe alternative to vaccination.

43. Explain to these concerned parents what a vaccine is and what it does in the body. Be sure to include the following:

♦ what is in a vaccine and how vaccination promotes immunity to various diseases
♦ the difference between antibiotics and vaccines in the prevention of bacterial diseases

44. State *one* advantage of using vaccinations in fighting bacterial diseases and *one* disadvantage of using antibiotics in fighting bacterial diseases.

Part C—Reading Comprehension

Base your answers to questions 45 to 47 on the information below and on your knowledge of biology. Use one or more complete sentences to answer each question.

Antibiotics are used to treat infections in people and animals. Due to the enormous success of antibiotics, their use is very common worldwide. When we are ill, we have come to expect quick, effective treatment with antibiotics. Physicians often prescribe antibiotics at the earliest sign of an infection.

One result of the widespread use of these medicines is a growing number of antibiotic-resistant strains of bacteria. Some scientists have warned about the alarming possibility of infections that will not be treatable by the antibiotics we have. Already, one disease, tuberculosis—which was largely under control—has reappeared in a strain that is much more difficult to treat with antibiotics.

Recently, scientists became alarmed to find bacteria, in the food that is given to chickens, that are resistant to the most powerful antibiotics. Even though those particular bacteria were harmless, the finding raised the disturbing possibility that these bacteria could pass on their antibiotic resistance to disease-causing bacteria in chickens and, ultimately, in humans. One reason it is thought that such drug-resistant bacteria are being found more frequently is the heavy, routine use of antibiotics in farm animals.

This is an issue for everyone to be aware of and concerned about. Science has provided us with a group of wonder drugs to treat diseases that once killed many people. However, we must be thoughtful and wise in our use of antibiotics. The laws of nature—in this case, the process of natural selection that produces resistance to antibiotics—can never be ignored.

45. How is the use of antibiotics a matter of both good news and bad news?

46. Why should people be concerned about the use of antibiotics in farm animals?

47. How is knowledge of the process of natural selection necessary in order to understand the problem of overuse of antibiotics?

15

How Cells Divide

▓ THE CONTINUITY OF LIFE

You have learned what it takes for an individual to remain alive in a constantly changing environment. Although staying alive is important for every individual organism, it is not sufficient to maintain life on Earth. No individual organism lives forever. Every organism has a typical life span—the length of time between when its life begins and when it ends. The continuity of life requires **reproduction**, the ability of individuals within a species to produce more of their own kind. Individuals are members of populations. It is reproduction within populations that allows species to survive. It is reproduction that allows life on Earth to continue.

▓ THE LIFE OF A CELL

Every cell has a life of its own. This is as true for single-celled organisms as it is for each of the billions of cells that make up the bodies of plants and animals, including ourselves. Each cell has a beginning (with a period of growth), a middle stage, and then an ending—a pro-

cess known as the *cell cycle*. In the first stage, the cell begins to grow in size. Organic materials, such as amino acids and sugars, and inorganic materials, such as water, are moved into the cell. The cell increases in size by adding these materials to itself. The cell also increases the number of its parts. For example, its mitochondria divide in two to make more mitochondria. If it is a plant cell, the same thing happens to its chloroplasts. (See Figure 15-1 on page 112.)

During the next stage in the cycle, the cell stops getting larger. Now, the genetic material in the cell—that is, the set of instructions received from the previous cell—duplicates. The genetic material is the building plan, similar in some ways to the set of blueprints used to build a house; it contains all the information about how the cell is to be built and how it functions. The genetic material is made up of the chemical called **DNA**, or **deoxyribonucleic acid**.

In reproducing cells, DNA is found in "packages" known as **chromosomes**. Bacterial cells may have just a single chromosome, a fruit fly has eight chromosomes in each body cell, a cabbage plant has 18, and a human has

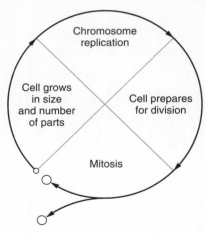

Figure 15-1 The cell cycle is the series of events that occurs in the life of a cell—from its beginning to its ending (in mitosis or death).

46. The number of chromosomes in other organisms varies. The chromosome number is specific for each type of organism. And the exact chromosome number must be maintained for the species to continue. This means that as cells reproduce, the new cells must have the same number of chromosomes as the original cells.

The duplication of the cell's genetic material, during this middle stage in the cell's life cycle, is called **replication**. This is the most important stage in preparation for reproduction of the cell. Following this stage, some additional cell growth occurs. What is growing here is material needed for the final stage in its life cycle, called *cell division*. This is how a single cell reproduces; it divides into two new cells. (See Figure 15-2.)

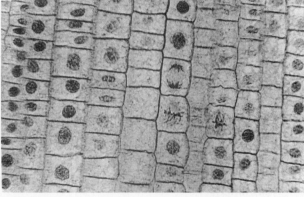

Figure 15-2 Several plant cells can be seen here preparing for cell division: chromosomes have replicated and are dividing into two groups.

▓ CELL DIVISION

During **cell division**, the genetic material must be equally divided. When a cell divides, it must send one copy of each of its chromosomes to each of the new cells. In addition, the cytoplasm and other cell parts must be divided between the two cells.

The division of the chromosomes occurs first. This division happens during a sequence of events called **mitosis**. During mitosis, the chromosomes of a cell are divided into two equal groupings. Following mitosis, the cytoplasm of the cell divides. After that occurs, each of the new cells has a complete set of chromosomes, just like the original cell. The two new cells are called *daughter cells*. The cell that they came from, which no longer exists, is called the *parent cell*. (See Figure 15-3.)

▓ MAKING NEW INDIVIDUALS: ASEXUAL REPRODUCTION

Cell division produces two new daughter cells from one parent cell. The daughter cells are identical. They are also genetically identical to the parent cell. But have new individuals really been produced?

The answer is *yes*, if the original parent was a single-celled organism. An ameba, through cell division, becomes two new, identical organisms. Reproduction in an ameba involves only one parent. (Refer to Figure 1-2 on page 2.) This is **asexual reproduction**. (For reproduction to be called sexual, it must involve two parents.)

Plants have a variety of types of asexual reproduction. In each type, a plant or a part of the plant reproduces itself through mitosis. As a result, the offspring are identical to the parent plant. For example, strawberry plants send out horizontal stems, called *runners*, above the ground. When these runners touch the surface of the soil at another spot, an entirely new, identical plant with roots and leaves begins to grow there. The production of identical genetic copies of a parent plant is called **cloning**. (See Figure 15-4.)

In some cases, animals also can reproduce asexually. For example, if an arm of a sea star is broken off, the arm can sometimes grow into a whole new sea star. The sea star that

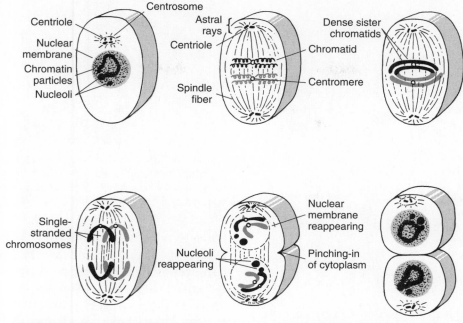

Figure 15-3 The stages of mitosis in an animal cell: the chromosomes have replicated; they pull apart; and the cytoplasm is divided to form two new cells.

lost the arm will grow another one. (See Figure 15-5.) Other organisms, such as yeasts, sponges, and hydra, can produce offspring by **budding**. During the process of budding, a new small individual begins to grow out of the side of the parent organism. The cells that form this new individual, or *bud*, result from mitotic cell division. The bud breaks free of the parent organism when it is large enough to live on its own. (See Figure 15-6.)

▓ THE RATE OF CELL DIVISION

When does a cell divide? How long does it take for one segment of cell division to begin and end? Do all types of cells divide at the same rate? The answers are very important to the

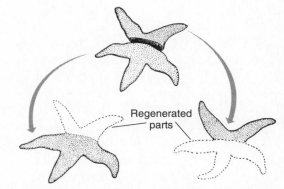

Figure 15-5 Some animals, such as the sea star, can reproduce asexually by regrowing lost body parts when cut in two.

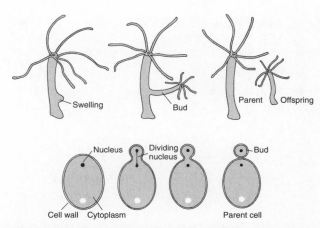

Figure 15-6 The hydra (top) and yeasts (bottom) can reproduce asexually by budding. The offspring, or bud, is smaller than the parent cell or organism.

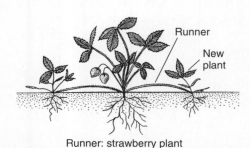

Runner: strawberry plant

Figure 15-4 Strawberry plants can reproduce asexually by means of runners; this method of producing identical offspring is called cloning.

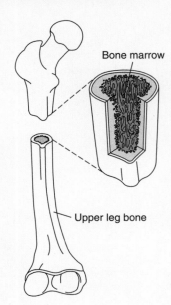

Bone marrow

Upper leg bone

Figure 15-7 The cells of different types of tissues divide at different rates, depending on their functions. Millions of red blood cells, which divide quickly, are made in the bone marrow each day. By contrast, the bone tissue cells divide much more slowly.

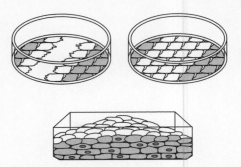

Figure 15-8 Normal cells grown in a lab dish stop dividing when they touch each other (top). Cells that continue to divide after they touch each other exhibit the kind of uncontrolled cell growth seen in cancer (bottom).

process of growth and development in organisms.

Every multicellular organism is made up of various types of tissues and cells. For example, the human body contains blood, skin, muscle, bone, and nerve tissues, among other tissues. Controlling the rate at which cells of each particular kind of tissue divide is a necessary part of homeostasis. Red blood cells have a relatively short life span, and we need millions of them, so the cells that develop into red blood cells divide quickly. (See Figure 15-7.) Bone cells, on the other hand, divide much more slowly. Skin cells normally take about 20 hours to complete one cell division; but their rate of division speeds up if you cut yourself.

▒ CANCER: CELL DIVISION OUT OF CONTROL

Cancer results from uncontrolled cell division. Cancer cells do not seem to follow the rules or recognize the signals that control normal cell division. Uncontrolled cell growth can occur in many different types of cells. As a result, there are different types of cancer, such as skin cancer, breast cancer, prostate cancer, lung cancer, and many others. There does not seem to be a single cause for all types of cancer. (See Figure 15-8.)

Even though there are differences, it is quite certain that all types of uncontrolled cell division involve the cells' genetic instructions, which are made of DNA. Factors that cause cancer do so by damaging or changing the DNA. These factors include the exposure of cells to certain chemicals and to radiation.

One of the most important ways to reduce the risk of cancer is to maintain good health habits. The immune system constantly attacks not only invading cells but also abnormal, cancerous cells from our own body. Much of the time, the immune system is successful. It destroys cancer cells before they can develop and cause problems. It is no surprise then that many people with AIDS actually die from some type of cancer. The patient's damaged immune system is not able to protect the person from cancerous cells. Therefore, a healthy immune system is one of the best protections against cancer.

Chapter 15 Review

Part A—Multiple Choice

1. An organism's typical life span is its

   1 body length, from head to tail
   2 time between birth and death
   3 average age when it reproduces
   4 normal population size

2. The survival of a species depends on

   1 an environment that never changes
   2 a continuously increasing life span
   3 reproduction within populations
   4 a limit of no more than two populations

3. During the first stage of the cell cycle, a cell

   1 grows in size
   2 divides in half
   3 duplicates its genetic material
   4 makes a copy of itself

4. What causes a cell to grow in size?

   1 The cell takes in other cells.
   2 The genetic material of the cell replicates.
   3 The cell divides into two parts.
   4 The cell takes in organic and inorganic materials.

5. The genetic material of the cell is most like

   1 the blueprints for a building
   2 the tracks for a train
   3 an advertisement for a store
   4 a fence for a house

6. A cell's DNA is located within structures known as

   1 mitochondria
   2 chromosomes
   3 chloroplasts
   4 cytoplasm

7. The number of chromosomes

   1 is specific for each type of organism
   2 is the same for every species of organism
   3 decreases from one generation to the next
   4 increases from one generation to the next

8. Before cell division, the genetic material must undergo a process called

   1 reduction
   2 disintegration
   3 replication
   4 reproduction

9. During the process of mitosis, the chromosomes

   1 are cut in half twice
   2 are equally divided
   3 form a circle in the cell
   4 spread through the cell

10. The diagram below represents the chromosomes in a cell. Which of the following diagrams best illustrates the daughter cells that result from the normal division of this cell?

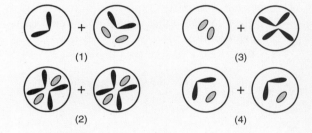

   1 diagram 1
   2 diagram 2
   3 diagram 3
   4 diagram 4

11. What happens *after* mitosis has occurred?

   1 The cell grows in size again.
   2 The genetic material replicates.
   3 The genetic material forms a parent cell.
   4 The cytoplasm of the cell divides in two.

12. Compared to the parent cell, each daughter cell that results from the normal mitotic division of the parent cell contains

   1 the same number of chromosomes, but different genes from those of the parent cell
   2 half the number of chromosomes, but different genes from those of the parent cell
   3 the same number of chromosomes and identical genes to those of the parent cell
   4 twice the number of chromosomes and identical genes to those of the parent cell

13. In asexual reproduction, the genetic material is supplied by

1 one daughter cell
2 one parent cell
3 two daughter cells
4 two parent cells

14. The diagram below represents a cell process. Which statement regarding this process is correct?

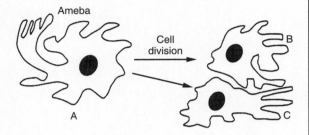

1 Cell B contains the same genetic information as cells A and C.
2 Cell A has DNA that is only 75 percent identical to cell B.
3 Cell C has DNA that is only 50 percent identical to cell B.
4 Cells A, B, and C each contain different genetic information.

15. The DNA of a plant produced by asexual reproduction would be

1 identical to that of the parent plant
2 similar, but not identical, to that of the parent plant
3 totally different from that of the parent plant
4 a combination of genetic information from several plants

16. A researcher determines that all the members of a certain population of plants on a lawn are genetically identical. The best explanation for this is that the plant

1 reproduces sexually, by cloning
2 reproduces sexually, by budding
3 reproduces asexually, by cloning
4 reproduces asexually, by budding

17. A new hydra can be produced from groups of cells that enlarge and stay attached to the parent hydra for a time before breaking off and becoming independent. This method of reproduction is called

1 sporulation
2 cloning by runners
3 binary fission
4 budding

18. One way to produce many genetically identical offspring is by

1 using radiation to change their genes
2 using chemicals to change their genes
3 cloning them, so they have the same genes
4 inserting a new DNA section into their genes

19. Which phrase does *not* describe cells cloned from a carrot?

1 they are genetically identical
2 they have the same DNA codes
3 they are reproduced sexually
4 they have identical chromosomes

20. Which statement about the rate of cell division is true?

1 All the cells of all organisms divide at the same rate.
2 The rate of cell division is related to a cell type's function.
3 All the cells within an organism divide at the same rate.
4 The rate of cell division is random in every organism.

21. Damage to a cell's DNA can cause cancer, which results from

1 a slower than normal cell division
2 a complete stop to all cell division
3 an uncontrolled type of cell division
4 no changes in the genetic instructions

Part B—Analysis and Open Ended

22. Why does the survival of a species depend more on populations than on the life spans of individual organisms?

23. Briefly explain the main purpose of the genetic material in a cell.

24. Use the following terms to replace the definitions given within the boxes in the following concept map: *Cell division; Cell growth; Mitosis; Replication.*

First Stage —— | Cell increases in size and number of parts. |

followed by

Middle Stage —— | Genetic material duplicates. |

leads to

| Chromosomes divide into two equal groupings. |

Final Stage <

results in

| Cytoplasm divides to form two daughter cells. |

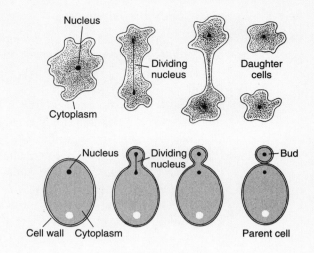

25. The best title for this concept map probably would be

1　The Functions of a Cell
2　The Life Cycle of a Cell
3　The Regeneration of a Cell
4　The Genetics of the Cell

26. How does the chromosome number of one species compare with that of another species?

27. Explain why the duplication of chromosomes is necessary for the process of cell division.

28. Why is cell division necessary for all living things? Your answer should include the following:

◆ why it is important for one-celled organisms

◆ why it is important for multi-celled organisms

◆ *one* example of each type of organism discussed

29. During cell division, both the genetic material and the cytoplasm have to be equally divided. Which process occurs first? Which process is called mitosis?

Refer to the two diagrams at the top of the next column to answer questions 30 and 31.

30. What type of reproduction is illustrated in these two diagrams?

1　sexual reproduction only
2　asexual reproduction only
3　sexual and asexual reproduction
4　neither type of reproduction

31. In what way is reproduction in the ameba (top) the same as reproduction in the yeast (bottom)? In what way is it different? What is the specific term for the yeast's method of reproducing?

32. Which of the following diagrams represents asexual reproduction (by mitosis)? Explain why. [*Note:* The "n" stands for number of chromosomes in each cell.]

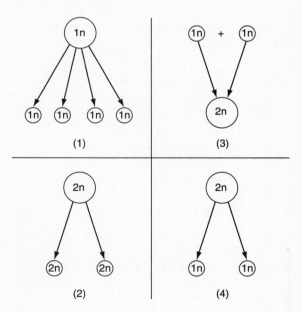

33. Does the rate of cell division differ from one tissue type to another within an organism? Why is this important for homeostasis? (Give *one* example.)

34. Explain why exposure to radiation and certain chemicals can cause uncontrolled cell division. What type of disease can this process lead to?

Part C—Reading Comprehension

Base your answers to questions 35 to 37 on the information below and on your knowledge of biology. Use one or more complete sentences to answer each question.

Non-Hodgkin's lymphoma (NHL) is a cancer of the lymph system. This system collects intercellular fluid from throughout the body, returning it in tubes to the bloodstream. The tiny lymph vessels join together to eventually form large ones that empty into veins in the neck. Enlargements along these lymph vessels are known as lymph nodes. These nodes, or glands, are involved in the body's defenses against diseases. However, the lymph system is also the site for NHL cancer—one of the few cancers that is occurring with greater frequency. No one knows why the incidence of NHL is increasing, but it now accounts for more than 4 percent of cancer deaths in this country.

Chemotherapy, the traditional use of drugs to treat cancers such as NHL, was developed during the twentieth century. These anti-cancer drugs use a variety of methods to attack cancer cells: by attacking DNA; by shutting down protein synthesis; or by stimulating the immune system. In the mid-1990s, trials began for the use of a very different type of drug—monoclonal antibodies. These drugs are actually designer-made antibodies that have been produced to find and attack cell-surface targets that exist only on cancer cells. The monoclonal antibody drugs are therefore referred to as *targeted drugs*; they search out the cancer cells. The cutting-edge capability of these twenty-first-century drugs is to attach radioactivity or some other cancer-fighting drug to the monoclonal antibody. The targeted drug will find the cancer cells, deliver the deadly payload, and then kill the cancer cells. This treatment is now being used against NHL with some success.

Doctors currently stress that the best approach is to use both methods—twentieth-century and twenty-first-century cancer treatments. Well-respected experts are optimistic about the chances for real progress in the years ahead. For example, Dr. Andrew Zelenetz, chief of the lymphoma services at Memorial Sloan Kettering Cancer Center in New York City has said, "This is a very exciting time. We didn't have new important agents for the treatment of lymphoma for many years. Now we're seeing the emergence of these targeted therapies that are very exciting, and in fact, we're starting to see the emergence of other chemotherapeutic agents that actually have activity in lymphoma. We're entering a new era where we have both the traditional tools as well as these new targeted tools, and we're going to be seeing more of them coming down the pike. There are a number of new agents that are in development that are being tested that I think have real promise." Hopefully these new cancer-fighting agents will be developed in time to fight the increase in incidence of NHL and other potentially deadly cancers.

35. Compare the lymph system with the circulatory system.

36. How do traditional anti-cancer drugs work?

37. How do targeted drugs work in fighting cancer?

16

Meiosis and Sexual Reproduction

⚒ SEXUAL REPRODUCTION: IT TAKES TWO

For almost all animals, it takes two to reproduce: a male and a female. This is *sexual reproduction*. Most plants use this method of reproduction to make more of their own kind, too. Sexual reproduction is very important in understanding living things. It also plays a significant role in the process of evolution. To understand why this is so, we must look at individual cells and closely examine the chromosomes within them. (See Figure 16-1.)

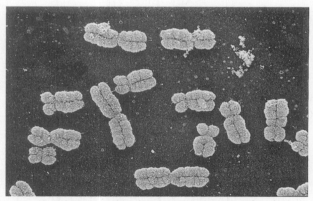

Figure 16-1 A photograph of human chromosomes.

⚒ IT'S ALL ABOUT CHROMOSOMES

Each of our cells contains chromosomes. The chromosomes contain the inherited information that has been passed along since the beginning of life on Earth. It is this information that determines an individual's characteristics. The chromosomes also contain the "know-how" that keeps our cells functioning correctly.

Why is sexual reproduction all about chromosomes? When a **sperm** cell and an **egg** cell unite during sexual reproduction, it is the nu-cleus from each cell that joins. What does the nucleus contain? The chromosomes. So sexual reproduction is about the combining of chromosomes from two individuals, a male (the father) and a female (the mother).

Each human body cell contains 46 chromosomes. The first cell from which each of us came—the cell that resulted from the combination of a sperm cell and an egg cell—had 46 chromosomes. Every body cell now in you still has 46 chromosomes. The question is, how did a sperm cell and an egg cell combine to make a new cell with just 46 chromosomes?

There is only one way. Both the sperm and the egg must have had only 23 chromosomes each, half the amount of chromosomes from the normal number of 46. And indeed this is the case. A special type of cell division produces sperm and egg cells, each with that reduced number of chromosomes.

⚒ GAMETES

The sperm and egg cells, or **sex cells**, are also called **gametes**. In the process of sexual reproduction, the nuclei of the gametes join together. This fusion of the nuclei is called **fertilization**. The resulting cell, a fertilized egg cell, is called a **zygote**. (See Figure 16-2 on page 120.) Each gamete, as we have said, has exactly half the normal number of chromosomes. The zygote and all body cells that come from the mitotic division of the zygote have two sets of chromosomes in them, one from each parent. Gametes are produced by a type of cell division that reduces the chromosome number by one-half, giving them just one set of chromosomes. Thus, when fertilization occurs, the normal number of chromosomes for

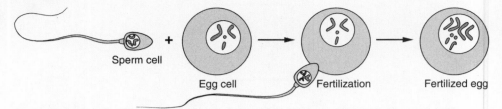

Figure 16-2 Sexual reproduction involves the joining of chromosomes from a sperm cell and an egg cell in the process called fertilization.

the species is maintained. This special type of cell division is called **meiosis**.

▚ A CLOSER LOOK AT CHROMOSOMES

Our chromosomes exist in pairs. Essentially, we have two chromosomes of each type. And where does each of these two chromosomes come from? The answer is: one from each parent. Beginning with a normal body cell, which has the double set of chromosomes, gametes must be produced through meiotic cell division. Each gamete contains a single set of chromosomes. And it must be an exact set, meaning one and only one from each of the pairs of chromosomes. (See Figure 16-3.)

▚ MEIOSIS: REDUCING THE CHROMOSOME NUMBER

Mitosis and meiosis take place during cell division, and in some ways these two processes are similar. The chromosomes replicate before either process begins. However, the results of mitosis and meiosis are very different. When mitosis is completed, the chromosome number remains the same as in the original parent cells. When meiosis is completed, the chromosome number is half the original number. Meiosis actually involves two separate cell divisions, which take place one after the other.

▚ MEIOSIS: THE SOURCE OF OUR DIFFERENCES

With the exception of identical twins, children in the same family are never exactly alike. Differences can occur in eye color, hair

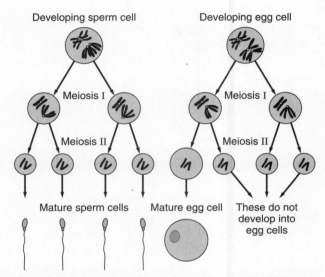

Figure 16-3 As a result of meiosis, sperm cells and egg cells have half the normal number of chromosomes for their species.

color, height, nose shape, ear size, and many other characteristics. Why is this so, if the children were born of the same parents? The explanation arises from one of the two important jobs of meiosis. The first job of meiosis, as we have said, is to maintain the normal species chromosome number by preparing gametes with single sets of chromosomes.

The second important job of meiosis is to increase genetic variability by **recombining** genes in the eggs and sperm. (See Figure 16-4.) Genes may get exchanged between

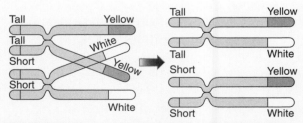

Figure 16-4 During meiosis, genetic recombination occurs when chromosomes overlap and exchange pieces.

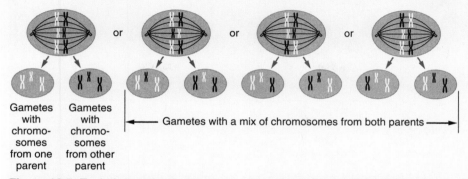

Figure 16-5 Each time meiosis occurs, the chromosomes line up in a different arrangement, resulting in variability among offspring.

chromosomes during meiosis. Also, chromosomes may get resorted into new groupings. Because of genetic recombination during meiosis, sexual reproduction results in offspring that are different from each other and from their parents. (See Figure 16-5.) This genetic variation is what natural selection acts on. A greater variety of characteristics in offspring increases the chances that some individuals will be better suited than others to survive in a particular place and time. As natural selection acts on the varied offspring in a population, generation after generation, the species evolves.

❖ UNUSUAL MEIOTIC EVENTS

The sorting of chromosomes that occurs in cell division—especially during meiosis—is a wonderfully complex sequence of events. However, it does not always proceed correctly. A gamete may have an extra chromosome because it receives both members of a pair of chromosomes, instead of only one. Or a gamete may be one chromosome short, having received neither member of a pair. If a gamete with such an abnormality fuses with another gamete, problems may occur. In most instances, the zygote fails to develop. However, in some cases, the zygote does develop into an individual with an abnormal chromosome number.

❖ THE SEX LIFE OF FLOWERING PLANTS

Of the many types of plants on Earth, flowering plants are the group that has evolved

most recently. Most types of plants reproduce sexually. As in animals, the male gamete (sperm) joins together with the female gamete (egg) to produce a zygote (the seed). The zygote then begins to grow into a new plant. What is special about flowering plants is that the place where all this happens is very visible. The location is often brightly colored, beautifully shaped, and sweet smelling—that is, flowers are the parts of plants where sexual reproduction occurs. In fact, the parts that make up flowers include the sex organs of the plants. (See Figure 16-6.)

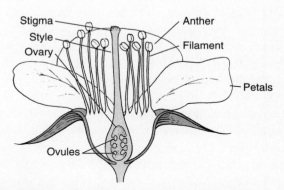

Figure 16-6 In flowering plants, sexual reproduction takes place in the flowers, which contain the reproductive structures.

❖ SEXUAL REPRODUCTION: INTERNAL OR EXTERNAL?

Various methods of sexual reproduction occur in plant and animal species. Yet whatever the method, sexual reproduction always involves fertilization (the fusion of nuclei from two gametes) and **development**, the growth of the zygote into a new individual. One of the main differences in the types of reproduction

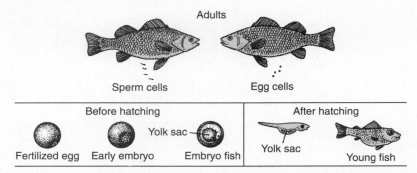

Figure 16-7 In most species of fish, both fertilization and development are external; many eggs are released to ensure that some offspring will survive.

involves the location of the events. Both fertilization and development may occur either inside or outside the bodies of the reproducing organisms.

For many aquatic plants, the sex cells meet in the open water; fertilization occurs and the zygote begins to develop. Many invertebrates simply release their gametes into the water, where fertilization and development then occur. Two groups of vertebrates, the fish and the amphibians, also reproduce in water. In most species of fish, the eggs and sperm are released directly into the water, where fertilization and development of the zygotes then occur. These events, which occur in the environment and not inside the organism, are known as *external fertilization* and *external development*. External fertilization and development are risky. Therefore, large numbers of eggs are released to increase the chances that some of the offspring will survive. This is natural selection at work. (See Figure 16-7.)

In the case of vertebrates that reproduce on land, the gametes still need moisture to meet and fuse. Reptiles and birds make use of the fluids inside their bodies for fertilization. The male and the female must mate for the sperm to be deposited inside the female, a process known as **internal fertilization**. Then the zygote is prepared for development on land: a watertight membrane and protective shell form around the zygote; the egg is laid (usually within a nest); and development of the new organism occurs externally. Internal fertilization increases the chances of reproductive success and survival. Fewer eggs are produced, but there is some parental care to

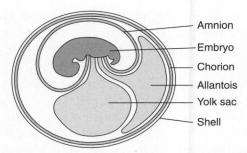

Figure 16-8 In birds, fertilization is internal. The embryo is surrounded by a watertight membrane (the amnion) and then covered by a protective shell. The egg is laid and development is external.

help protect the developing zygote. (See Figure 16-8.)

One final pattern of sexual reproduction takes place mainly in mammals. Fertilization occurs internally, but the big difference from most other animal groups is that development of the zygote occurs within the female's body, too. Thus, mammals have **internal development**. The food for the developing **embryo** comes entirely from the body of the mother. A structure called the **placenta** has evolved to bring nutrients to the developing baby and to remove its wastes. (*Note:* The exceptions to this are the marsupial mammals, which complete their embryonic development within a protective pouch, and the egg-laying mammals.) After birth, continuing nourishment of the baby mammal occurs through its nursing on milk provided by the mother's mammary glands. Although the embryos are fewer in number, they have the most complete form of protection, since they develop within the body of the female and then receive more parental care after their birth.

Chapter 16 Review

Part A—Multiple Choice

1. During sexual reproduction, the chromosomes of
   1 two separate individuals are combined together
   2 one individual are transferred to another
   3 one parent only are copied for its offspring
   4 two separate individuals are split apart

2. If each human body cell has 46 chromosomes, how many were in your very first body cell?
   1 23   3 92
   2 46   4 100

3. Most cells in the body of a fruit fly contain eight chromosomes. How many of these chromosomes were contributed by each parent of the fruit fly?
   1 8   3 2
   2 16   4 4

4. Sperm cells of the Russian dwarf hamster contain 14 chromosomes. What is the total number of chromosomes that would be found in each cell of a normal, newly formed zygote of this species?
   1 7   3 14
   2 28   4 42

5. The gamete (sex cell) for any species should *always* contain
   1 an even number of chromosomes
   2 the normal number of chromosomes
   3 twice the normal number of chromosomes
   4 half the normal number of chromosomes

6. The following diagram represents some events in a cell undergoing normal meiotic cell division.

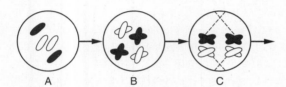

A   B   C

Which of the following diagrams most likely represents the next cell that would result from the process shown in the diagram above?

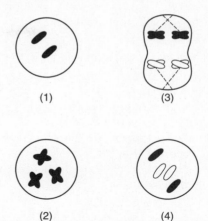

(1)   (3)

(2)   (4)

7. Compared to human cells resulting from mitotic cell division, human cells resulting from meiotic cell division should have
   1 twice as many chromosomes
   2 one-half as many chromosomes
   3 the same number of chromosomes
   4 one-quarter as many chromosomes

8. During fertilization, the parts of the sex cells that join are the
   1 membranes
   2 nuclei
   3 ribosomes
   4 vacuoles

9. Which of these is formed during fertilization?
   1 an egg cell
   2 a sperm cell
   3 a zygote
   4 a gamete

10. Most cells in the body of a fruit fly contain eight chromosomes. In some cells, only four chromosomes are present, a condition that is a direct result of
    1 mitotic cell division
    2 embryonic differentiation
    3 meiotic cell division
    4 internal fertilization

11. Which statement best explains the significance of meiosis in the evolution of a species?
    1 Meiosis produces egg cells and sperm cells that are completely alike.
    2 Meiosis ensures the continuation of a species by asexual reproduction.

3 Meiosis produces equal numbers of egg cells and sperm cells in animals.

4 Meiosis results in genetic variation among the gametes that are produced.

12. Mitosis and meiosis are similar in that

1 the chromosomes are replicated before either process starts

2 the chromosome number is the same when each process is completed

3 two separate cell divisions occur during each process

4 each process combines genetic material from two individuals

13. Which diagram correctly represents part of the process of sperm formation in an organism that has a normal (species) chromosome number of eight?

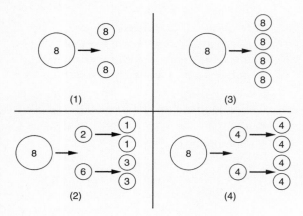

14. The great variety of possible gene combinations in a sexually reproducing species is due in part to the

1 sorting of genes as a result of gene replication

2 pairing of genes as a result of differentiation

3 sorting of genes as a result of meiosis

4 pairing of genes as a result of mitosis

15. During meiosis, recombining (gene exchange between chromosomes) may occur. Genetic recombination usually results in

1 overproduction of gametes

2 variation within the species

3 fertilization and development

4 formation of identical offspring

16. The following diagram shows a process that can occur during meiosis.

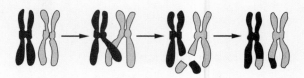

The most likely result of this process is

1 a new combination of inheritable traits that can appear in the offspring

2 a loss of genetic information that will produce a genetic disorder in the offspring

3 an inability to pass either of these chromosomes along to future offspring

4 an increase in the chromosome number of the organism in which this occurs

17. Mitosis produces new body cells and meiosis produces

1 new body cells, too

2 body cells and sex cells

3 sex cells, only

4 red blood cells

18. Which of the following is a characteristic found *only* in sexual (*not* asexual) reproduction?

1 cell division

2 cell growth

3 fertilization

4 chromosomes

19. Sexual reproduction in flowering plants occurs within the

1 roots

2 stems

3 leaves

4 flowers

20. An animal that has external fertilization will produce more eggs than an animal that has internal fertilization, because

1 the siblings help raise each other without any parental involvement

2 an animal can reproduce externally only once during its lifetime

3 it increases the chances that some of the offspring will survive

4 that way the parent animal can locate some eggs after they are fertilized

21. In terms of reproduction, how do mammals *differ* from most other animals?

1 The gametes are formed internally.

2 Fertilization takes place internally.

3 The zygote is formed externally.
4 The embryo develops internally.

21. What is the role of the placenta in the embryonic development of a mammal?
1 It forms a protective barrier around the developing baby.
2 It brings nutrients to and removes wastes from the developing baby.
3 It provides the location for fertilization of the egg to occur.
4 It provides a method of nourishing the baby after it is born.

Part B—Analysis and Open Ended

23. Briefly state what chromosomes contain and what they determine in organisms.

24. The diagram below represents an incomplete process of meiosis in an animal's ovary. In your notebook, copy and complete the diagram by drawing in the chromosomes of cell A. Your drawing should show the usual result of meiosis in the formation of an egg cell.

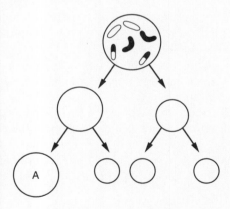

25. In what way is sexual reproduction basically all about chromosomes?

26. Use the words *gametes*, *zygote*, and *fertilization* in one sentence to explain sexual reproduction.

Base your answers to questions 27 and 28 on the diagram at the top of the next column and on your knowledge of biology.

27. State why process 2 is necessary in sexual reproduction.

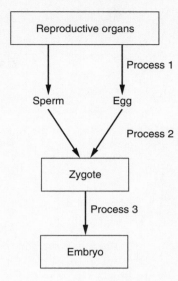

28. State *one* difference between the cells produced by process 1 and the cells produced by process 3.

29. Why are gametes essential to sexual reproduction, in terms of their chromosome number?

30. Explain why, in a mammal, a mutation in a gamete may contribute to evolution while a mutation in a body cell will not.

31. Although paramecia (single-celled organisms) usually reproduce asexually, some have developed a method by which they exchange genetic material with each other in a simple form of sexual reproduction. State *one* advantage this simple form of sexual reproduction would provide over asexual reproduction for the survival of these single-celled organisms.

Refer to the diagram below to answer questions 32 to 34. [Note: The "n" stands for the number of chromosomes in each cell.]

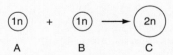

32. What reproductive process does the diagram represent?
1 gamete formation
2 cell division
3 fertilization
4 recombination

33. Which of the structures in the diagram represents a gamete?

1 A only
2 B only
3 A and B
4 C only

34. Which of the structures in the diagram represents a zygote?

1 A only
2 B only
3 C only
4 none of them

35. Which one of the following diagrams represents meiosis? [*Note:* The "n" stands for the number of chromosomes in each cell.]

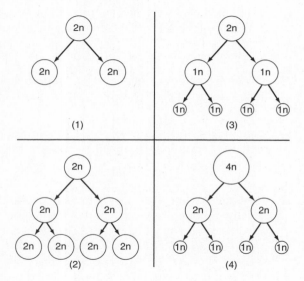

(1)   (2)   (3)   (4)

36. In terms of chromosome number, what is the main difference between the results of mitosis and meiosis?

37. Why are offspring of organisms that reproduce sexually *not* genetically identical to their parents?

38. Why is meiosis important in terms of evolution? Your answer should explain *how* it affects the following:

◆ the chromosome number in a species
◆ genetic variability within a population
◆ natural selection and survival of offspring

39. Compare asexual reproduction to sexual reproduction. In your comparison, be sure to include:

◆ which type of reproduction results in off-spring that are usually genetically identical to the previous generation and explain why this occurs
◆ one other way these methods of reproduction differ

Refer to the illustration below, which shows an important event that occurs during meiosis, to answer questions 40 and 41.

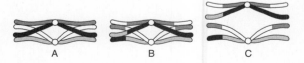

A   B   C

40. What is occurring in this process from steps A to B to C? How are the chromosomes in step C different from those in step A?

41. Why is this process significant in terms of the offspring that are produced?

42. Briefly describe the process of reproduction in flowering plants.

43. Which characteristic of sexual reproduction has specifically favored the survival of animals that live on land?

1 fusion of gametes in the outside environment
2 male gametes that may be carried by the wind
3 fertilization within the body of the female
4 female gametes that develop within ovaries

44. Which process normally occurs at the placenta?

1 Oxygen diffuses from fetal blood to maternal blood.
2 Maternal blood is converted into fetal blood.
3 Materials are exchanged between fetal and maternal blood.
4 Digestive enzymes pass from maternal blood to fetal blood.

Base your answers to questions 45 to 47 on the paragraph below and on your knowledge of biology.

Three groups of animals in which most species lay eggs for reproduction are amphibians, reptiles, and birds. Most female amphibians lay hundreds of eggs in water, which are then fertilized by sperm from the male. Many reptiles lay up to 200 eggs at a time, often in nests on land. The eggs have a leathery shell. Birds usually lay between one and four eggs at a time in nests on land. Wild bird eggs usually have shells similar to those of the domestic chicken. Most mammals bear live young. Some of these mammals, such as humans, usually give birth to just one live offspring at a time.

45. State *one* reason that individuals of some species must lay hundreds of eggs in order for the species to survive.

46. Explain why fertilization in reptiles and birds must be internal.

47. State *two* reasons (related to reproduction) that the human species has been able to survive, even though only one offspring is usually produced at a time.

Part C – Reading Comprehension

Base your answers to questions 48 to 50 on the information below and on your knowledge of biology. Use one or more complete sentences to answer each question.

It seems impossible to imagine all organisms of a single species reproducing at the same time. But this is exactly what happens when all trees of a species simultaneously release pollen into the wind to fertilize the female flowers of that species. Still, in animal species, simultaneous reproduction of many individuals is a rare occurrence. However, one of the most spectacular underwater events involves the mass spawning of the millions of small organisms that make up a coral reef. A coral reef is a stony structure made of minerals removed from the water, over a long period of time, by tiny coral animals that live on the outer edges of the reef. Reefs are found only in the clear, warm, shallow waters of the tropics. During their lifetime, coral organisms remain in one place. They catch and remove food particles from the water that surrounds them. One good place to observe the mass spawning of coral animals is the Flower Gardens National Marine Sanctuary in the Gulf of Mexico.

Scientists think that mass spawning increases the chances of successful fertilization in three ways. First, with so many eggs and sperm in the water at the same time, fertilization is more likely to occur. Second, with gametes from different colonies of one species being released at the same time, cross-fertilization between different colonies is more likely. This increases the genetic variation among the offspring. Finally, with so many fertilized gametes in the water at once, the amount lost to predation is limited.

48. Compare the process of simultaneous reproduction in plants to that in animals such as coral.

49. Why is cross-fertilization between different coral colonies beneficial?

50. What advantage does mass spawning provide for the fertilized eggs?

Human Reproduction

▓ THE MALE REPRODUCTIVE SYSTEM

The male reproductive system in humans has two main functions: first, to produce male gametes (the sperm cells); second, to deposit the sperm cells it produces inside the female. In addition, the male reproductive system provides a pathway for the removal of urine. (See Figure 17-1.)

The sperm cell formation occurs in the two **testes**. The formation of sperm requires a temperature that is a few degrees cooler than that of the rest of the body. This lower temperature occurs in the testes because they are not located within the body cavity. Instead, the testes are suspended in a sac called the *scrotum*. The scrotum is an adaptation that has evolved to increase the chances of producing healthy sperm.

Inside the testes are a great many tiny tubes, or tubules. As cells move through these tubules, they undergo the meiotic cell division that leads to formation of the gametes. Nowhere else in the male's body does meiotic cell division occur.

Sperm cells are highly specialized cells that are able to move. Each sperm cell must be able to deliver a single set of chromosomes from the male to an egg cell in the female. The structure of a mature sperm cell is well adapted to its function. Almost the entire head of the sperm is the nucleus, the all-important genetic information that is delivered to the egg. Attached to the head of the sperm is a long tail that propels the cell along. Also present are large numbers of mitochondria that produce ATP, which supplies the energy that sperm use to propel themselves to the egg. (See Figure 17-2.)

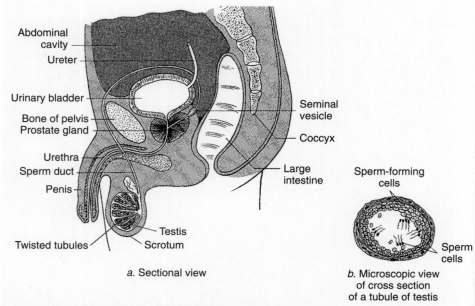

a. Sectional view

b. Microscopic view of cross section of a tubule of testis

Figure 17-1 The human male reproductive system.

Figure 17-2 The mature sperm cell is well adapted to deliver its single set of chromosomes to the egg cell in the female—the head contains the genetic information and the mitochondria supply the energy used by the tail to propel the sperm cell.

After sperm move from the testes, a number of glands add fluids. The sperm and these fluids make up the *semen*. In fact, most of the semen consists of fructose, a sugar that provides an additional source of energy for the sperm.

The male reproductive system is adapted for internal fertilization. The penis is a structure that has evolved to deposit sperm safely within the female's reproductive tract. This occurs when the semen is forced from the body during ejaculation.

❖ THE FEMALE REPRODUCTIVE SYSTEM

Three important functions are performed by the female reproductive system: first, to produce the female gametes (the egg cells); second, to provide a pathway for sperm cells to reach the egg; and third, to provide a temporary home for the developing embryo. (See Figure 17-3.)

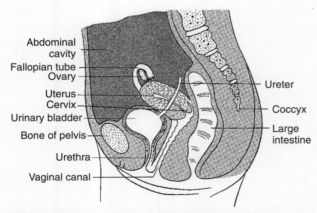

Figure 17-3 The human female reproductive system (side view): egg cells are produced in the ovaries; sperm cells join the egg cells in the oviduct (fallopian tube); and the embryo develops in the uterus.

In females, egg cells are produced in the **ovaries**, a pair of reproductive organs. In an adult male, sperm production occurs all the time; an average of 30 million sperm are produced each day. In a female, all potential eggs are already present when she is born. Throughout her reproductive life, a female releases only a few hundred of these eggs. Usually only a single egg matures and is released each month, packed with the nutrients needed to nourish an embryo right after fertilization.

This development of egg cells occurs (in a sexually mature female) within the ovaries every month. One mature egg cell is released from one of the ovaries. This event is called *ovulation*. The egg cell gets swept into a long tubular structure called the *oviduct,* or Fallopian tube, found next to each ovary. If fertilization occurs, the sperm usually joins the egg in the oviduct. The egg continues to move along the oviduct to the **uterus**, a pear-shaped organ with thick muscular walls. If the egg cell has been fertilized, the embryo becomes attached to the inside wall of the uterus and continues to develop. If fertilization did not occur, the egg cell breaks down within 24 hours of ovulation and is passed from the body. (See Figure 17-4.)

At the lower end of the uterus is the cervix, a narrow opening through which the sperm travel on their way to the egg cell. Connecting the cervix to the outside of the body is the vagina, which is made up of muscular tissue. It is into the vagina that sperm are ejaculated from the penis. Also, the vagina is the birth canal, through which the infant passes as it leaves the mother's body during birth.

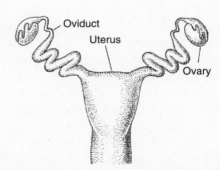

Figure 17-4 Front view of the female reproductive system (simplified).

Unlike in the male, the reproductive pathway in females is not combined with a pathway for excretion. Instead, the urine passes through an opening near the vagina.

⚏ HORMONES AND SEXUAL REPRODUCTION

During one's life, many changes and events occur in the body to make sexual reproduction possible. Hormones coordinate these changes. The main endocrine gland in charge of producing these hormones is the pituitary gland in the brain. The pituitary gland is controlled by the hypothalamus, a part of the brain.

The effects of a hormone depend not on the hormone itself but on its target tissue. The testes produce **testosterone**, the main male sex hormone. The effects of testosterone include the development of the male sex organs before and after birth. Around the age of 11, the level of testosterone suddenly increases in a boy's body. As a result, sperm production begins. This event is the beginning of *puberty*, or sexual maturation. During puberty, the penis and the testes begin to mature.

Testosterone also affects various other tissues in the male, causing the growth of facial and body hair, changes in body proportions, deepening of the voice, and other changes. These developments are called *secondary sex characteristics* because they are not directly related to sexual reproduction. In males, the level of testosterone in the body remains much the same for about 40 years after puberty, after which it gradually begins to decrease.

In females, the major sex hormones, **estrogen** and **progesterone**, are produced and released from the ovaries. The onset of puberty in females occurs somewhat earlier than in males. At about age 10, the levels of estrogen and progesterone increase dramatically, causing the uterus, vagina, and ovaries to mature. Secondary sex characteristics, such as the growth of body hair and breast (mammary gland) development, also are influenced by estrogen and progesterone. In addition, a monthly cycle of events known as the *menstrual cycle* begins. Remember, in males, sperm production occurs all the time after

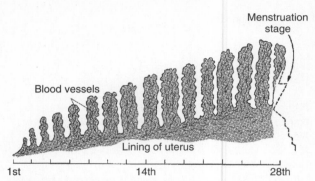

Figure 17-5 Ovulation—the release of a mature egg from the ovary—occurs in the middle of the menstrual cycle, as hormones cause the lining of the uterus to thicken.

puberty. In females, the menstrual cycle occurs every month after puberty. Part of this cycle includes the release of an egg cell from the ovaries. (See Figure 17-5.)

In addition to ovulation, another critical function occurs during the menstrual cycle. The woman's body must be prepared in case fertilization occurs. Everything must be ready to nurture the developing embryo. So during the first two weeks of the cycle, estrogen causes the lining of the uterus to thicken. There is also an increase in the amount of blood flow to that area.

During the second half of the cycle, after ovulation has occurred, progesterone, the pregnancy hormone, prepares the uterus for an embryo. If **pregnancy** occurs, the embryo becomes attached to the inner lining of the uterus. The **placenta** develops between the fetus and the uterus for the exchange of materials. The growing tissue then begins to release hormones to keep everything in the right condition. However, if fertilization does not occur, the continued preparations in the uterus are unnecessary. The body realizes this near the end of the four-week period of the menstrual cycle. As the level of progesterone decreases, the uterine lining no longer remains intact, so it breaks down. The built-up tissue along with some blood and the unfertilized egg are released from the body. This flow of blood, called *menstruation*, lasts for about four days. Then the cycle begins again. (See Figure 17-6.)

In women, the menstrual cycle continues

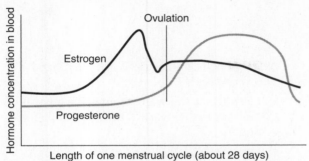

Figure 17-6 The menstrual cycle and ovulation, which occur every month after puberty, are controlled by the release of hormones.

for about 40 years from puberty. Between the ages of 45 and 55, the levels of hormones change; the menstrual cycle becomes less regular and eventually stops. This stage, called *menopause*, marks the point at which a female is no longer capable of reproducing. Menopause is a normal occurrence in all women. However, the effects of menopause vary widely from one woman to another. In men, by contrast, sperm production continues throughout life, although the number of healthy sperm likely declines with age.

Chapter 17 Review

Part A—Multiple Choice

1. The reproductive system of the human male produces gametes and

   1 transfers gametes to the female for internal fertilization
   2 releases hormones involved in external fertilization
   3 produces enzymes that prevent fertilization
   4 provides an area for fertilization of the gamete

2. The testes are adapted to produce

   1 body cells involved in embryo formation
   2 immature gametes that undergo mitosis only
   3 sperm cells that may be involved in fertilization
   4 gametes with large food supplies that nourish a developing embryo

3. The scrotum is located outside the body cavity, enabling healthy sperm to form because

   1 blood does not flow to this region
   2 the cells in the testes do not divide
   3 its temperature is higher than that of the body
   4 its temperature is lower than that of the body

4. In the male, meiotic cell division (that is, sperm cell formation) occurs within the

   1 penis
   2 bladder
   3 testes
   4 semen

5. The shape of a sperm cell can best be described as

   1 an oval with four limbs
   2 a head with a long tail
   3 a tree with branches
   4 a long, twisted tunnel

6. ATP is important to sperm cells because it

   1 supplies the energy they need to move
   2 enables the cells to replicate and divide
   3 reduces their chromosome number
   4 doubles their chromosome number

7. Most of the semen consists of

   1 sperm
   2 lipids
   3 sugar
   4 protein

8. The development of a human female's egg cells occurs within her

   1 ovaries
   2 oviduct
   3 cervix
   4 uterus

9. How does the production of male gametes differ from that of female gametes?

   1 A male is born with all of his potential gametes, whereas a mature female produces them every day.
   2 A female is born with all of her potential gametes, whereas a mature male produces them every day.

3 Female gametes are produced during fertilization, whereas male gametes are produced in advance.

4 Males produce one gamete at a time, whereas females produce millions each month.

10. The diagrams below represent cells that transport chromosomes. These cells are specialized for

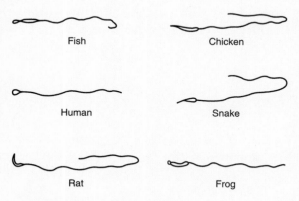

Fish

Chicken

Human

Snake

Rat

Frog

1 anaerobic respiration
2 sexual reproduction
3 chemical transmission
4 antibody production

11. Which statement does *not* correctly describe an adaptation of the human female reproductive system?

1 It produces gametes within the ovaries.
2 It provides for external fertilization of an egg.
3 It provides for internal development of the embryo.
4 It removes the wastes produced by the fetus.

12. Structures (in side view) in the human female are represented in the diagram below. Fertilization of the egg cell normally occurs within the part labeled

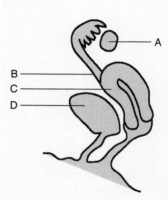

1 A   3 C
2 B   4 D

13. What happens if an egg cell is *not* fertilized after it is released?

1 It attaches to the wall of the uterus anyway.
2 It returns to the ovary to be released the next month.
3 It breaks down and is passed out of the female's body.
4 It remains in the oviduct until new sperm are introduced.

14. Regulation of sexual reproductive cycles of human males is related most directly to the presence of the hormone

1 estrogen
2 testosterone
3 progesterone
4 insulin

15. Estrogen and progesterone are examples of

1 organelles
2 tissues
3 hormones
4 enzymes

16. Secondary sex characteristics are traits that are

1 caused by sex hormones but are not directly related to reproduction
2 caused by the female sex hormones only
3 caused by the male sex hormones only
4 involved in reproduction, but only for the second child

17. Which process normally occurs at the placenta?

1 Oxygen diffuses from fetal blood to maternal blood.
2 Maternal blood is converted into fetal blood.
3 Materials are exchanged between fetal and maternal blood.
4 Digestive enzymes pass from maternal blood to fetal blood.

18. Why does the lining of the uterus thicken during the first half of the menstrual cycle?

1 to provide a place for sperm cells to attach
2 to produce a mature egg cell
3 to prepare to nurture an embryo
4 to rid the body of unfertilized egg cells

19. Why does menstruation occur?

    1 to produce an egg in one of the ovaries
    2 to release an egg from one of the ovaries
    3 to allow an egg to be fertilized in the uterus
    4 to remove an unfertilized egg from the uterus

20. If fertilization of the egg does *not* occur after ovulation, the

    1 level of progesterone then increases
    2 level of progesterone then decreases
    3 level of testosterone then increases
    4 level of estrogen then decreases

21. The diagram below represents a side view of the human male reproductive system. If structure X were to be tied and cut off at the arrow, which change would occur in this system?

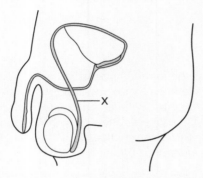

    1 Sperm cells would no longer be produced.
    2 Hormones would no longer be produced.
    3 Sperm would be produced but no longer released from the body.
    4 Urine would be produced but no longer released from the bladder.

22. After the onset of menopause, women normally

    1 release more eggs than they did before
    2 are no longer capable of reproducing
    3 menstruate for longer periods of time
    4 produce more estrogen and progesterone

Part B—Analysis and Open Ended

23. Describe the three functions of the human male reproductive system.

24. In what way is the cell division that occurs in the testes different from that which occurs elsewhere in the male body?

Base your answers to questions 25 and 26 on the diagram below, which represents the human male reproductive system.

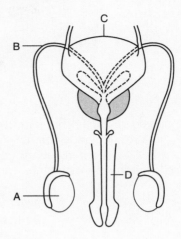

25. The hormone produced in structure A most directly brings about a change in the male's

    1 blood sugar concentration
    2 rate of digestion
    3 physical characteristics
    4 rate of respiration

26. Which pair of letters indicates both a structure that produces gametes and a structure that makes possible the delivery of those gametes for internal fertilization?

    1 A and D   3 C and A
    2 B and D   4 C and B

27. Explain how the structure of a mature sperm cell is related to its function. Your answer should mention the following features:

    ♦ the head of a sperm cell
    ♦ the tail of a sperm cell
    ♦ the mitochondria of sperm

28. List the three main functions of the female reproductive system. Include the specific structures involved in each of these functions.

29. What is the main difference between males and females in terms of the formation of their sex cells?

Base your answers to questions 30 and 31 on the following diagram, which represents the human female reproductive system.

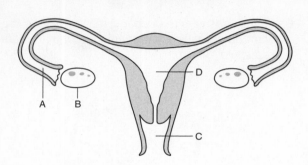

30. New, inherited characteristics may appear in offspring as a result of new combinations of existing genes—or from mutations in genes—in the cells produced by structure

1 A   3 C
2 B   4 D

31. A fetus normally develops within the structure labeled

1 A   3 C
2 B   4 D

32. Explain the meaning of the term *puberty*. How is the onset of puberty related to hormones and secondary sex characteristics?

The diagrams below represent the reproductive organs of two individuals—one male and one female. The diagrams are followed by a list of sentences. For each phrase in questions 33 to 35, select the sentence from the list that best applies to that phrase.

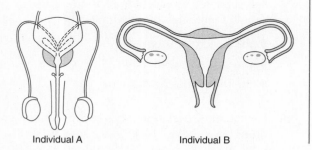

Individual A   Individual B

Sentences:
A The phrase is correct for both Individual A and Individual B.
B The phrase is *not* correct for either Individual A or Individual B.
C The phrase is correct for Individual A, only.
D The phrase is correct for Individual B, only.

33. Contains organs that produce gametes (sex cells).

1 A   3 C
2 B   4 D

34. Contains organs involved in internal fertilization.

1 A   3 C
2 B   4 D

35. Contains the structure in which a zygote develops.

1 A   3 C
2 B   4 D

36. Use the following terms to complete the boxes in the flowchart below: *facial hair; testosterone; wider hips; broader shoulders; estrogen; deeper voice; progesterone; ovaries and uterus mature; breasts develop; menstrual cycle begins; sperm production starts.*

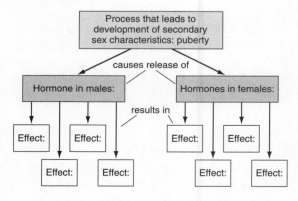

37. What is the main purpose of the menstrual cycle?

Part C—Reading Comprehension

Base your answers to questions 38 to 40 on the information below and on your knowledge of biology. Use one or more complete sentences to answer each question.

On December 8, 1998, Nkem Chukwu gave birth to a 1-pound, 6-ounce girl. At the time of her baby's birth, doctors discovered that the mother was carrying seven additional fetuses. At this point, the doctors were able to stop her labor for two more weeks. They wanted to give the tiny fetuses more time to develop. On December 20, the remaining seven infants were born by cesarean section. For the first time in recorded history, a woman had given birth to eight surviving babies.

It was reported that Mrs. Chukwu had taken two hormone-based fertility drugs—follicle-stimulating hormone and human chorionic gonadotropic hormone—to produce multiple eggs. The occurrence of this woman giving birth to eight babies raises some serious questions about the use of fertility drugs. In this case, the eight babies were born very prematurely. They spent only six and one-half months developing in the uterus, instead of the normal nine months. The beginning of their lives involved intravenous feeding, living in closely monitored incubators, and breathing through a tube attached to a respirator. Once the eight premature babies got past the grave health risks that could occur in the first few days, they had to face the risks of life-threatening infections due to their almost nonexistent immunity at birth.

Fertility experts frequently express concerns about multiple births such as this. Dr. Alan DeCherney, chairman of the Department of Obstetrics and Gynecology at UCLA, says, "The risk here is tremendous with these kids, of death as well as of severe neurological impairment. So there's just no reason to do this."

Dr. DeCherney pointed out that doctors are able to limit the number of eggs that become fertilized in the womb. Another specialist in reproductive medicine, Dr. Jirain Konialian, said it was "reckless" for a doctor not to monitor egg production more closely to reduce the chances for such a high multiple birth. Clearly, there are some people who see the birth of octuplets as a wondrous event, and others who view this situation as a consequence of reckless medical practice that can endanger the lives of the babies and even that of the mother, too.

38. Why did doctors stop the labor process of Mrs. Chukwu after the birth of her first baby girl?

39. What did Mrs. Chukwu take that caused her to produce multiple eggs?

40. List four ways in which the use of fertility drugs may involve health risks for the resulting newborn babies.

18

Growth and Development

EMBRYONIC DEVELOPMENT: FROM THE BEGINNING

Embryonic development is the sequence of events that gradually changes a zygote into a functional organism. Most of the instructions that control this series of events are in the genetic material (chromosomes) of the zygote. In addition, the environment that surrounds a zygote can have profound effects on its development. The process of embryonic development is mostly the same for all animals, whether vertebrate or invertebrate. (See Figure 18-1.)

The beginning of embryonic development occurs as soon as the egg is fertilized. A zygote forms at the moment the cell membrane of the egg and the sperm join. This event makes it possible for the nuclei of the two cells to fuse.

The changes that begin to occur after fertilization, and which continue throughout life, are known as **development**. The most dramatic developmental changes—growth and differentiation—occur early in the life of an organism. Through growth, the organism becomes larger as its number of cells increases; **differentiation** occurs as these cells begin to develop their own specific structures and functions. We increase in size because our bodies are made up of many cells. However,

we stay alive because our cells differentiate into more than 200 types, such as blood, skin, muscle, and bone cells.

All divisions of the zygote after fertilization are mitotic cell divisions. The number of chromosomes is maintained at each division. Therefore, all cells in the body still have the same complete set of chromosomes found in the fertilized egg cell. After the first series of mitotic cell divisions, the zygote becomes a hollow ball of cells. Although this ball of cells appears to have little organization, experiments have shown that each of the cells already "knows" which part of the organism it will become. (See Figure 18-2.)

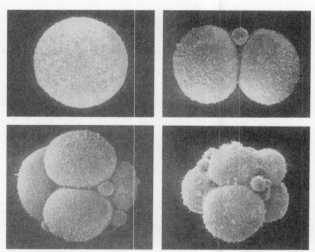

Figure 18-2 This series of scanning electron micrographs shows the mitotic cell division of a fertilized egg, or zygote, up to the eight-cell stage.

DIFFERENTIATION OF CELLS

During the next stage of embryonic development, cells begin to move, changing position in a highly regulated fashion until a three-layered structure is formed. Each layer gives

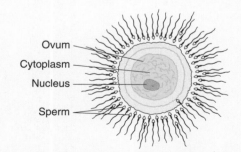

Ovum
Cytoplasm
Nucleus
Sperm

Figure 18-1 Embryonic development begins after the egg is fertilized by a sperm.

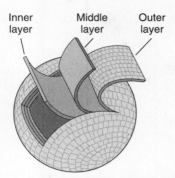

Inner layer   Middle layer   Outer layer

Figure 18-3 In vertebrates, the three embryonic cell layers form in a regulated fashion from the zygote's hollow ball of cells. These cell layers give rise to all body parts and structures.

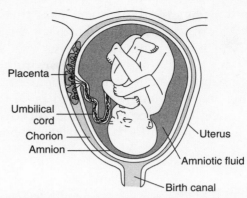

Placenta

Umbilical cord

Chorion

Amnion

Uterus

Amniotic fluid

Birth canal

Figure 18-5 At the final stage of development, the fetus grows in size until it is ready for birth.

rise to particular body parts and body systems of the developing embryo. (See Figure 18-3.)

All eggs develop in a fluid environment. For animals that reproduce in the water, such as frogs, this presents no problem. However, for animals that reproduce on land, the need for a watery environment poses an important problem. The solution to this problem evolved millions of years ago. A series of *membranes* forms around the developing embryo. The most important membrane, the *amnion*, has three main functions: it surrounds the embryo, protects the embryo, and holds in a fluid. Because of the amnion, the embryos of land animals develop in water just like those of aquatic animals do. (See Figure 18-4.)

At this point in development, the rest of the embryo's cells are still quite similar to each other. However, the process of cellular changes now begins. As the cells start to develop into muscle cells, skin cells, blood cells, or other tis-

sue types, they begin to join to form organs. Cell differentiation is occurring. All of the cells contain the same genetic information, but each type of cell uses this information differently.

In most vertebrates, by the time birth occurs, all the major structures of the animal have been formed. The development that occurs after birth is mostly confined to an increase in size as the animal develops into its adult form. (See Figure 18-5.)

❖ THE DANGERS THAT FACE A FETUS

The **fetus** (an embryo after the first three months of development) is surrounded by a watery cushion, in which it is kept warm and nourished. It has its safety provided by the mother's womb, or *uterus*. Most infections that may make the mother ill cannot cross over the placenta and into the fetus. However, the fetus is not entirely safe. Dangers can intrude into its small world.

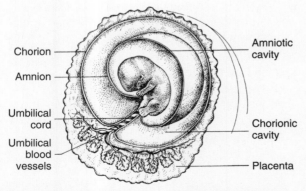

Chorion

Amnion

Umbilical cord

Umbilical blood vessels

Amniotic cavity

Chorionic cavity

Placenta

Figure 18-4 In mammals, the umbilical cord connects the placenta to the fetus. Nutrients and wastes are exchanged between the mother and fetus through the blood vessels of the umbilical cord and placenta.

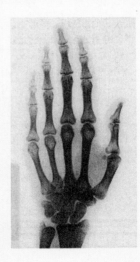

Figure 18-6 During embryonic development, the cells are dividing and growing. Dividing cells are easily damaged by X rays, which can penetrate through soft tissues to show bones. For this reason, X rays are one of the things that can be harmful to a fetus and should be avoided during pregnancy.

Some forms of radiation can pass through the tissues of the mother and into the fetus. X rays, for example, can affect a fetus. (See Figure 18-6.) Cells are dividing, growing, and changing during embryonic development; and dividing cells are easily damaged by X rays. Some infectious microorganisms within the mother can enter a fetus. Cigarette (tobacco) smoking by the mother, as well as chemicals, or **toxins**, taken in by her during pregnancy, can also harm a developing fetus. Use of heroin, LSD, cocaine, and alcohol can endanger a fetus, and many serious effects result. Babies can be born addicted to these drugs, and mental retardation can occur due to the alcohol that may pass from the mother's blood into the fetus.

Chapter 18 Review

Part A—Multiple Choice

1. What happens during embryonic development?

    1 An egg cell is released from an ovary.
    2 An egg cell is fertilized in the female.
    3 A zygote changes into a functional organism.
    4 A developed organism leaves its mother's body.

2. Which event does *not* occur between stages 2 and 11 in the process represented in the diagrams below?

    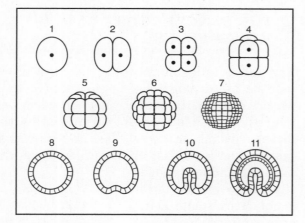

    1 a decrease in cell size
    2 the development of embryonic layers
    3 DNA replication
    4 fertilization

3. Embryonic development begins when an egg cell is

    1 produced in the ovary
    2 released from the ovary
    3 fertilized by the sperm cell
    4 removed from the body

4. During the process of growth, the

    1 number of cells in an organism increases
    2 number of cells in an organism decreases
    3 cells begin to develop specific structures
    4 cells undergo mitosis and meiosis

5. The diagram below represents part of the human female reproductive system. Most of an embryo's development normally occurs within

    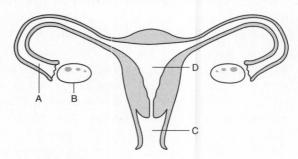

    1 structure A
    2 structure B
    3 structure C
    4 structure D

6. Compared with the number of chromosomes in a fertilized egg cell, each body cell of an adult organism has

    1 half as many chromosomes
    2 the same number of chromosomes
    3 twice as many chromosomes
    4 varying numbers, depending on function

7. Which phrase best describes the process represented in the diagram below?

    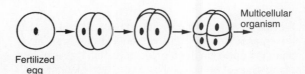

1 a zygote dividing by mitosis
2 a gamete dividing by mitosis
3 a zygote dividing by meiosis
4 a gamete dividing by meiosis

8. Cells develop into skin cells, muscle cells, bone cells, and so on, during the process of

1 fertilization
2 cell growth
3 differentiation
4 classification

9. In animals, the normal development of an embryo is dependent upon

1 fertilization of a mature egg by many sperm cells
2 production of body cells having half the number of chromosomes as the zygote
3 production of new cells having twice the number of chromosomes as the zygote
4 mitosis and the differentiation of cells after fertilization has occurred

10. The diagram below represents a developing bird embryo within an egg. What is the primary function of the egg?

1 to serve as a food supply for wild predators
2 to ensure survival of all the genes of that species
3 to protect and nourish the developing bird embryo
4 to give the parent birds freedom of movement

11. Heavy cigarette smoking and the use of alcohol throughout pregnancy usually increase the likelihood of the birth of

1 identical twins
2 a male baby
3 a baby with a viral infection
4 a baby with medical problems

12. Whether from frogs, snakes, cows, or humans, all fertilized eggs

1 have half the chromosome number of body cells
2 form a hard outer shell while developing
3 develop within a fluid environment
4 are ready to hatch within a few months

13. The role of the amnion is to

1 protect the embryo within a fluid
2 provide a food source for the embryo
3 control the differentiation of cells
4 exchange gases and remove wastes

14. Cells undergo differentiation because they

1 originate from different egg cells
2 have very different genetic information
3 each receive a different chromosome pair
4 use the same genetic information differently

15. By the time most vertebrates are born,

1 their three germ layers are still being formed
2 all their major structures have been formed
3 only half their major structures have been formed
4 they no longer have to increase their body size

16. The term that describes an embryo in the uterus, after its first three months of development, is

1 placenta
2 zygote
3 fetus
4 newborn

17. During the last months of pregnancy, the brain of a human embryo undergoes a "growth spurt." Which action by the mother would most likely pose the greatest threat to the normal development of the fetus's nervous system at this time?

1 spraying pesticides in the garden
2 taking prescribed vitamins on a daily basis
3 maintaining a diet high in fiber and low in fat
4 not exercising anymore

18. When a pregnant woman ingests toxins such as alcohol and nicotine, the embryo is put at risk because these toxins can

1 diffuse from the mother's blood to the embryo's blood at the placenta
2 enter the embryo when it opens its mouth
3 transfer to the embryo through the mother's mammary glands
4 enter the uterus through the mother's navel

Part B—Analysis and Open Ended

19. Some stages in the development of an organism are listed below. Which sequence represents the correct order of these stages?

(A) differentiation of cells into tissues
(B) fertilization of egg by sperm
(C) development of organs
(D) mitotic cell division of zygote

1 A–B–C–D
2 B–C–A–D
3 D–B–C–A
4 B–D–A–C

20. The sequence of diagrams below represents some events in the reproductive process. To regulate these cellular events, what adaptation is required?

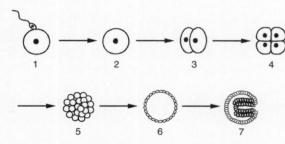

1 the presence of active genes in each cell for stages 1 to 7
2 the removal of all enzymes from the cells in stage 6
3 an increase in the number of genes in each cell for stages 3 to 5
4 a decrease in mitotic activity in all cells after stage 5

21. What are the two main factors that have an effect on embryonic development?

22. Distinguish among the terms *development, growth,* and *differentiation.* Which one of these terms includes the other two?

23. Arrange the following terms in the correct sequence in which they occur: *fertilization, growth, gamete formation, differentiation,* and *zygote.*

24. Which process is illustrated in the following diagram?

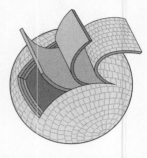

1 Three new egg cells are being produced.
2 Three new organisms are being produced.
3 The zygote is splitting into three separate pieces.
4 The three embryonic cell layers are being formed.

25. How does the single-layered zygote change to give rise to all body parts?

26. Use the following terms to complete the definitions in the flowchart below: *three-layered structure; fetus; zygote; hollow ball of cells;* and *differentiation.*

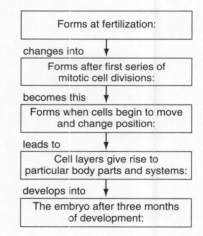

Refer to the diagram below to answer questions 27 and 28.

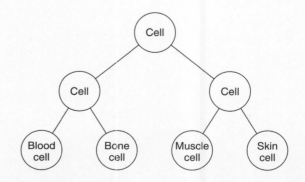

27. Which process is represented by the diagram?

    1 fertilization
    2 differentiation
    3 meiosis
    4 mutation

28. Why is this process vital to the embryonic development of a fetus?

29. Which of the diagrams shown below represents development and growth in a fetus?

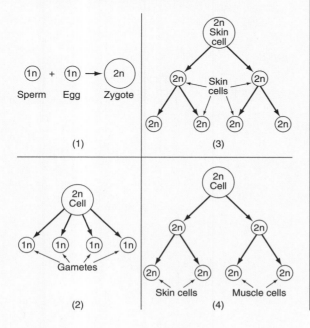

    1 diagram 1
    2 diagram 2
    3 diagram 3
    4 diagram 4

30. List the three functions of the amnion membrane. Explain why the amnion is so important for the embryonic development of land animals.

31. Why is the environment that surrounds an embryo important for its normal cell growth and development?

32. Use a specific example to identify *one* action taken by a mother that could have a negative effect on the embryonic development of her baby.

33. Identify four dangers that a fetus can face while inside the womb. Explain why they can be harmful to its development.

34. Write a brief essay that explains why someone who smokes cigarettes and has become pregnant should stop smoking cigarettes at this time.

Part C—Reading Comprehension

Base your answers to questions 35 to 39 on the information below and on your knowledge of biology. Use one or more complete sentences to answer each question.

Scientists have learned that the end section of a chromosome—called the telomere—plays an important role in the life of a cell. Normal human cells divide only a limited number of times. Each time a cell divides, it gets a little older. Research has suggested that when a cell divides, the telomere becomes a little shorter. The telomere is now thought of as a kind of molecular clock that keeps track of, or controls, the age of cells. It did not take long for researchers to begin to think that if the telomere could somehow be kept from getting shorter, a cell could continue to divide forever. Because their cells would never get older, the person who had such cells would also never age. In effect, an ageless telomere would become a molecular fountain of youth.

Normally, the telomeres become shorter and shorter with each cycle of cell division. It is thought that a short telomere tells a cell to stop dividing. A key enzyme that can change this shortening process is telomerase, which reverses the process by adding DNA to the telomeres at the chromosome ends.

In support of this hypothesis, researchers found that telomerase remains active in most immortal cell lines, such as cancerous cells that keep dividing in an uncontrolled manner. The importance of telomerase therefore has become even greater. Not only may the absence of telomerase lead to cell aging, but its presence may lead to cancer. Telomerase is also normally active in human cells that give rise to sperm and egg cells, which have to replicate.

Will further research provide a means to keep cells from aging through the action of telomerase—yet without also dividing uncontrollably? These are significant questions to answer and an important area for future research.

35. Why has the end section of chromosomes attracted a great deal of scientific attention?

36. What events seem to occur when a telomere gets reduced in length to a certain point?

37. How is the enzyme telomerase involved in cell aging?

38. Why is there great interest in the activity of the enzyme telomerase?

39. Why is there great interest in the results of the absence of telomerase?

Genetics and Molecular Biology

19

DNA Structure and Function

⊞ GENETIC MATERIAL: A JOB DESCRIPTION

During the 1940s and 1950s, several scientists conducted research to determine if it was the protein or the DNA within a cell's chromosomes that contained the genetic material. As a result of careful experiments and chemical analyses, they discovered that **DNA** (**d**eoxyribo**n**ucleic **a**cid) contains the information on which all life depends; that is, DNA *is* the genetic material. A substance that serves as the genetic material has the most significant job in the world: to carry on life itself. In order to carry out this job, the genetic material must do the following:

♦ It must be able to store information that can be passed on from one generation of cells to the next. It must be able to store enough information to make an organism like a tree or like you.

♦ It must be able to make a copy of itself in order to pass its information on again and again.

♦ It must be strong and stable so that it does not easily fall apart and cause per-

haps harmful changes to its store of information.

♦ It must be able to mutate, or change, slightly from time to time. These changes allow a species to produce the variations on which natural selection acts, which can lead to the evolution of new species.

We can now look at how DNA is built and how it functions in order to do these jobs.

⊞ THE WORLD LEARNS OF THE DOUBLE HELIX

DNA is made up of smaller **subunits**. These subunits, or **nucleotides**, include four types of bases, which occur in two pairs. The amounts of adenine (A) and thymine (T) are always the same (A pairs with T). The amounts of guanine (G) and cytosine (C) are always the same (G pairs with C). In 1953, scientists James Watson and Francis Crick described the structure of DNA for the first time, as a double helix. (See Figure 19-1.)

To understand the double helix structure of DNA, picture a ladder that has been twisted.

Figure 19-1 Scientists James Watson (left) and Francis Crick (right) shown in 1953 with their model of part of a DNA molecule.

The two sides of the ladder are parallel to each other, and the steps of the ladder link the two sides together. The sides of the ladder are the backbone of the DNA molecule (composed of sugar and phosphate molecules). Stretching between the two sides are the pairs of bases. The Watson-Crick model showed that the only possible way all the parts could fit was for each large adenine base to be matched opposite a smaller thymine base. Similarly, the large guanine had to be opposite a smaller cytosine. (See Figure 19-2.)

So, a molecule of DNA consists of two strands, opposite each other, connected by matching base pairs. If we look at one strand, we can describe it in terms of the order, or sequence, of its subunits. Because the subunits are in a long line, the order of the subunits is called a *linear sequence*. This linear sequence of nucleotides builds the DNA molecule, which may be very long. (Recall that long molecules such as DNA, which can contain thousands of nucleotides in a sequence, are called *polymers*.)

Imagine walking along a single strand of DNA. The bases in the subunits may occur in any order. The linear sequence on a short molecule of DNA might be A-T-T-G-A-C-C-G. Now imagine walking along the opposite strand, starting at the same place. Opposite the A in the first strand is a T. Because we know the sequence of bases in the first strand we automatically know the sequence of bases in the other strand. In this example, beginning with the T, the sequence must be T-A-A-C-T-G-G-C. This is the key to how the DNA molecule copies itself. The process by which DNA copies itself depends on the matching base pairs in the subunits of each strand. What is so important about the order of the subunits in a strand of DNA? The sequence of bases in the subunits *is* the genetic information that the strand of DNA contains. (See Figure 19-3.)

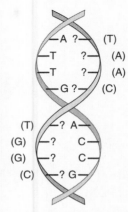

Figure 19-3 From the sequence of bases on one strand of DNA, we can determine the sequence on the opposite strand: A pairs with T, and C pairs with G.

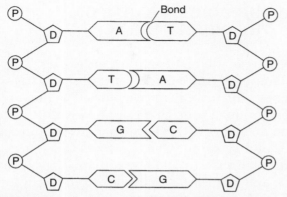

Figure 19-2 The structure of a DNA molecule—the nucleotide subunits include four types of bases (A, T, G, and C).

DNA: A LIBRARY OF INFORMATION

In some ways, the bases in DNA are like the letters of an alphabet, only the DNA "letters" are chemical letters. Because there are only four letters (A, T, G, and C) in the DNA alphabet, scientists thought that DNA was too simple to contain the complex genetic information of life. But what is also significant in DNA is the sequence of the letters, not just the letters themselves. Using these four letters in long sequences, nature can create an almost unlimited variety of genetic messages.

When you realize that human DNA consists of three billion pairs of bases, you can begin to imagine how much information can be stored in the DNA of our cells. All of the information for constructing our bodies, determining all of our character-istics or traits, and keeping our bodies func-tioning is stored in the linear sequences of bases in our DNA. The same is true for all other organisms on Earth. The evolutionary relationship between two organisms can be learned by comparing their DNA. The more similar their sequences of bases, the more re-cently the two organ-isms evolved from a common ancestor. (See Figure 19-4.)

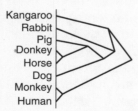

Figure 19-4 Evolution-ary relationships are confirmed by DNA closeness—the more similar their base se-quences, the more re-cently two organisms evolved from a common ancestor. For example, the donkey and the horse are more closely related than are the pig and the horse.

To make use of the genetic information stored in DNA, organisms must change that information into proteins. Proteins are made up of amino acids, subunits that—like nu-cleotide bases—are joined in a linear se-quence. The sequence of DNA subunits is used to direct the synthesis of proteins that have the correct sequence of amino acid subunits. In other words, through a chemical process, the order of the nucleotides determines the or-der of the amino acids in the proteins that are built.

▪▪ DNA REPLICATION: PASSING IT ON

To qualify as genetic material, DNA has to be able to **replicate**, or make a copy of, itself. This process of DNA replication occurs during the middle of the cell cycle. What we already know about its structure is enough to explain how DNA replicates.

To make a copy, you need an original, some-times called a **template**. Because DNA is a double helix, it has templates built into it. To begin the process, the double helix unwinds.

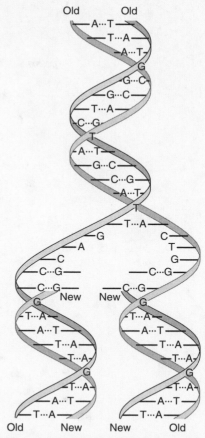

Figure 19-5 During DNA replication, the double helix unwinds, the strands separate, and the new strands form opposite each of the original DNA strands.

As with all metabolic activities, **enzymes** are needed for this process. Once the double-stranded molecule is untwisted, it begins to unzip, just like a zipper. Through the activity of an enzyme, the bonds between bases begin to break apart. (See Figure 19-5.)

As the bonds break, each strand of the DNA molecule becomes separate. Many free sub-units float around in the cell. Specific enzymes match up these free subunits with the exist-ing subunits in each DNA strand. Wherever a T is located on a strand, an A pairs to it; wherever a C is located, a G joins up, and so on. One by one, new subunits are joined to-gether to make a new strand opposite each old strand. The sequence of bases in the old strands determines the linear sequence of subunits in the new strands. When replication is complete, two double-stranded DNA mole-cules are formed. Each molecule is made up of one old strand joined to a newly synthesized

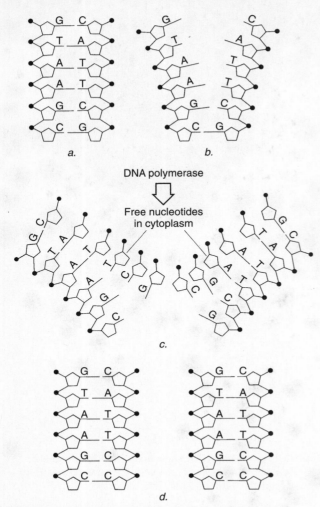

a.   b.

DNA polymerase

⇩

Free nucleotides
in cytoplasm

c.

d.

Figure 19-6 Through the process of DNA replication, two identical double-stranded DNA molecules are formed.

strand. How do the two new DNA molecules compare to the original one? They are identical. DNA replication has occurred. (See Figure 19-6.)

■ ERRORS IN DNA REPLICATION

In life, nothing is perfect. This is true about DNA replication, too. The enzymes responsible for directing the correct pairing of subunits during DNA replication occasionally make mistakes. A nucleotide base may be left out. Or the wrong base may be matched up. Sometimes an extra one is added. These mistakes produce errors in the linear sequence in one strand of the DNA molecule. Such an error is called a genetic **mutation**. From what we know about the replication process, once an error occurs in a DNA strand, it may be copied again and again. Thus, a mutation in the genetic material of one cell can easily be passed on to future cells.

A mutation is simply a change. However, many changes in the genetic material are harmful and may make it impossible for future cells, or even the entire organism, to survive. Other mutations cause an unnoticeable change; rather than harming the organism, the mutation seems to produce no effect. And sometimes a mutation gives the organism a sudden advantage that other similar organisms lack. Not only can mutations in DNA be good, but they are actually an important source of the genetic variation that is necessary for natural selection to occur. Much of the evolution of different life forms on Earth has depended on the chance occurrence of these mutations. (Remember: Only mutations within the DNA of gametes can be passed along to offspring; mutations within the DNA of body cells cannot.)

Chapter 19 Review

Part A—Multiple Choice

1. Which is *not* a necessary characteristic of the genetic material?

   1 It must be able to make a copy of itself.
   2 It must be weak so that it can fall apart easily.
   3 It must be able to mutate from time to time.
   4 It must be able to store information.

2. If a set of instructions that determines all of the characteristics of an organism is compared to a book, and a chromosome is compared to a chapter in the book, then what might be compared to a paragraph in the book?

   1 a starch molecule
   2 an amino acid
   3 a protein polymer
   4 a DNA molecule

3. A portion of a molecule is shown in the diagram below. Which statement best describes the main function of this type of molecule?

   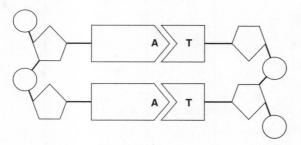

   1 It is a structural part of the cell wall.
   2 It determines what traits may be inherited.
   3 It stores energy for metabolic processes.
   4 It transports materials across the cell membrane.

4. The subunits of proteins are

   1 simple sugars
   2 phosphates
   3 amino acids
   4 enzymes

5. Watson and Crick contributed to the study of DNA by

   1 experimenting with pea plants
   2 recognizing that traits are inherited
   3 discovering the double helix structure of DNA
   4 mapping the entire human genome

6. The genetic code of a DNA molecule is determined by its specific sequence of

   1 ATP molecules
   2 carbohydrates
   3 sugar molecules
   4 nucleotide bases

7. The DNA molecule is formed from subunits arranged in a

   1 sequence with three kinds of bases
   2 circle with four kinds of bases
   3 sequence with four kinds of bases
   4 sequence with four kinds of acids

8. The base pairs in DNA are similar in arrangement to the

   1 sides of a ladder
   2 steps of a ladder
   3 railing of a staircase
   4 surface of a ramp

9. The order of the subunits in a strand of DNA is called a

   1 subunit sequence
   2 linear sequence
   3 strand sequence
   4 nucleotide sequence

10. If one strand of a DNA molecule is G-A-T-C-C-A-T, the sequence of the opposite strand is

    1 G-A-T-C-C-A-T
    2 C-T-A-G-G-T-A
    3 A-T-G-G-A-T-G
    4 T-A-C-C-T-A-G

11. The organization of bases in DNA can best be likened to the

    1 arrangement of letters in a word
    2 kinds of tools in a garage
    3 number of books in a library
    4 colors in a rainbow

12. When DNA separates into two strands, the DNA would most likely be directly involved in

    1 replication
    2 differentiation
    3 fertilization
    4 evolution

13. The sequence of subunits in a protein is most directly dependent upon the

1 region in the cell where enzymes are produced
2 type of cell in which starch is found
3 DNA in the chromosomes in a cell
4 kinds of materials in the cell membrane

14. In the diagram below, strands I and II represent sections of a DNA molecule. Strand II would normally include (top to bottom)

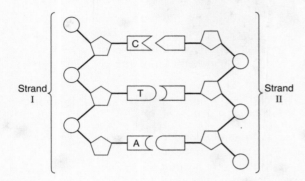

1 AGC
2 TAC
3 TCG
4 GAT

15. During the first step in the replication of DNA, the

1 double helix unwinds
2 base template is created
3 subunits of DNA form pairs
4 double helix rewinds itself

16. What causes the base pairs of DNA to break apart?

1 a mutation during replication
2 the activity of an enzyme
3 the production of new bases
4 the introduction of a fifth base

17. After DNA replication, the new DNA molecules are

1 the reverse of the original
2 the mirror image of the original
3 identical to the original
4 totally different from the original

18. Which statement is true regarding an alteration or change in DNA?

1 It is always referred to as a mutation.
2 It is always passed on to the offspring.
3 It is always advantageous to an individual.
4 It is always detected by chromatography.

19. A mutation occurs in a cell. Which sequence best represents the order of events for this mutation to affect traits expressed by the cell?

1 amino acids joining in sequence → a change in the sequence of DNA bases → appearance of characteristic
2 a change in the sequence of DNA bases → amino acids joining in sequence → appearance of new characteristic
3 appearance of new characteristic → amino acids joining in sequence → a change in the sequence of DNA bases
4 a change in the sequence of DNA bases → appearance of new characteristic → amino acids joining in sequence

20. A mutation is considered positive when it

1 makes it hard for the organism to survive
2 has absolutely no effect on the organism
3 changes the organism in an undetectable way
4 provides a sudden advantage that aids survival

Part B—Analysis and Open Ended

21. What four qualities must the genetic material have in order to do its job?

22. List the four bases of the DNA nucleotides and tell which bases pair together.

23. Explain the basic structure of DNA as described by Watson and Crick.

24. Why did scientists once think that DNA was too simple to contain the genetic information of living things? Explain why their reason was not correct.

25. Molecule 1 represents a section of hereditary information, and molecule 2 represents part of the substance that is determined by the information in molecule 1. What will most likely happen if there is a change in the first three subunits on the upper strand of molecule 1 shown below?

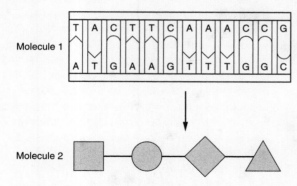

1 The remaining subunits in molecule 1 will also change.
2 Molecule 1 will split apart, triggering an immune response.
3 A portion of molecule 2 may be formed differently.
4 Molecule 2 may form two strands rather than one.

26. In an experiment, DNA from dead pathogenic bacteria was transferred into living bacteria that were, normally, not pathogenic. These altered bacteria were then injected into healthy mice. The mice died of the same disease caused by the original pathogens. Based on this information, which statement would be a valid conclusion?

1 DNA is present only in living organisms.
2 DNA functions only in the original organism from which it comes.
3 DNA changes the organism receiving an injection into another organism.
4 DNA from a dead organism can become active in another organism.

27. Briefly explain how the genetic information is arranged within a DNA molecule.

28. You see a photograph of a famous man and his teenaged son. You notice that they look very much alike, and that they even wear similar eyeglasses. What conclusion can you draw from this observation?

1 The DNA present in their body cells is identical.
2 Their percentage of having the same proteins is high.

3 The base sequences of their genes are all identical.
4 The mutation rate is the same in their body cells.

Refer to the figure below to answer questions 29 to 31.

29. The diagram at right represents a molecule of
1 ATP
2 RNA
3 DNA
4 FSH

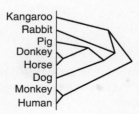

30. The structures labeled G, C, T, and A all represent
1 acids
2 sugars
3 bases
4 phosphates

31. Starting from the top of the diagram, what would be the letters of the missing units on the matching strand?

32. Complete the analogy: Nucleotide bases are to DNA as amino acids are to
1 sugars
2 proteins
3 lipids
4 nucleic acids

33. Use data from the diagram at right to explain why DNA nucleotide sequencing is important to the study of evolution.

Kangaroo
Rabbit
Pig
Donkey
Horse
Dog
Monkey
Human

34. How do the nucleotides of the DNA molecule allow it to replicate?

35. Briefly describe the process of DNA replication. Your answer should include the following terms (but not necessarily in this order):

♦ template
♦ enzymes
♦ subunits

Base your answers to questions 36 and 37 on the following chart, which provides information about heredity, and on your knowledge of biology.

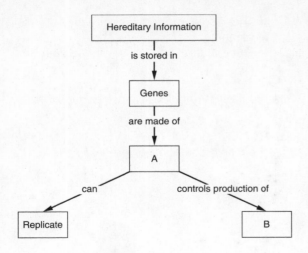

36. The molecule that is represented by box A serves as a template. Identify the type of molecule and explain how it is a template.

37. Which types of molecules are represented by box B?
    1 bases
    2 proteins
    3 lipids
    4 sugars

38. How would you explain to someone who has never heard of DNA why it is such an important molecule?

39. Mutations can be helpful to organisms; yet people fear the effects of substances that can cause mutations. Explain how mutations can be both helpful and harmful.

Base your answers to questions 40 through 42 on the passage below and on your knowledge of biology.

When making movies about dinosaurs, film producers have sometimes used ordinary lizards and enlarged their images thousands of times. We all know, however, that while they may look like dinosaurs and be related to dinosaurs, modern lizards are not actually dinosaurs.

Recently, some scientists have developed a hypothesis that challenges this view. These scientists suggest that some dinosaurs were actually the same species as some modern lizards that had grown to unbelievable sizes. They think that such growth might be due to a special type of DNA called *repetitive DNA*, often referred to as "junk" DNA because scientists do not understand its functions.

The scientists studied pumpkins that can reach sizes of nearly 1000 pounds and found them to contain large amounts of repetitive DNA. Other pumpkins that grow to only a few pounds in weight have very little of this kind of DNA. In addition, cells that reproduce uncontrollably have almost always been found to contain large amounts of repetitive DNA.

40. State *one* reason why scientists formerly thought of repetitive DNA as "junk."

41. Which kind of cells would most likely contain large amounts of repetitive DNA?
    1 red blood cells
    2 cancer cells
    3 nerve cells
    4 skin cells

42. Which fact best supports the hypothesis that large amounts of repetitive DNA are responsible for increased sizes of organisms?
    1 Lizards look very much like little dinosaurs.
    2 Modern lizards may be related to dinosaurs.
    3 Large pumpkins contain a lot of repetitive DNA.
    4 Another term for repetitive DNA is "junk" DNA.

Part C—Reading Comprehension

Base your answers to questions 43 to 45 on the information below and on your knowledge of biology. Use one or more complete sentences to answer each question.

Every time a prisoner awaiting a death sentence is proven innocent by DNA evidence and released, it makes the news. And it should. Nothing demonstrates the power of DNA technology better. Ray Krone owes his freedom, and probably his life, to this technology. In 2002, he was released from an Arizona prison after serving ten years. During that time, Mr. Krone, who had served in the U.S. Air Force and worked as a letter carrier with no criminal record, was tried twice for the sexual assault and stabbing murder of a bartender in 1991. Mr. Krone was in the bar where the victim worked the night of the murder. The only evidence used to convict him was the similarity between the pattern of tooth marks on the victim, where she had been bitten, and Mr. Krone's teeth.

The first trial sentenced Mr. Krone to death, the second trial to a life sentence. Finally, after 10 years, DNA testing was done on saliva from bite marks found on the victim's clothing. Not only did the DNA *not* match that of Mr. Krone, but it *did* match that of a person serving time in another Arizona prison for an unrelated sex crime. The odds were 1.3 quadrillion (1,300,000,000,000,000) to 1 that it was this other man's DNA on the victim and not that of Mr. Krone or anyone else. A judge ordered the immediate release of Ray Krone when the DNA test results were announced.

The DNA match was made possible because Arizona now has a database that contains a DNA profile of every prison inmate. In fact, every state in America now has such a database; and a national system, the National DNA Index System (NDIS), was started in 1998. By 2002, the one-millionth DNA profile had been entered into the computerized system. DNA evidence collected from any crime scene can now be quickly compared to that of any one of the million convicted offenders in the NDIS database. The system is quickly growing and the technology of DNA testing is rapidly improving. For example, a portable DNA testing kit is under development in Britain. It will be smaller than a suitcase and will be linked to the national DNA database of that country. It is expected that the crime scene evidence will be put in a solution and then placed inside the mobile unit. Silicon chip technology in the testing kit will then extract a DNA profile that will be sent to the national database via a laptop computer. The results may be returned in under an hour to the detective's palm-held computer. Saliva on discarded cigarette butts at crime scenes has already been used successfully to provide DNA profiles of suspects.

It is hoped that someday, thanks to this kind of technology, there will be no more wrongful convictions such as that of Mr. Krone, and more positive identifications of those who do deserve the jail time.

43. Compare the evidence used to convict Ray Krone in 1991 with the evidence used to release him in 2001.

44. Describe the system that has been put in place in the United States to use DNA technology to solve crimes.

45. How is the technology of DNA testing being improved for use at crime scenes?

Genes and Gene Action

GENES AND PROTEINS

Now that it has been shown that DNA is what makes up the genetic material, it is time to look more closely at genes. What is a gene? **Genes** are really packages of information that tell a cell how to make proteins. Proteins are polymers, or long chains, of amino acids. As you learned already, there are 20 different types of amino acids. The order in which the amino acids are joined determines which protein is made. Every different protein has a unique sequence of amino acids. This sequence determines the shape of a protein molecule. It is the shape of the protein that allows the molecule to do its work in the cell.

Genes are specific sections of DNA molecules that are made up of linear sequences of subunits. Proteins are linear sequences of amino acids. How do cells use a linear sequence of subunits in DNA to build a linear sequence of amino acids for a protein? In all cells, except for bacteria, DNA is stored in the nucleus. Yet protein synthesis occurs outside the nuclear membrane, at the **ribosomes**. These small organelles are distributed throughout the cytoplasm. How does the genetic information in DNA within the nucleus get to the ribosomes? A third type of molecule, *ribonucleic acid*, or *RNA*, works as a helper to transfer the information. That is, the genetic information flows from the DNA to the RNA to a protein. (See Figure 20-1.)

FROM DNA TO RNA

Each gene is a portion of a chromosome, in effect a portion of the DNA chain. An RNA molecule called *messenger RNA* does the job of moving the information in the base sequence

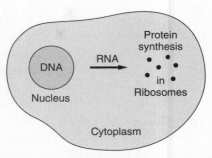

Figure 20-1 The flow of genetic information in a cell—from DNA in the nucleus to RNA to amino acids at the ribosomes.

out to the ribosomes. DNA is copied into RNA by a process that is similar to DNA replication. The DNA double helix opens up where a particular gene is located. Special enzymes begin to match up RNA subunits with the correct DNA subunits. The new RNA molecule has the same base sequence as one strand of the original DNA. This RNA molecule then goes out of the nucleus through pores in the nuclear membrane to ribosomes in the cytoplasm. (See Figure 20-2.)

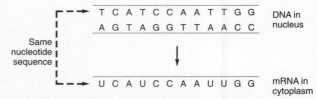

Figure 20-2 The DNA sequence is copied into messenger RNA, which goes out to the ribosomes in the cytoplasm. *Note:* In RNA, the base uracil (U) substitutes for the DNA base thymine (T).

FROM RNA TO PROTEIN

So far, the genetic information, stored as a base sequence, has moved from the nucleus to the cytoplasm by using RNA. Another problem

		Second Position			

First Position		U	C	A	G	Third Position
	U	UUU ⎫ Phe UUC ⎭ UUA ⎫ Leu UUG ⎭	UCU ⎫ UCC ⎬ Ser UCA ⎪ UCG ⎭	UAU ⎫ Tyr UAC ⎭ UAA Stop UAG Stop	UGU ⎫ Cys UGC ⎭ UGA Stop UGG Trp	U C A G
	C	CUU ⎫ CUC ⎬ Leu CUA ⎪ CUG ⎭	CCU ⎫ CCC ⎬ Pro CCA ⎪ CCG ⎭	CAU ⎫ His CAC ⎭ CAA ⎫ Gln CAG ⎭	CGU ⎫ CGC ⎬ Arg CGA ⎪ CGG ⎭	U C A G
	A	AUU ⎫ AUC ⎬ Ile AUA ⎪ AUG Met	ACU ⎫ ACC ⎬ Thr ACA ⎪ ACG ⎭	AAU ⎫ Asn AAC ⎭ AAA ⎫ Lys AAG ⎭	AGU ⎫ Ser AGC ⎭ AGA ⎫ Arg AGG ⎭	U C A G
	G	GUU ⎫ GUC ⎬ Val GUA ⎪ GUG ⎭	GCU ⎫ GCC ⎬ Ala GCA ⎪ GCG ⎭	GAU ⎫ Asp GAC ⎭ GAA ⎫ Glu GAG ⎭	GGU ⎫ GGC ⎬ Gly GGA ⎪ GGG ⎭	U C A G

Figure 20-3 The amino acid triplet codes. Note that most amino acids are represented by more than one codon.

remains: how to use the nucleotide base sequence in the RNA to build a protein with the correct sequence of amino acids. This problem involves a change of "language," from the base sequence language of RNA into the amino acid language of proteins. This process is called **translation**, and it occurs at the ribosome.

Built into every living cell in the world is a genetic **code**. It is called the *triplet code*. Each different combination of three bases makes up a word, called a *codon*. Each codon represents a specific amino acid. Each of the 20 amino acids has at least one codon, and most have more than one. This genetic code is universal; in other words, all organisms on Earth use the same genetic code. For example, the codon GCA stands for the amino acid alanine in all life-forms, from bacteria to trees to humans. This similarity among living things is good evidence that all organisms evolved from a common ancestral life-form in Earth's distant past. (See Figure 20-3.)

MUTATIONS: A CLOSER LOOK

In Chapter 19, a mutation was defined as a change in the base sequence of a DNA molecule. The possible effects of a mutation can now be explained in terms of what you know about protein synthesis.

The order of bases in DNA determines the order of amino acids in proteins. In certain cases, a mutation in one subunit will change the triplet code, which in turn may make a change in an amino acid. If this change occurs in a body cell, then all other cells in the organism's body that reproduced (through mitosis) from that cell will have the same change. It is more important, however, if the mutation occurs in the DNA of a gamete. If that gamete fuses with another gamete in sexual reproduction, then the mutation will be inherited. The change in the DNA will be passed on to succeeding generations. The new organism will have the mutation, as will all offspring of that organism. This will be an inherited condition. If the mutation is harmful, the individual and its offspring will have a genetic disease.

GENE EXPRESSION AND CELL DIFFERENTIATION

Chromosomes contain extremely long DNA molecules. Many genes are stretched out along these molecules. For example, it is estimated that there are 20,000 to 30,000 different genes in human cells. After fertilization, every cell of a growing organism arises from the mitotic cell division of other cells. Through mitosis, every cell in our body has the same 46 chromosomes with the same DNA as the original fertilized egg cell.

You learned in Chapter 18 that there are different types of cells in our bodies. We have skin cells, muscle cells, bone cells, nerve cells, blood cells, and so on. If all of these cells have the same DNA, why are they so different from each other? The answer is that only certain genes are used in certain cells. The use of specific information from a gene is called gene **expression**. Proteins are synthesized only from genes that are being expressed, or "turned on." All other genes in the cell are kept silent, or "turned off." This gives the cell its own structure, enzymes, functions, and physical characteristics. A muscle cell contracts, a nerve cell transmits an impulse, and a skin cell helps form a flat, protective layer. The process by which special types of cells are formed through controlled gene expression is called cell **differentiation**. This is an essential process of life. Without cell differentiation, we could not survive, because our bodies would be made up of only one type of cell. While the exact process is not known for certain, it is thought that environmental factors—both outside and inside each cell—influence gene expression. (See Figure 20-4.)

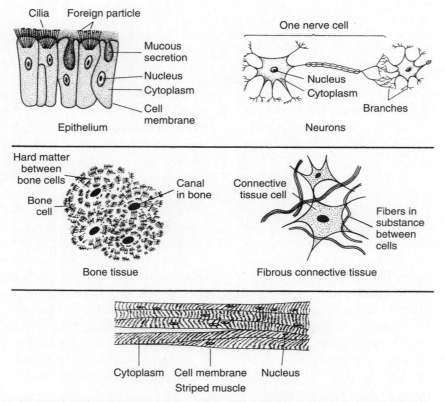

Figure 20-4 Many different types of cells make up the human body. This cell differentiation results from differences in gene expression—only some genes are "turned on" to make the specific proteins needed for each cell type.

Chapter 20 Review

Part A—Multiple Choice

1. Genes can best be described as
   1. directions for making DNA
   2. directions for making proteins
   3. subunits of proteins
   4. molecules that transfer information out of the nucleus

2. Which path correctly describes the flow of information in cells?
   1. DNA → RNA → protein
   2. protein → RNA → DNA
   3. protein → DNA → RNA
   4. RNA → DNA → protein

3. The kinds of genes that an organism has is determined by the
   1. type of amino acids in its cells
   2. size of simple sugar molecules in its organs
   3. sequence of the subunits A, T, C, and G in its DNA
   4. shape of the protein molecules in its organelles

4. A change in the order of DNA bases that code for a respiratory protein will most likely cause
   1. the production of a starch that has a similar function
   2. a change in the sequence of amino acids determined by the gene
   3. the digestion of the altered gene by enzymes
   4. the release of antibodies by certain cells to correct the error

5. The role of messenger RNA is to
   1. prevent mutations during DNA replications
   2. match ribose-containing subunits to subunits of DNA
   3. move the information in a base sequence out to the ribosomes
   4. translate the base sequence at the ribosomes

6. RNA receives information from DNA by
   1. binding with a double helix as a third strand
   2. matching with subunits of a single strand of DNA
   3. making an exact copy of the DNA molecule
   4. accepting proteins through pores in the nuclear membrane

7. What happens at the ribosome?
   1. The DNA strands separate.
   2. RNA matches up with DNA strands.
   3. Genetic information is mutated.
   4. RNA is translated into amino acids.

8. The diagram below represents a process that occurs within a cell in the human pancreas. This process is known as

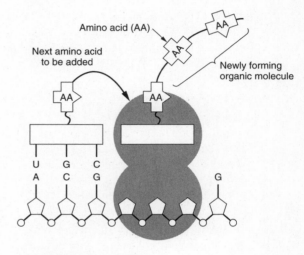

   1. digestion by enzymes
   2. energy production
   3. protein synthesis
   4. replication of DNA

9. How many bases make up a codon?
   1. one
   2. two
   3. three
   4. four

10. What does a codon represent?
    1. a specific amino acid
    2. a specific base
    3. an RNA molecule
    4. an enzyme

11. The genetic code is
    1. different for every organism
    2. the same for all organisms
    3. constantly changing
    4. impossible to identify

12. The sequence of amino acids in a protein is determined by the

    1 speed at which translation occurs
    2 size of the cell involved
    3 number of ribosomes in a cell
    4 order of bases in the DNA

13. The diagram below provides some information concerning proteins. Which phrase does the letter A represent?

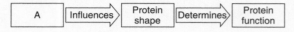

    1 Sequence of amino acids
    2 Sequence of starch molecules
    3 Sequence of simple sugars
    4 Sequence of ATP molecules

14. A mutation is inherited if it

    1 occurs in a gamete used in sexual reproduction
    2 occurs in a cell that undergoes mitosis
    3 gives the organism a better chance for survival
    4 endangers the organism's chance for survival

15. People with cystic fibrosis inherit defective genetic information and cannot produce normal CFTR proteins. Scientists have used gene therapy to insert normal DNA segments that code for the missing CFTR protein into the lung cells of people with cystic fibrosis. Which statement does *not* describe a result of this therapy?

    1 Altered lung cells can produce the normal CFTR protein.
    2 The normal CFTR gene may be expressed in altered lung cells.
    3 Altered lung cells can divide to produce other lung cells with the normal CFTR gene.
    4 Offspring of someone with altered lung cells will inherit the normal CFTR gene.

16. About how many genes are contained on the 46 chromosomes in each human body cell?

    1 200–1000
    2 5000–10,000
    3 20,000–30,000
    4 50,000–100,000

17. The cells that make up a person's skin have some functions that are different from those of the cells that make up the person's liver. This is because

    1 all types of cells have a common ancestor
    2 environment and past history have no influence on cell function
    3 different cell types have completely different genetic material
    4 different cell types use different parts of the genetic instructions

18. Gene expression means that

    1 different genes are found in different cells
    2 genes are passed through the nuclear membrane
    3 only some genes are turned on in each type of cell
    4 some cells have genes and some cells do not

19. Scientific studies have shown that identical twins who were separated at birth and raised in different homes may vary in height, weight, and intelligence. The most probable explanation for these differences is that

    1 the original genes of each twin increased in number as they developed
    2 the environments in which they were raised were different enough to affect the expression of their genes
    3 one twin received genes only from the mother while the other twin received genes only from the father
    4 the environments in which they were raised were different enough to change the genetic makeup of both individuals

20. After a series of cell divisions, an embryo develops different types of body cells, such as muscle cells, nerve cells, and blood cells. This development occurs because

    1 the genetic code changes as the cells divide
    2 different genetic instructions are created to meet the needs of new types of cells
    3 different segments of the genetic instructions are used to produce different cell types
    4 some sections of the genetic material are lost as a result of fertilization

Part B—Analysis and Open Ended

21. What is so important about the order in which amino acids are joined?

22. The diagram below shows two different structures, 1 and 2, that are present in many single-celled organisms. Structure 1 contains protein A, but not protein B; and structure 2 contains protein B, but not protein A. Which statement is correct concerning protein A and protein B?

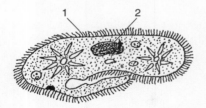

1 Proteins A and B have different functions and different amino acid chains.
2 Proteins A and B have the same function but a different sequence of bases (A, C, T, and G).
3 Proteins A and B have different functions but the same amino acid chains.
4 Proteins A and B have the same function and the same sequence of bases (A, C, T, and G).

23. The letters in the diagram at right represent genes on a particular chromosome. Gene B contains the code for an enzyme that cannot be synthesized unless gene A is also active. Which statement best explains why this can occur?

1 A hereditary trait can be determined by more than one gene.
2 All the genes on a chromosome act to produce a single trait.
3 Genes are made up of double-stranded segments of DNA.
4 The first gene on each chromosome controls all the other genes.

24. Hemoglobin is a complex protein molecule found in red blood cells. Hemoglobin with the normal amino acid sequence can effectively carry oxygen to body cells. In the disorder known as sickle-cell anemia, one amino acid is substituted for another in the hemoglobin. One characteristic of this disorder is poor distribution of oxygen to the body cells. Explain how one change in the amino acid sequence of this protein could cause the results described.

25. In what way are the structures of DNA and protein similar?

26. Why is the location of DNA within the nucleus a potential problem for protein synthesis? How does RNA solve this problem for the cell?

27. Arrange the following structures from the largest to the smallest: DNA molecule; chromosome; nucleus; gene.

28. Explain how the process of copying DNA into messenger RNA is similar to DNA replication. Your answer should include the following terms:

♦ double helix
♦ enzymes
♦ subunits
♦ base sequence

29. Why must the bases be grouped into triplets in order to represent amino acids?

Base your answers to questions 30 through 32 on the table below, which provides the DNA codes for several amino acids.

Amino Acid	DNA Code Sequence
Cysteine	ACA or ACG
Tryptophan	ACC
Valine	CAA or CAC or CAG or CAT
Proline	GGA or GGC or GGG or GGT
Asparagine	TTA or TTG
Methionine	TAC

30. A certain DNA strand has the following base sequence: TACACACAAACGGGG. What is the sequence of amino acids that would be synthesized from this code (if it is read from left to right)?

31. Suppose the DNA sequence undergoes the following change: TACACACAAACGGGG → TACACCCAAACGGGG. How would the sequence of amino acids be changed as a result of this mutation?

32. The original DNA sequence undergoes the following change: TACACACAAACGGGG → TACACACAAACGGGT. State *one* reason why this mutation produces *no change* in the action of the final molecule that will be synthesized from this code.

Refer to the flowchart below to answer questions 33 and 34.

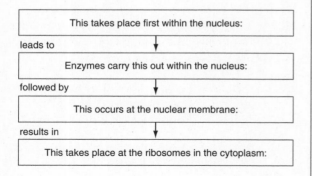

33. Use the following phrases to complete the steps listed in the flowchart boxes: a new RNA molecule moves out through pores; RNA base sequences translate into amino acid sequences; the DNA double helix opens up and unwinds; RNA subunits match up with DNA subunits.

34. The *best* title for this flowchart probably would be

1 How RNA Is Made
2 How DNA Is Made
3 How Proteins Are Made
4 How Ribosomes Are Made

35. How does RNA allow for translation from the genetic base sequences to the amino acid sequences?

36. Explain why the genetic code is called universal. What is the evolutionary significance of this fact?

37. In what way do ribosomes help in the process of translation?

38. How might a mutation in a DNA molecule result in a different protein?

39. Use the diagram below to answer the following questions. Suppose these cells are from the same body. How is gene expression related to cell differentiation? Do they have the same proteins? Do they have the same DNA?

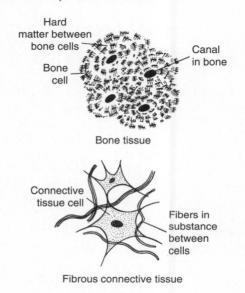

Part C—Reading Comprehension

Base your answers to questions 40 to 42 on the information below and on your knowledge of biology. Use one or more complete sentences to answer each question.

The first big surprise to arise from the decoding of the human genome was a matter of numbers. Where were all the genes? Rather than 100,000 or more genes, as scientists had predicted for years, the Human Genome Project has revealed that humans have perhaps only 30,000 genes. This can be compared to the fruit fly's 12,371 genes or the 19,098 genes of the tiny roundworm *Caenorhabditis elegans*. But it is not just a matter of human pride for people to consider themselves more complex. We *are* more complex. The little worm *C. elegans*, with over 19,000 genes, has a body of only 959 cells, of which 302 are neurons—the worm's "brain". The human body, built by perhaps only 50 percent more genes than the worm, has 100 trillion cells, with the brain alone containing 100 billion cells. So where does the complexity come from?

There are two main ways that scientists think human complexity has arisen. The first way concerns proteins. Proteins are the working parts of every cell and it turns out that proteins themselves have different sections or domains in them. Ninety-three percent of the protein domains in humans are also in the worm and the fly. However, it seems that a lot of mixing and matching of these domains has occurred. Dr. Francis S. Collins, director of the genome institute at the National Institutes of Health said, "Maybe evolution designed most of the basic folds that proteins could use a long time ago, and the major advances in the last 400 million years have been to figure out how to shuffle those in interesting ways." The second ingenious way that evolution seems to have increased complexity is by dividing the genes themselves into several different segments, and using these segments in different arrangements to make different proteins.

There are many different ways of thinking about human complexity. One scientist has compared people to the machines created by them. Dr. Jean-Michel Claverie of the French National Research Center writes, "In fact, with 30,000 genes, each directly interacting with four or five others on average, the human genome is not significantly more complex than a modern jet airplane, which contains more than 200,000 unique parts, each of them interacting with three or four others on average."

40. What is the first unexpected finding from the decoding of the human genome?

41. How do proteins contribute to the complexity of the human organism?

42. How do genes contribute to the complexity of the human organism?

Patterns of Inheritance

GREGOR MENDEL: THE FOUNDER OF GENETICS

How characteristics are passed from parents to offspring was a question that puzzled people for thousands of years. A set of experiments completed more than 100 years ago by Gregor Mendel helped our understanding of **inheritance**, how traits are passed from one generation to the next. Mendel, an Austrian monk, conducted hundreds of experiments using thousands of pea plants. By applying careful mathematical analysis to his work—something that had rarely been done in biology before—Mendel discovered much about the way heredity works. The study of inheritance really began with Mendel; he can rightly be called the "founder of the science of genetics." (See Figure 21-1.)

Mendel discovered that hereditary information is passed from parents to offspring in individual units that he called *factors*, which we now call genes. He also discovered that the factors were passed on in specific, predictable patterns from both parents. These patterns of inheritance are described in Mendel's laws.

MENDEL'S IDEAS: THE LAWS OF INHERITANCE

Mendel proposed several ideas that explained his results. These ideas were correct, even though Mendel knew nothing about chromosomes, genes, or DNA. Since that time, scientists have been able to combine Mendel's ideas with what has been learned about genetics.

Mendel's first main idea was that each characteristic, or trait, exists in two versions. These

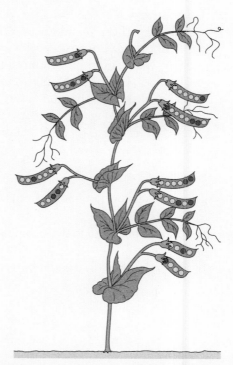

Figure 21-1 Gregor Mendel investigated the inheritance of traits from one generation to the next by carefully studying characteristics of the common pea plant.

two versions of a gene are called **alleles**. For example, in pea plants, there are two alleles for the gene for flower color: one allele for purple flowers and one allele for white flowers. Genes exist at specific locations on chromosomes. So one chromosome may have an allele for purple at the location for flower color, while another chromosome may have an allele for white at the same place. The DNA at these locations consists of a sequence of subunits. At one allele, the DNA subunits code for proteins that result in the color purple. At the other allele, a different sequence of DNA subunits codes for proteins that make the color white.

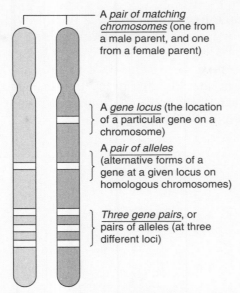

A *pair of matching chromosomes* (one from a male parent, and one from a female parent)

A *gene locus* (the location of a particular gene on a chromosome)

A *pair of alleles* (alternative forms of a gene at a given locus on homologous chromosomes)

Three gene pairs, or pairs of alleles (at three different loci)

Figure 21-2 The gene for one trait, such as flower color, exists at the same place on each member of a pair of matching chromosomes. Each gene has two versions, or alleles, of that trait, for example, either purple or white for flowers.

Mendel's second main idea was that for each characteristic, an individual inherits two alleles, one from each parent. All offspring—plant or animal—that result from sexual reproduction have a double set of chromosomes made up of chromosome pairs. Each pair of chromosomes consists of one chromosome from the mother and one chromosome from the father. Since corresponding genes (such as for flower color) occur at the same place on each chromosome in a pair, any particular gene exists twice in each cell. Thus, every cell has two alleles for each gene. (See Figure 21-2.)

❖ GENE EXPRESSION

You already know that the genes of an organism determine its characteristics. Yet the environment in which an organism lives can also affect the way its genes are expressed. The fur coloration pattern of Siamese cats is an example of this interaction of genes and the environment. Siamese cats have a gene that codes for an enzyme that produces dark fur. However, this enzyme works only at cool temperatures. Most of a cat's body is too warm for this enzyme to work. But you can easily

Figure 21-3 The expression of the gene that codes for dark-colored fur in Siamese cats depends on temperature—the parts of the body that have a lower (cooler) temperature allow that gene to be expressed. The rest of the body, which is warmer, has light fur.

identify the cooler areas where the enzyme does work—the dark ears, paws, face, and tail that are typical of a Siamese cat. Even though all parts of a cat's body have the same combination of genes, the way that the genes are expressed differs from one part to another because of different environmental influences. Thus, the environment frequently affects the final appearance of an organism. (See Figure 21-3.)

❖ PLANT AND ANIMAL BREEDING

For centuries, people have chosen to breed plants and animals that had desirable traits. This is called **selective breeding**. By allowing only those organisms to reproduce, traits such as size, shape, and color have been altered over time.

The breeding of plants and animals has been greatly helped by the discoveries of Mendel and later geneticists. Often, breeders can identify the exact traits in which they are interested and then prepare the best genetic crosses. For example, plant breeders have produced new crops that are resistant to disease, can live in new environments, grow more plentifully, and look and taste better. (See Figure 21-4 on page 162.)

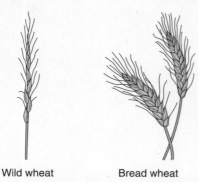

Wild wheat   Bread wheat

Figure 21-4 As a result of centuries of selective breeding, the small kernels of wild wheat have been transformed into the large kernels of bread wheat, which is a more useful crop for people.

People have bred animals for many different purposes, too. For example, sheep have been bred to produce more wool of better quality, turkeys have been bred to produce more white meat, chickens to lay larger eggs, cows for more milk, and pigs to produce meat that contains less fat.

Chapter 21 Review

Part A—Multiple Choice

1. Mendel studied inheritance patterns in

1 pink roses
2 fruit flies
3 Siamese cats
4 pea plants

2. Mendel is credited with

1 discovering the structure of DNA
2 beginning the science of genetics
3 recognizing the role of RNA
4 beginning animal-breeding programs

3. An allele is a

1 version of a gene
2 specialized enzyme
3 subunit of DNA
4 three-base code

4. One of Mendel's ideas was that

1 alleles are responsible for mitosis
2 alleles can cause mutations
3 there are two alleles for each trait
4 each gene exists as only one allele

5. In the plants Mendel studied, one allele produces purple flowers while the other produces white flowers because

1 it depends on how the sunlight reflects off the flowers' petals
2 the DNA subunits at those alleles code for two different proteins

3 the DNA subunits at those alleles code for different carbohydrates
4 it depends on how hot or cold the environment is when they bloom

6. According to Mendel, for each trait inherited, offspring receive

1 just one allele per cell
2 one allele from each parent
3 two alleles from each parent
4 several pairs of alleles

7. How are an organism's traits related to the environment?

1 An organism inherits different genes depending on the environment.
2 Genetic information is never affected by the environment.
3 The environment can affect the expression of genetic traits.
4 The environment affects genetic traits only in wild organisms.

8. In a particular variety of corn, the kernels turn red when exposed to sunlight. In the absence of sunlight, the kernels remain yellow. Based on this information, it can be concluded that the color of these corn kernels is due to

1 a different type of DNA that is produced when sunlight is present
2 the effect of sunlight on the number of chromosomes inherited
3 a different species of corn that is produced only in sunlight

4 the effect of the environment on gene expression in the corn

9. In Siamese cats, the fur on the ears, paws, tail, and face (that is, the extremities) is usually black or brown, while the rest of the body fur is almost white. If a Siamese cat stays indoors, where it is warm, it may grow fur that is almost white on its extremities. In contrast, if a Siamese cat mostly stays outside, where it is cold, it will grow fur that is quite dark on its extremities. The best explanation for these changes in fur color is that

1 an environmental factor influences the expression of this inherited trait
2 skin cells that produce pigments have a higher mutation rate than other cells
3 the location of pigment-producing cells determines the DNA code of the genes
4 the alleles for fur color are mutated by interactions with the environment

10. The diagram below represents the change in a sprouting onion bulb when sunlight is present and then when sunlight is no longer present. Which statement best explains this change?

1 Plants need carbon dioxide to survive.
2 Plants produce hormones for growth.
3 Environmental conditions never affect genetic traits.
4 The environment can influence the expression of traits.

11. Fruit flies with the curly-wing trait will develop straight wings if kept at a temperature of 16°C during development and curly wings if kept at 25°C. The best explanation for this change in the shape of wings is that the

1 genes for curly wings and for straight wings are found on different chromosomes
2 outside environment affects the expression of the genes for this trait
3 type of gene present in the fruit fly is dependent on environmental temperature
4 lower outside temperature always produces the same genetic mutation

12. To produce large tomatoes that are resistant to cracking and splitting, some seed companies use the pollen from one variety of tomato plant to fertilize a different variety of tomato plant. This process is an example of

1 selective breeding
2 direct harvesting
3 DNA sequencing
4 plant cloning

13. Mendel experimented by carrying out selective breeding and

1 natural selection
2 mathematical analysis
3 molecular selection
4 animal husbandry

14. Research applications of the basic principles of genetics have contributed greatly to the rapid production of new varieties of plants and animals. Which activity is an example of such an application?

1 testing new chemical fertilizers on food crops
2 developing new irrigation methods to conserve water
3 selective breeding of crops that show a resistance to disease
4 using natural predators to control insect pests

15. Which process has been used by farmers for hundreds of years to develop new animal varieties?

1 genetic cloning
2 DNA splicing
3 genetic engineering
4 selective breeding

16. When humans first domesticated dogs, there was very little physical diversity in the species. Today there are many varieties, such as the German shepherd and the Boston terrier. This increase in diversity is most closely associated with

1 cloning of selected body cells
2 years of mitotic cell division
3 selective breeding for desirable traits
4 environmental influences on inherited traits

Part B—Analysis and Open Ended

17. Even though Mendel did not know about genes, he can be called the "founder" of the science of genetics. Why?

18. In one sentence, tell what Mendel noticed about the "factors" that were passed from parents to offspring among pea plants.

Refer to the figure below to answer questions 19 and 20.

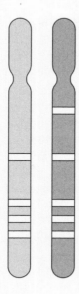

19. The two main structures represent
1 the pods of two different pea plants
2 one pair of genes from a pea plant
3 a pair of matching chromosomes
4 mutations in the genes of a trait

20. The figure illustrates that
1 every chromosome has at least four or five genes on it
2 an allele for a trait is on either chromosome, but not on both
3 several pairs of alleles are necessary to determine any trait
4 the gene for a trait is at the same place on each matching chromosome

21. How are Mendel's ideas about "factors" explained by what we now know about genes and chromosomes?

22. Explain how fur color in Siamese cats demonstrates an important fact about the expression of genes. In what way might this be an adaptive feature?

Refer to the following figures and paragraph to answer question 23.

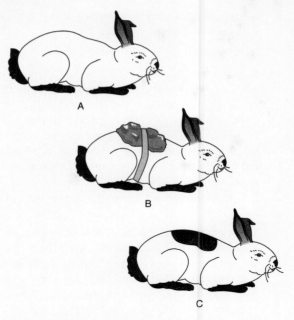

The normal color pattern of a Himalayan rabbit's fur is shown in Figure A. In Figure B, the rabbit is shown with the fur shaved from an area on its back and an ice pack applied to that area. Figure C shows the same rabbit after new fur has grown in the shaved area.

23. The best explanation for this change in the rabbit's color pattern is that the low temperature of the ice pack
1 caused a genetic mutation in the fur
2 deleted the gene for white-colored fur
3 allowed the gene for dark fur to be expressed
4 had no impact; it was due to a change in diet

24. List five examples in which humans have altered the characteristics of plants or animals by preparing particular genetic crosses.

25. For many years, people have used a variety of techniques to influence the genetic makeup of organisms. These techniques have led to the production of new varieties of organisms that possess characteristics that are useful to humans. Identify *one* technique presently being used to alter the genetic makeup of an organism, and explain how people can benefit from this change. Your answer must include:

◆ the name of the technique used to alter the genetic makeup

◆ a brief description of what is involved in this technique

♦ *one* specific example of how this technique has been used

♦ how humans have benefited from the production of this new variety of organism

26. When people carry out specific genetic crosses (such as those done by Mendel), are they working with organisms that reproduce sexually or asexually? Explain.

Part C—Reading Comprehension

Base your answers to questions 27 to 29 on the information below and on your knowledge of biology. Use one or more complete sentences to answer each question.

It is not good for most organisms to breed with close relatives, a process called *inbreeding*. In corn, a very important crop, inbreeding causes the plant to become shorter, less strong, less productive, and more likely to develop diseases. The first person known to experiment with developing new strains of corn was the Pennsylvania farmer John Lorain. In 1812, he described experiments in which he crossed two different types of corn to make a hybrid that produced greater yields than either parent plant.

The American botanist and geneticist George H. Shull is well known for his work on hybrid corn. Due to his research, corn yields increased 25 to 50 percent. In 1917, at Harvard University, the chemist Edward East and his student, Donald Jones, successfully combined two single-cross hybrid corn varieties to produce the first highly productive corn variety that could be grown commercially. Since the 1930s, corn hybrids have been used in countries throughout the world. Now molecular geneticists have produced even better breeds of corn by inserting genes to make the corn naturally resistant to pests. A controversy has arisen since some of this new "super corn" has apparently appeared, uninvited, in Mexico where the native corn stills grows. Nevertheless, the study of genetics has helped provide more and better food for people all over the world.

27. Prepare a chart that has two columns. In the first column, list the name of each of the researchers who helped develop a better corn, in the order in which they did their research. In the second column, next to each researcher's name, describe the contribution that the person made to this research.

28. Explain the controversy surrounding genetically altered "super corn."

29. How has the study of genetics, in this particular case, had a direct impact on people's lives?

22

Human Genetics

◼◼ GENETIC COMPLEXITIES: BEYOND THE PEA PLANTS

Mendel's experiments provided the foundation for the study of inheritance. Much of Mendel's success was due to the traits he studied. The pea plants he worked with were either tall or short, and their seeds were either yellow or green, smooth or wrinkled, and so on. But are people either tall or short? Are there only two skin colors? Do people have only blond or black hair? Of course not. In humans, and in most organisms, almost all traits are not as clearly defined as the traits Mendel studied in pea plants. Yet Mendel was correct in his explanations.

The traits Mendel picked to study were, luckily for him, each determined by single genes. The height of pea plants is due to a single gene that occurs in two different versions, or alleles. Human height is a different story. Height in humans is due to several genes. Pieces of DNA on different chromosomes code for the proteins that affect human height. If a large group of people were arranged according to height, those with average height would be the most numerous. There would be fewer extremely short and extremely tall people. Grouping individuals in this way produces a bell-shaped curve. Likewise, the many variations in human skin color show that multiple genes are involved in this trait, too. (See Figure 22-1.)

◼◼ LINKED TRAITS

In some cases, the genes for one type of trait are inherited along with the genes for another particular type of trait. Such traits are said to

Figure 22-1 When a group of men are arranged by height, a bell-shaped curve is produced, which shows variation within a population.

be *linked* because they are located on the same chromosome. For example, why do people with red hair also have freckles? The reason is that the genes for red hair and for freckles are on the same chromosome. These genes are linked, so they tend to be inherited together. Now we will examine more closely the types of chromosomes found in human body cells.

▚ IS IT A GIRL OR IS IT A BOY?

Scientists can take a photograph of the chromosomes in a human cell by using a camera attached to a microscope. The chromosomes are paired up and then numbered from largest to smallest. The picture that results is called a *karyotype*. If we examine a human karyotype, we find 22 perfectly matched pairs of chromosomes. These chromosome pairs, numbered from 1 to 22, are called *autosomes*. This accounts for 44 of the 46 chromosomes found in the cell. However, the last two chromosomes do not always make a matched pair. It depends on whether the cell came from a male or a female.

If the cell came from a male, the last two chromosomes will not be an identical match. Cells from males contain an X and a much smaller Y chromosome. If the cell came from a female, the last two chromosomes will both be X chromosomes, a matched pair. Differences in the last pair of chromosomes—whether a person has an X and a Y chromosome or two X chromosomes—determine the person's sex. Therefore, these last two chromosomes are called the *sex chromosomes*. (See Figure 22-2.)

Egg cells are produced in the mother via meiosis. Her cells have 22 pairs of autosomes and one pair of X's (her sex chromosomes), for a total of 46 chromosomes. Through meiosis, every egg cell gets one of each of the 22 autosomes and one X chromosome. Now consider

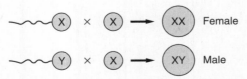

Figure 22-2 The X and Y chromosomes are called the sex chromosomes. The sex of the offspring is determined by the sperm cell.

sperm cells. Every cell in the father's body has 22 pairs of autosomes and an X and a Y chromosome. During meiosis, 50 percent of his sperm cells get 22 autosomes and an X chromosome, and 50 percent get 22 autosomes and a Y chromosome. Egg cells have only X chromosomes in addition to the autosomes. So who determines the sex of the child? The father does. If a sperm with a Y chromosome fertilizes the egg, the zygote is XY, and a male develops. If a sperm with an X chromosome fertilizes the egg, the zygote is XX, and a female develops.

▚ FAMILY HISTORY AND GENETIC DISORDERS

Genetic disorders occur in some infants. These disorders are not caused by infectious microorganisms. Instead, genetic disorders result from inborn errors that are caused by defects in genes. Now that the human genome has been decoded, new and powerful information processing tools are being used to find the genes that cause human genetic disorders. Is it possible to learn if the baby has a genetic disorder before birth? Can we predict how often genetic disorders will occur?

As we learn more about genetic disorders, we realize that there are risks in having children. To assess the risks, people must have information about their family's medical history. A man and a woman who want to become parents can go to a trained genetics counselor for help in determining their risks of having a child with a genetic disorder. A genetics counselor prepares a chart showing the occurrence of any genetic disorders in past generations of the couple's families. Such a chart showing a person's family history for a particular trait can be analyzed, and patterns of inheritance can be determined. This information helps prospective parents make informed decisions about having children.

▚ OTHER GENETIC DISORDERS

Genetic disorders can result from an abnormal number of chromosomes. As described in Chapter 16, unusual events can happen during

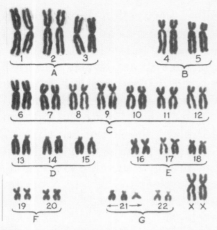

Figure 22-3 The karyotype of a person with Down syndrome shows that the disorder is caused by inheritance of three copies of chromosome 21.

meiosis as the paired chromosomes separate. Abnormal chromosome numbers cause disorders that can be detected prior to or at birth. For example, Down syndrome is due to an extra chromosome number 21. (See Figure 22-3.)

Sickle-cell anemia is a disorder that is due to a single gene defect, not an abnormal chromosome number. A mutation in the DNA base sequence of the gene for hemoglobin causes this disease, which reduces the ability of red blood cells to carry oxygen. (See Figure 22-4.)

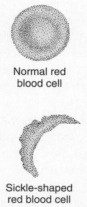

Normal red
blood cell

Sickle-shaped
red blood cell

Figure 22-4 The sickle-shaped red blood cell is characteristic of sickle-cell anemia—a genetic disorder caused by a mutation in the base sequence of the gene for hemoglobin.

Phenylketonuria is one of the most studied genetic disorders. The allele for this condition prevents a newborn from producing the enzyme that breaks down the amino acid phenylalanine. As a result, phenylalanine builds up in the baby's blood, which interferes with the development of the brain, causing mental retardation. Fortunately, some genetic disorders can be treated. Routine tests for this disease are now done on newborns. With such early detection, the baby's diet can be changed to prevent the disease's effects from developing, and the baby can lead a normal life.

GENETIC SCREENING

The presence of some genetic disorders can be determined before birth. Biochemical tests can be done to show the presence of a genetic disorder. More often today, the DNA of a fetus is studied directly to see if something is abnormal in its genes. Other prenatal tests are also possible. Sound waves can be used to make images of the fetus (while it remains safely in its mother's uterus) in order to see if it has any physical abnormalities. (See Figure 22-5.)

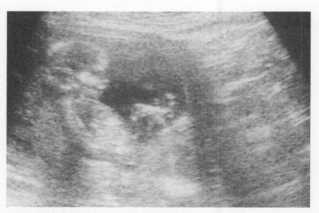

Figure 22-5 This ultrasonograph shows the outline of a developing fetus. (Note: The fetus is facing the center of the image.) This prenatal test is used to see if the fetus has any physical abnormalities.

Chapter 22 Review

Part A—Multiple Choice

1. The traits Mendel studied were particularly useful because they were

    1 inherited on one chromosome
    2 determined by single genes
    3 carried by just a few genes
    4 carried by several genes

2. Unlike height in pea plants, human height is determined by

    1 a single gene
    2 several genes
    3 two alleles
    4 proteins instead of genes

3. A bell-shaped curve for height indicates that most people are

    1 short
    2 tall
    3 either very short or very tall
    4 average height

4. People with red hair also have freckles because these traits are

    1 very common in people
    2 found in all chromosomes
    3 on the same chromosome
    4 on two linked chromosomes

5. The term *karyotype* refers to a

    1 group of similar alleles
    2 photograph of chromosome pairs
    3 cross between two plants or animals
    4 pair of traits that are linked

6. Autosomes differ from sex chromosomes in that

    1 autosomes do not occur in pairs
    2 there are 22 sex chromosomes
    3 autosomes are always perfectly matched
    4 sex chromosomes are always perfectly matched

7. In humans, a cell that has an X chromosome and a Y chromosome

    1 comes from a male
    2 comes from a female
    3 causes Down syndrome
    4 is missing a chromosome

8. The diagram below represents the organization of genetic information within a cell nucleus. The circle labeled Z most likely represents the

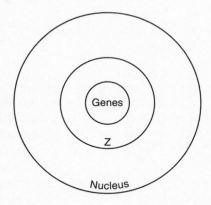

    1 amino acids
    2 vacuoles
    3 chromosomes
    4 molecular bases

9. In humans, a zygote will develop into a male if

    1 a sperm with an X chromosome has fertilized an egg cell
    2 a sperm with a Y chromosome has fertilized an egg cell
    3 an egg with a Y chromosome has been fertilized by any sperm cell
    4 the egg cell and sperm cell both have Y chromosomes

10. Couples may consult with a genetics counselor to

    1 map the genes for their future children
    2 determine the exact genes that their children will have
    3 determine patterns of inheritance and risks of disorders
    4 correct all genetic disorders before their child is born

11. Inborn genetic disorders can result from all of the following *except*

    1 an abnormal chromosome number
    2 a single gene defect
    3 an infectious microorganism
    4 a mutation in a DNA base sequence

Part B—Analysis and Open Ended

12. How did Mendel's choice of pea plants contribute to his scientific success?

13. Why do traits such as human height show a very wide range of variations?

14. Explain why red hair and freckles are most often inherited together.

15. The sex chromosomes are identified for each cell in the figure below. Copy the figure and correctly label each cell using the following terms: *Sperm cell*; *Egg cell*; *Male zygote*; and *Female zygote*.

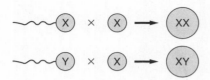

16. Discuss the importance of the sex chromosomes. Your answer should explain the following:

    ♦ In what way are the two sex chromosomes different from the autosomes?

    ♦ How do chromosomal differences determine whether a person is male or female?

    ♦ In what way is the father responsible for determining the sex of a child?

17. How can a chart that shows a family's medical history be used to assist a couple that is thinking about having children?

18. Sickle-cell anemia is due to a single gene defect that affects hemoglobin. Use your knowledge of genetics to explain how one defect in a DNA base sequence could upset production of this protein.

19. Why is early detection important for treating the disease phenylketonuria?

20. Refer to Figure 22-5 on page 168 to answer this question. The image of the fetus was formed by a method that uses

    1 X rays
    2 light rays
    3 sound waves
    4 chemical tests

21. Why is prenatal diagnosis important for a couple that has a family history of a genetic disorder?

22. How can sound waves be used for prenatal diagnosis? What can they show?

Part C—Reading Comprehension

Base your answers to questions 23 to 25 on the information below and on your knowledge of biology. Use one or more complete sentences to answer each question.

Joseph and his parents had been referred by their family doctor to the genetics clinic at a local hospital. The doctor had made this suggestion after testing the levels of sex hormones in Joseph's blood and ordering a chromosome analysis. Joseph, a healthy 16-year-old, was doing well in school. However, even though he was almost two meters tall, he showed no signs of sexual maturity. He had not developed additional body hair, and his voice retained the high pitch of a much younger person. A long talk occurred in the geneticist's office.

The counselor explained that Joseph has one more X chromosome than is usually found in a male. This is due to an error that occurred during meiosis when the egg or sperm from which he developed formed. Joseph and his parents learned from the counselor that about one in every thousand males has this XXY chromosome makeup—a total of 47 chromosomes instead of the normal 46.

The fact that this condition is called Klinefelter syndrome was not important to Joseph and his family. However, learning about the characteristics of the syndrome was very important. The counselor told them that the characteristics of men with this syndrome usually included tall stature, some minor birth defects, small testicles, and sterility. Sometimes, but not in Joseph's case, mental retardation begins at birth.

As a result of the genetics counseling, Joseph realized that he could most likely look forward to a normal life. Indeed, in a few months, his beard began to grow. More important, Joseph learned about possible options for having a family. His parents also learned that amniocentesis might be a good idea if they decided to have another child, since some families have shown a tendency to produce other errors in meiosis.

23. Why did Joseph's doctor order hormone and chromosome tests for him?

24. Describe the genetic explanation for the symptoms experienced by Joseph.

25. How did the genetics counseling that Joseph and his family received benefit them?

23

Biotechnology

RECOMBINANT DNA TECHNOLOGY: A BRIEF DESCRIPTION

A revolution in biology began with the discovery of the structure and function of DNA, the molecule of life. This revolution has increased in importance through advances in *recombinant DNA technology*.

We know that genes are made of DNA and that they determine the characteristics of every organism on Earth. Now scientists have learned how to identify and find individual genes. Once found, these pieces of DNA can be removed and put together, or **recombined**, with other pieces of DNA. The genes can then be moved from one cell into another. The methods for doing this make up recombinant DNA technology.

For thousands of years, people have selectively bred certain characteristics in plants and animals to produce different crops and breeds. Now, recombinant DNA technology makes it possible to put "new" genes into organisms—that is, actually to change the genetic makeup of plants and animals.

Through recombinant DNA technology—also called **biotechnology** or **genetic engineering**—human genes can be inserted into the genetic material of bacteria. These altered bacteria then become tiny "factories" that produce human proteins. Many other types of genes can be inserted into the genetic material of bacteria and other organisms. For example, agricultural scientists improve crops by inserting genes that make them disease-resistant, and improve livestock by inserting genes that make them grow faster or produce more milk. Perhaps most important, through biotechnology, human gene therapy may be used to treat some genetic disorders. Although

progress has been slow, someday it may be possible to remove defective genes that cause certain disorders. In their place, healthy genes may be inserted that will prevent people from developing the disorder.

BASIC TOOLS OF RECOMBINANT DNA TECHNOLOGY

As you read in Chapter 20, scientists now estimate that humans have about 20,000 to 30,000 different genes in each cell. Although this figure is much lower than the 100,000 genes that had been estimated years earlier, it is still a very large number of genes. From these genes, after considerable processing, come a much larger number of different proteins in our cells. Scientists wanted to know where the genes for specific proteins were located in the DNA. They wanted to be able to

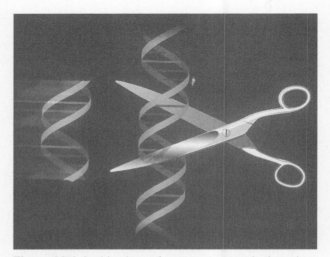

Figure 23-1 In this piece of computer artwork, the scissors represent the restriction enzyme that is used to cut a piece of DNA. Another piece of DNA, perhaps the gene to produce human insulin, would be inserted where the cut is made.

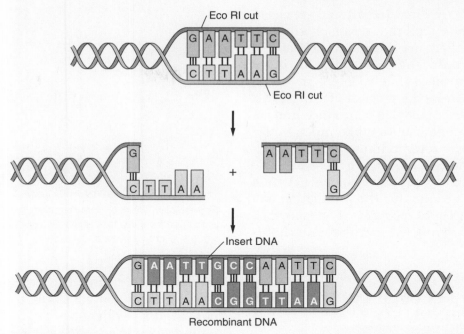

Figure 23-2 The action of a restriction enzyme called Eco RI: The original double strand of DNA is cut, and another piece of DNA that has been cut (perhaps from a different organism) is inserted to form recombinant DNA.

move the genes from one organism and place them into another. The task seemed hopelessly complex until the 1970s, when restriction enzymes were discovered. A *restriction enzyme* recognizes a sequence of between four and six base pairs. Whenever it finds this specific sequence, the restriction enzyme cuts the DNA. The place where the cut occurs is called a *restriction site*. (See Figure 23-1.)

For a scientist, restriction enzymes are very powerful tools. With them, DNA can be cut at precise locations. In addition, the same restriction enzyme can be used to cut DNA from two completely different organisms, such as a frog and a bacterium. Pieces of DNA from one organism can then be inserted, or spliced, into the DNA of another organism. Remember, those "pieces of DNA" *are* the genes themselves. (See Figure 23-2.)

Yet, restriction enzymes are not enough. Scientists also need a way to move pieces of DNA from the cell of one organism to a cell of another organism, where the foreign DNA can replicate. So scientists have now developed special molecules, called *vectors*, which can move pieces of DNA from one organism to another. (See Figure 23-3.)

Bacterial cells are used most often to receive the piece of DNA from a vector. Bacteria

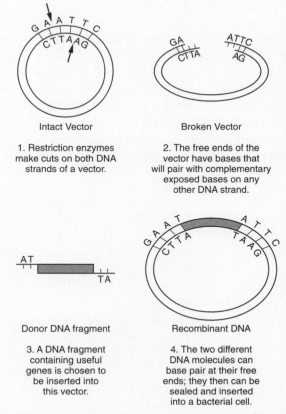

1. Restriction enzymes make cuts on both DNA strands of a vector.

2. The free ends of the vector have bases that will pair with complementary exposed bases on any other DNA strand.

3. A DNA fragment containing useful genes is chosen to be inserted into this vector.

4. The two different DNA molecules can base pair at their free ends; they then can be sealed and inserted into a bacterial cell.

Figure 23-3 Scientists use vectors, such as circular pieces of bacterial DNA, to insert and move genes from one organism to another. The vector is usually placed within a bacterial cell, because it can reproduce quickly and make many more copies of the recombinant DNA.

reproduce quickly. Since the altered genes are passed on to every cell that develops (through mitosis), soon there are thousands of bacterial cells that contain the new DNA. In other words, once the piece of new DNA is in the reproducing bacteria, the amount of it increases as the number of bacteria reproduce and increase. This is one way to make much larger quantities of the recombinant DNA.

There is another reason why the bacteria are encouraged to reproduce. The new DNA in the bacteria is coding for the production of a protein of interest to scientists. The more bacterial cells there are, the more protein that is produced.

⠿ USES OF RECOMBINANT DNA TECHNOLOGY

Medicine is benefiting from the use of biotechnology. Dozens of human genetic disorders can now be diagnosed with recombinant DNA technology. Often, individuals can be diagnosed with a disease even before they show any symptoms. This is because the gene that causes the disease can be identified in their DNA. This identification can even be performed on a fetus in the mother's womb.

Some medicines are now being produced in bacteria through biotechnology. For example, genetically engineered bacteria use a human gene to produce insulin. It is, therefore, pure

Figure 23-5 Results from the analysis of DNA fingerprints can be used to argue the guilt or innocence of individuals in criminal trials.

human insulin. (See Figure 23-4.) A type of biotechnology called *DNA fingerprinting* is now used as evidence in legal cases to determine the guilt or innocence of individuals in criminal trials. (See Figure 23-5.) Biotechnology has been enlisted to help clean up the environment, too. For example, genes are inserted into bacteria to give them the ability to remove hazardous substances from the environment. As mentioned before, animals that are raised for food are receiving genes from other animals to make them grow faster and produce more food in a shorter time. In addition, genes that make agricultural plants resistant to disease can be added to a crop plant's genes.

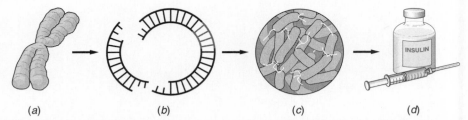

(a)   (b)   (c)   (d)

Figure 23-4 The insulin gene from a human chromosome *(a)* is inserted into a vector *(b)*, a circular piece of bacterial DNA. The bacteria that receive this vector then contain the gene to produce insulin *(c)*, which is collected for use by people who have diabetes *(d)*.

Chapter 23 Review

Part A—Multiple Choice

1. Recombinant DNA technology involves
   1. creating new DNA from their molecular subunits
   2. inbreeding plants or animals with similar DNA
   3. interbreeding plants or animals with different DNA
   4. splicing pieces of DNA into other sections of DNA

2. Terms that describe the methods by which scientists change the genetics of organisms include all of the following *except*
   1. biotechnology
   2. genetic engineering
   3. agricultural engineering
   4. recombinant DNA technology

3. Humans produce more than 30,000 different proteins in their bodies. To make these proteins, we must have between
   1. 100 and 500 genes
   2. 5,000 and 15,000 genes
   3. 25,000 and 30,000 genes
   4. 50,000 and 100,000 genes

4. Scientists use restriction enzymes to
   1. limit the length of DNA molecules
   2. stop parts of DNA from replicating
   3. prevent certain genes from being expressed
   4. cut specific base-pair sequences out of DNA

5. The diagram below represents some steps in a procedure used in biotechnology. Letters X and Y represent the

Bacterial DNA

Foreign DNA

   1. hormones that stimulate the replication of bacterial DNA
   2. hormones that trigger rapid mutation of genetic information
   3. enzymes involved in the insertion of genes into a new organism
   4. energy required for the manipulation of genetic information

6. The molecules that can move cut pieces of DNA from one organism to another are called
   1. vectors
   2. splicers
   3. transformers
   4. combiners

7. Genetic engineering has been used to improve crop varieties by
   1. reproducing old genes for wild characteristics
   2. removing genes that cause them to get diseases
   3. inserting genes that make them disease resistant
   4. adding animal genes that make them grow faster

8. Why do scientists insert human genes into bacteria?
   1. to give bacteria some human traits
   2. to make large amounts of human proteins
   3. to dispose of our defective genes
   4. to find out what the bacteria will do

9. When a human gene is inserted into a bacterial cell to become part of its DNA, the process is an example of
   1. DNA fingerprinting
   2. biotechnology
   3. karyotyping
   4. reproduction

10. Why are bacterial cells useful in recombinant DNA technology?
    1. They reproduce very quickly.
    2. They reproduce very slowly.
    3. They are almost identical to human cells.
    4. They can be placed within the human body.

11. The production of certain human hormones by genetically engineered bacteria results from
    1. inserting a specific group of amino acids into the bacteria
    2. interbreeding two different species of harmless bacteria
    3. splicing a piece of human DNA into a vector and then inserting it into bacteria
    4. deleting a specific amino acid from human DNA and inserting it into bacterial DNA

Part B—Analysis and Open Ended

12. How is recombinant DNA technology different from the traditional practice of selective breeding of plants and animals?

13. Describe three examples of how recombinant DNA technology is being used today. Your answer must include at least:

♦ *one* example for plants

♦ *one* example for animals

♦ *one* example for humans

Base your answers to questions 14 and 15 on the passage below and on your knowledge of biology.

For a number of years, scientists at Cold Spring Harbor Laboratory in New York have been working on the Human Genome Project to map every known human gene. By mapping, researchers mean that they are trying to find out on which of the 46 chromosomes each gene is located and exactly where on the chromosome the gene is located. The scientists want to be able to improve the health of people. By locating the exact positions of defective genes, scientists hope to cure diseases by replacing defective genes with normal ones, a technique known as *gene therapy*. Scientists can use specific enzymes to cut out the defective genes and insert the normal genes. They must be careful to use the enzyme that will splice out only the target gene, since different enzymes will cut DNA at different locations.

14. Using one specific example, state why the Human Genome Project is considered important.

15. Explain why scientists must use only certain enzymes when inserting genes or removing genes from a cell.

16. What discovery was made in the 1970s that advanced genetic engineering?

17. Explain why this discovery was so important for recombinant DNA technology.

18. What are vectors and how are they used in genetic engineering?

19. The following diagram represents a procedure used in biotechnology. Name a specific substance that can be produced by this technique and state how humans have benefited from the production of this substance.

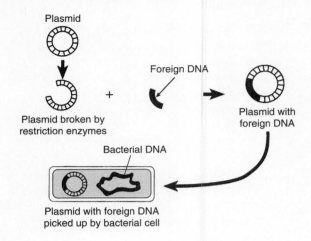

20. The diagrams below illustrate a process used in biotechnology. For each step (*a* through *d*) shown, use one of the following phrases to label what the structure represents: DNA fragment with useful genes is chosen; DNA fragment has been inserted into bacterial DNA; Free ends of bacterial DNA are exposed; Restriction enzyme makes cuts in bacterial DNA.

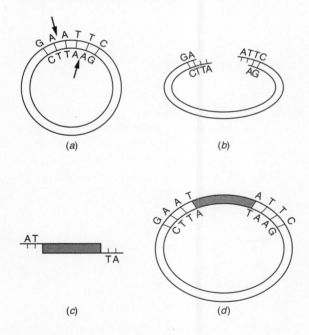

21. Biological research has generated knowledge used to diagnose genetic disorders in humans. Explain how a specific genetic disorder can be diagnosed. Your answer must include at least:

♦ a name of *one* genetic disorder that can be diagnosed

♦ a name or description of *one* technique used to diagnose the disorder

♦ a description of *one* characteristic of the disorder

22. How is biotechnology used to detect a disease even before its symptoms appear?

Refer to the set of diagrams below to answer questions 23 and 24.

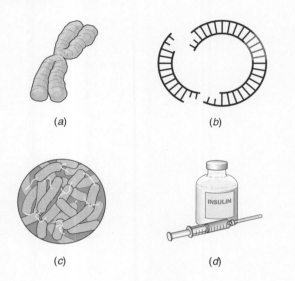

(a)

(b)

(c)

(d)

23. For each step (*a* through *d*) illustrated, write one sentence to explain which part of the genetic engineering process it represents.

24. Which one of the following titles would best describe the set of diagrams?
1 Restriction Enzymes and Genetic Advances
2 Agricultural Uses for Genetic Engineering
3 Genetic Engineering and Medicine Production
4 Genetic Engineering and DNA Fingerprinting

25. The bacteria that are used in recombinant DNA experiments have been changed so that they can survive only under special conditions in the laboratory. Why do you think scientists have included this safety precaution in their work?

Part C—Reading Comprehension

Base your answers to questions 26 to 28 on the information below and on your knowledge of biology. Use one or more complete sentences to answer each question.

The horse has played a major role in people's lives throughout history. Horses are important to people as a means of transportation, sport, recreation, and even companionship. People have fought wars, chased cattle, entered competitions, and won races—all while riding on the backs of horses. And headlines that concerned horses abounded in 2003, with the attempt at the Triple Crown by the up-and-coming New York racehorse Funnycide, and with the release of both a documentary and a Hollywood film (based on the book) about Seabiscuit, the unlikely champion racehorse of the 1930s. Indeed, the question of what makes a great racehorse has intrigued people for decades.

In 1990, the U.S. Department of Agriculture invited veterinary scientists to begin mapping the genes of domestic animals, just as researchers were doing for human genes. Work began quickly on the DNA of cattle, sheep, pigs, and chickens—but not horses. In spite of people's interest in horses, there did not seem to be enough money for that work at the time.

All that changed in 1995 when geneticists working with horses decided to begin to share information. By 2001, the horse genome project involved 25 laboratories in 15 countries. Already, genes have been found on horse chromosomes for a number of bone and muscle diseases. Breeders of champion racehorses are paying close attention to the progress of the horse genome project. Certainly, it would be important to know in advance if a racehorse had a gene for a particular disease. However, will genes for exceptional racing performance ever be found? In other words, will DNA technology help build a better racehorse? For now, the outcome is uncertain. But it's probably a safe bet that it is just a matter of time!

26. What project in genetic research began in 1995?

27. What progress has been made in the genetic research on horses to date?

28. Why is the racehorse industry so interested in the horse genome project?

Interaction and Interdependence

An Introduction to Ecology

✛ ECOLOGY AND ECOSYSTEMS

An aquarium is a self-contained miniature world of life. Like the living things in an aquarium, every organism on Earth lives within its surroundings, or **environment**. All living things interact—they affect other living things and their environment; and all living things depend on each other and their environment—they are interdependent. These relationships of *interaction* and *interdependence* between organisms and their environment are studied in the branch of biology known as **ecology**. (See Figure 24-1.)

Every organism has to live somewhere. The environment in which an organism lives is molded by many different *factors*, such as availability of food and water, amount of sunlight, temperature, and type of soil. Conditions that involve other living organisms are known as **biotic** factors. For a fish in the aquarium, the biotic factors could include other fish, snails, algae, and plants. Conditions that involve nonliving things are known as **abiotic** factors. For that same fish, the abiotic factors could include the water, air

Figure 24-1 An aquarium is like a miniature ecosystem —the living things interact with one another and with the nonliving parts of their environment.

bubbles, gravel, **acidity**, temperature, and light.

The interaction and interdependence of organisms and their environment can be understood by examining specific places. All the living and nonliving factors that interact in one specific place are the parts of an **ecosystem**. A pond is an ecosystem. A forest is an ecosystem. Even the little aquarium is an ecosystem that can be used to study ecology.

Figure 24-2 Organisms have adaptations that help them survive in particular environments. The camel's wide, two-toed feet enable it to walk in the desert sand without sinking.

Figure 24-3 Many rain-forest trees have wide, woody supports at the base of their trunks to keep them upright in the shallow tropical soil.

✖ ADAPTATIONS AND EVOLUTION

Camels have extremely wide, two-toed feet to avoid sinking into the desert sand. (See Figure 24-2.) Many rain-forest trees have wide supports at the base of their trunks to keep them upright in the shallow tropical soil. (See Figure 24-3.) Wherever we look on this planet, we see many examples of the features that enable organisms to survive Earth's great variety of living conditions. How did all this happen?

No individual organism intentionally changed to survive in a particular environment. What did happen is that there have always been some differences among individuals in populations. Two dogs from the same litter, for example, might be very different in size or color. Sometimes the differences are not obvious. Yet a slight difference in the biological makeup of an organism, such as the ability to make a particular enzyme, might provide it with a survival advantage over other organisms. These variations are due to inherited differences. Because of these differences, some individuals are better suited than others to certain environmental factors.

It is through the process of natural selection that species, not individual organisms,

evolve. This is the basis of evolution. Over time, a species' traits make a remarkable fit with its environment. If they do not, the species will probably not survive in that particular environment. The characteristics of an organism that give it this fit are called **adaptations**. Adaptations can be in physical traits, such as the size, shape, or color of the organism. They also can be in the behavior of an organism, such as the building of nests by birds or the release of seeds once a year by plants. (See Figure 24-4.)

✖ AN ORGANISM'S HABITAT AND NICHE

Every organism is adapted through evolution to live in a particular place. Each species of organism is adapted to a specific set of conditions. The place where an organism lives is its **habitat**. The habitat of a bullfrog is a pond. The habitat of a giant anteater is open grassland. An organism's habitat is its "address."

To understand an organism's relationship to its environment, we must know more than

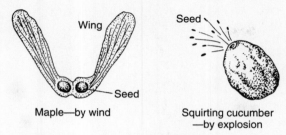

Maple—by wind

Seed
Squirting cucumber
—by explosion

Figure 24-4 Plants have many different adaptations for releasing and dispersing their seeds.

its address, or *where* it lives. We must also know its "occupation," or what it does and *how* it lives. The occupation of an organism is called its **niche**. The niche of an organism includes how it gets food, reproduces, avoids predators, and so on. The behavioral adaptations of an organism make up its niche. These adaptations are the result of evolution just as are its physical adaptations. The niche of an organism determines its habitat. In other words, the ways that an organism has evolved to survive will also determine where it can live. For example, a woodpecker cannot live in a grassland; its niche involves finding its insect food in the trunks of trees. So woodpeckers need a habitat that has trees, and the insects that live in them, in order to survive. (See Figure 24-5.)

Figure 24-5 The niche of an organism determines where it can live. For example, the woodpecker needs to live in a forested habitat because its niche involves finding insects that live in trees.

ENVIRONMENTAL FACTORS

Every type of organism is adapted to the specific conditions in its environment. Since organisms do have specific habitat requirements, different environmental conditions can limit where they live. These conditions are called *limiting* factors.

For example, the availability of sunlight is an important limiting factor for plants. The amount of sunlight in oceans and lakes varies but does not reach below a certain depth. Be-

low that depth, it is too dark for aquatic plants to grow. Temperature, of course, is another major limiting factor. Each species of plant or animal has a fairly narrow temperature range that it prefers. In other words, the organism can tolerate temperatures only within this range. Other environmental factors, such as chemical nutrients in the habitat, may be less obvious, but are still very important for a species' survival. In general, organisms have a tolerance range for a variety of environmental factors, sometimes a very narrow one, usually neither too low nor too high. The tolerance range determines the best conditions for a specific type of organism in a specific location.

All life depends on water, and organisms have a variety of adaptations that enable them to survive in very specific ranges of available moisture. For example, water constantly moves out of openings on the surface of leaves. So some species of trees, such as pines, have evolved ways to save water—they have narrow leaves, or needles, from which little moisture is lost. In areas with abundant rainfall, water loss is not a problem, so the trees have large, flat leaves.

AQUATIC ECOSYSTEMS

Because water is so essential to life, we will first look at ecosystems that occur in the water. A map of the world shows individual oceans; however, all the world's oceans are actually connected. Some ecologists consider this world ocean to be one tremendously large ecosystem. Three main limiting factors in the ocean are saltiness, temperature, and sunlight. As the amount of salt in ocean water varies, the density of the water also changes. The temperature of ocean water also differs from place to place and also affects water density. Cold water is denser than warm water, so the colder water sinks. The amount of sunlight also varies over different parts of the ocean, and it penetrates only to a certain depth. Because they need light to carry out photosynthesis, all plants and **algae** in the ocean live only in the top (photic) zone of the water. (See Figure 24-6 on page 182.)

There are two main types of freshwater

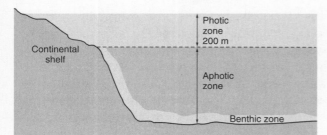

Figure 24-6 Sunlight is a limiting factor for the growth of plants. Because sunlight penetrates only about 200 meters below the ocean's surface, aquatic plant life is restricted to that top (photic) zone.

Figure 24-7 Water is another important limiting factor for plants. In areas of abundant rainfall, plants such as the maple (left) have broad, flat leaves; in areas of limited rainfall, plants such as the white pine (right) have narrow, needle-shaped leaves to reduce water loss.

ecosystems on the surface of Earth: lakes and ponds, which are still bodies of water; and rivers and streams, which are running water. In the still-water ecosystems, temperature and light are the main limiting factors. In the running-water ecosystems, the temperature and light are fairly constant at any given point but vary along the length of the river or stream. For example, the water flows faster and colder at the start of a river than at its end. So different kinds of plants and animals are adapted to survive in different parts of the river. In general, conditions in water are fairly constant over wide areas. They change little over time and, when they do, they change very slowly.

⚟ LAND ECOSYSTEMS

Conditions on land are very different from those in the water. From season to season, and from one part of the day to another, temperatures on land can vary widely. Variations in temperature, moisture, soil type, length of days and nights, seasons, and altitude all work together to produce many different land ecosystems.

There are three main forest ecosystems on Earth, each with its characteristic types of trees: tropical rain forests, broad-leaved forests, and needle-leaved forests. Tropical rain forests exist in a wide band north and south of Earth's equator. They have large amounts of rainfall, warm temperatures, and a stable length of daylight throughout the year. Abundant life of all kinds exists in the tropical rain forests.

Farther north and south of the equator, cli-

mate patterns change. Definite seasons occur, with variations in temperature and rainfall. This creates the forest environment of broad-leaved trees, such as maple and oak. Here, fewer tree species are found, and the seasonal dropping of leaves is typical. Still farther north, where it is colder, there are the great forests of needle-leaved evergreen trees, such as spruce and pine. (See Figure 24-7.)

Farthest from the equator, the temperatures and amount of rainfall are too low for

Figure 24-8 The giant saguaro cactus, which lives in the desert, stores water in its stem.

trees to grow. Only one or two months of the year are warm enough to support plant growth. Then, a thick layer of mosses, lichens, grasses, and low shrubs covers the surface.

In regions where precipitation, but not temperature, decreases, there are grasslands, because the limited moisture prevents the growth of trees. Finally, the driest places on Earth are the deserts. In a typical desert, plants include water-storing cactuses, shrubs with roots that grow deep to reach water, and wildflowers and grasses that flourish briefly after the infrequent rains. (See Figure 24-8.) Animals also show adaptations to the desert conditions. For example, the kangaroo rat lives underground for much of the day to avoid the heat, and desert predators such as coyotes and foxes are also more active at night.

Chapter 24 Review

Part A—Multiple Choice

1. Ecology can best be described as the study of
   1. all the plants in a certain environment
   2. living organisms and their environment
   3. living factors that affect an organism
   4. nonliving factors that affect an organism

2. A *biotic* factor in a snake's environment would be
   1. sunlight
   2. water
   3. sand
   4. a mouse

3. An *abiotic* factor in an eagle's environment would be
   1. a tree
   2. water
   3. a snake
   4. an insect

4. Which event illustrates the interaction of a biotic factor with an abiotic factor in the environment?
   1. Water temperature affects water density in the ocean.
   2. The lamprey eel survives as a parasite on other fish.
   3. Shorter daylight hours cause maple trees to lose leaves.
   4. A gypsy moth caterpillar eats the leaves of an oak tree.

5. An ecosystem is best described as the
   1. type of food that an organism eats
   2. type of home an organism builds
   3. group of organisms in a particular place
   4. living and nonliving factors in one place

6. Which of the following can be considered an ecosystem?
   1. a large rock
   2. a bird's nest
   3. a rain forest
   4. a rain cloud

7. Natural selection is important because it
   1. allows an individual organism to evolve
   2. enables a species to adapt and evolve
   3. allows humans to control animal traits
   4. changes an environment to fit the organisms

8. An adaptation is a body part or behavior that
   1. is no longer used by an organism
   2. prevents an organism from surviving
   3. helps an organism survive and be fit
   4. is useful to only one organism at a time

9. A habitat can be described as an organism's
   1. average size
   2. main function
   3. wild behavior
   4. natural address

10. A frog's habitat would be the
    1. pond it lives in
    2. sounds it makes
    3. insects it eats
    4. color of its skin

11. An organism's niche is most similar to a person's
    1. character
    2. occupation
    3. address
    4. personality

12. In a forest community, a shelf fungus and a slug live on the side of a decaying tree trunk. The fungus digests and absorbs materials from the tree, while the slug eats algae growing on the outside of the trunk. These organisms do not compete with one another because they occupy

1 the same habitat, but different niches
2 the same niche and the same habitat
3 the same niche, but different habitats
4 different habitats and different niches

13. In a pond, a carp eats decaying matter from the base of an underwater plant, while a snail scrapes algae from the leaves and stems of the same plant. These animals can both survive in the same pond because they occupy

1 the same habitat and the same niche
2 the same niche, but different habitats
3 the same habitat, but different niches
4 different habitats and different niches

14. Light, temperature, and water are examples of environmental

1 habitats
2 niches
3 limiting factors
4 adaptations

15. Why do plants live only in the top zone of the ocean?

1 There is too much salt in deeper water.
2 They automatically float to the top.
3 Plants cannot grow underwater.
4 Plants need sunlight to make food.

16. Which factor has the greatest influence on the variety of species that live in different regions of a marine habitat?

1 depth of light penetration
2 size of predators
3 daily fluctuations in temperature
4 average annual rainfall

17. The main limiting factors in freshwater ecosystems are

1 temperature and light
2 salt and temperature
3 density and light
4 altitude and depth

18. Ecosystems vary more on land than in water because

1 there is more land than water on Earth's surface

2 conditions on land vary more than they do in water
3 evolution of new species does not occur in water
4 organisms that live in water become extinct sooner

19. Earth's three main forest ecosystems vary because they

1 experience different temperatures and rainfall
2 are located near different major cities
3 differ greatly in size
4 are in different stages of development

Part B—Analysis and Open Ended

20. Identify five conditions that may be included as part of an organism's environment.

21. What is the difference between biotic and abiotic factors in an ecosystem? Your answer should include the following:

♦ the definition of *biotic*
♦ the definition of *abiotic*
♦ *two* examples of biotic factors
♦ *two* examples of abiotic factors

22. Provide one example of an interaction between a biotic factor and an abiotic factor that you have observed in your local environment.

Refer to Figure 24-1 on page 179 to answer questions 23 and 24.

23. Why is the aquarium a good example to use when studying about ecosystems?

24. List at least two biotic and two abiotic factors that are present in this aquarium.

25. Identify one *abiotic* factor that would directly affect the survival of organism A shown in the diagram below.

26. How does natural selection explain the close fit of organisms to their environment?

27. Why does the niche of an organism determine its habitat? Give one example.

28. Using an example such as sunlight or moisture, explain the idea of a limiting factor. How are such factors related to the diversity of environmental conditions and life forms on Earth?

Base your answers to questions 29 to 32 on the information below and on your knowledge of biology.

Duckweed is one of the smallest flowering aquatic plants. It grows floating on still or slow-moving fresh water. Duckweed species are found throughout the world, except in very cold regions, and are the subject of much scientific research. The plants are used to study basic plant biochemistry, plant development, and photosynthesis.

Environmental scientists are using duckweed plants to remove hazardous substances from water. Fish farmers use them as an inexpensive food source for the fish they raise. As with other aquatic plants, duckweed grows best in water containing high levels of nitrates (nitrogen compounds) and phosphates. The level of iron-containing compounds is often a limiting factor. A cover of duckweed on a pond shades the water below and reduces the growth of algae. A key for identifying duckweed species is shown below.

Duckweed Identification Key

Description of Duckweed	Plant Has No Roots	Plant Has One or More Roots
Plant body is flat	*Wolffiella*	
Plant body is oval (less than 1 mm)	*Wolffia*	
Plant has one mid-sized root		*Lemma*
Plant has two or more large-sized roots		*Spirodela*

29. Describe the value that duckweed has for heterotrophic organisms living in ponds where it grows.

30. Explain what is meant by the statement, "The level of iron-containing compounds is often a limiting factor."

31. State *one* way in which shading the water below it causes duckweed to affect the growth of algae.

32. Explain why *Spirodela* would most likely absorb more hazardous substances from the water than would the other duckweed species identified in the key.

33. Describe the three main limiting factors in the ocean. Why do they have an impact on life in the sea?

34. Suppose a dam is constructed near the headwaters (that is, the point of origin) of a river that flows from a mountain to the sea. How might the river change both upstream and downstream from the dam? How would the changes affect the biotic and abiotic factors in the river?

35. Use the following terms to fill in the table below, which outlines the three types of forest ecosystems: Broad leaves that drop seasonally; Equatorial regions; Colder temperatures and less rainfall; North and south of the equator; Wide bases that support trunks in shallow soil; Warmer temperatures and more rainfall; Needle-shaped leaves that conserve moisture; Variations in temperature and rainfall; Northern regions.

Characteristics	Tropical Rain Forests	Maple/ Oak Forests	Spruce/ Pine Forests
Type of Trees			
Type of Climate			
Geog. Location			

Part C—Reading Comprehension

Base your answers to questions 36 and 37 on the information below and on your knowledge of biology. Use one or more complete sentences to answer each question.

> Will the risk of a major hurricane in Florida be higher than normal this summer? Will Buffalo, New York, normally one of the snowiest cities in America, have another winter with no snowfall? Will intense Pacific storms cause destructive mudslides along the California coast?
>
> Questions like these have been asked in the United States and in other countries for years. The answers to these questions focus on the effects produced by two atmospheric events: "the little boy" El Niño and "the little girl" La Niña. Both El Niño and La Niña events occur regularly in the Pacific Ocean.
>
> El Niño is a periodic warming of the water in the eastern Pacific Ocean near the equator. La Niña, sometimes called the cooler sister, is marked by cool waters near the equator in the Pacific. Both of these events change the atmosphere above the ocean. A change in the wind causes a change in the movement of the ocean's surface waters. Changes in water temperature change the amount of moisture that leaves the ocean's surface and enters the atmosphere.
>
> El Niño events tend to transfer heat from the ocean to the atmosphere. This increases the strength and speed of the jet stream—the high winds that move from west to east across the United States—and contributes to relatively predictable climatic patterns. During La Niña, the colder Pacific water near the equator weakens the jet stream that helps to keep our weather constant. As a result, La Niña events cause unusual weather patterns in the United States. El Niño events occur every three to seven years. Because the effects of El Niño and La Niña are opposite each other, we are always living in one event and heading toward the other. Without a doubt, we will continue to learn more about how changes in water temperatures in the Pacific Ocean are related to changing weather patterns that occur in distant parts of the world and right at home.

36. Give two reasons why it seems surprising that El Niño and La Niña have effects on North American weather patterns.

37. Write a short paragraph in which you compare and contrast El Niño and La Niña. Be sure to describe both their similarities and their differences.

Populations and Communities

THE STUDY OF POPULATIONS

Ecologists are more interested in groups of organisms than in individuals. All the organisms of one species that live in one place at a particular time make up a **population**. No population ever lives alone. Other organisms —plants, animals, fungi, and microorganisms—are also present. For example, a field has a population of mice, but it also has populations of wildflowers, insects, mushrooms, birds, and snakes. All of the populations that interact with each other in a particular place make up a **community**. In large part, the study of ecology is about populations and communities.

FACTORS THAT AFFECT POPULATION GROWTH

Most organisms are able to reproduce rapidly. Even if a pair of individuals produced only two offspring each year, the growth rate would be enormous if all offspring survived. Suppose a population was founded by two individuals. By doubling each year, after only 10 years there would be a population of more than 1000 individuals. Many organisms produce even greater numbers of offspring, such as fish that lay thousands of eggs per year. If these eggs all hatched and survived to reproduce, the number of resulting fish would be enormous. Trees also produce many thousands of seeds per year. With these growth rates, Earth could quickly be overcrowded with organisms. Of course, this does not occur. Thus, an important area in ecology is the study of what controls population growth.

A key factor in population growth is *density*, which is the number of individuals in a population in a given area. A population with a small number of individuals in a particular area has a low density. When a population's density is low, there is usually sufficient food and space for existing organisms. The birth rate increases, while the death rate drops. As a result, the density of the population begins to increase at a faster rate. But this rate of increase cannot last forever. At some point, the population gets too crowded. (See Figure 25-1.)

The most basic needs of organisms from their habitats are food and space. However, every habitat has limits. When population density increases too much, the available food and space decrease. At that point, the population density has reached a maximum for the particular habitat. The death rate increases, while the birth rate drops. The size of a population that can be supported by any ecosystem is called the **carrying capacity**. Population growth slows and may reach zero growth as the population size approaches an area's carrying capacity. Zero growth means the population size is no longer increasing—the birth rate and death rate are about equal. The rate

Figure 25-1 You can visualize the population density of New York City from this photograph, taken at night by the crew of a space shuttle. The brighter areas are those where population density is higher.

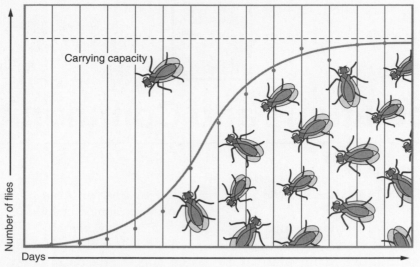

Figure 25-2 A typical population growth curve. A new population increases slowly at first, because there are few individuals. Then, as the population increases, the growth rate increases rapidly as more individuals reproduce. After some time, it levels off to zero growth as the population approaches the area's carrying capacity.

at which a population grows is shown in Figure 25-2.

Factors that limit the size of a population include **competition** for food and space. This competition increases as population density increases. Another problem with high population density is that the more crowded a population is, the easier it is for predators to find prey. In a crowded population, there is also a greater chance for disease to spread among individuals.

▓ COMMUNITY INTERACTIONS

Organisms that live in the same environment are always interacting with each other. The survival of many plants and animals often depends on the relationships they have with other organisms. (See Figure 25-3.) Some important types of relationships are discussed below. Competition is one of the main interactions between organisms. For example, if two different species of birds ate the same species of butterflies from treetops in the same forest, there would be competition between them. This competition between two different species would be intense, in this case, because of the overlap of the birds' niches. However, if the two bird species fed on different insects in the same treetops, there would be less overlap in their niches and less competition between them.

The greatest competition usually occurs between members of the same species, because

Figure 25-3 A community, such as that illustrated living in and around this pond, is made up of many different organisms. The survival of most plants and animals depends on the relationships they have with the other organisms in their environment.

Figure 25-4a The mouse is considered the predator when it preys on plants.

Figure 25-4b The snake is the predator when it preys on the mouse, which is then the prey.

such individuals most likely share identical niches. In other words, they live in exactly the same way and compete most intensely for the same limited resources in their area. Competition results in natural selection. Recall the thousands of eggs produced by the fish—most will not survive to maturity. Only the most fit individuals live to pass on their genes to offspring. Therefore, competition is an important force in the process of evolution.

Predation is one of the most basic relationships that occurs between organisms. In nature, living things either "eat or are eaten." That is, they are either the **predator** or the **prey**. In predation, members of one population are the food source for members of another population. Even grass seeds can be considered prey when a mouse—the predator, in this case—eats them. Of course, the mouse becomes the prey when a snake, another predator, eats it. (See Figures 25-4 *a* and *b*.)

▉ TYPES OF SYMBIOTIC RELATIONSHIPS

As stated above, almost no organism lives entirely alone. In fact, most organisms have close relationships with at least one other type of organism. This close relationship is called *symbiosis*. In this relationship, one type of organism can live near, on, or even in another organism. Each partner in the relationship can either help, harm, or have no effect on the other partner.

A **parasite** lives on or in another organism that it uses for food and, sometimes, for shelter. The organism the parasite uses is called

the **host**. In this type of interaction, one organism is helped while the other is harmed. This relationship is different from the **predator-prey** relationship. In that relationship, the prey is usually killed right away. In parasitism, the host organism usually continues to live, but it is harmed. (See Figure 25-5 on page 190.) Parasites usually evolve together with their host and have characteristics that make them specifically adapted to it. For example, the human tapeworm is a parasite that is adapted to live inside the human intestines.

In many symbiotic relationships, one organism benefits while the other organism remains unaffected. For example, when the Cape buffalo walks through the African plains to browse, it disturbs insects in the grass. Birds called cattle egrets gather around the buffalo and feed on the insects. The birds are helped, and the Cape buffalo is unaffected.

Finally, there are symbiotic relationships in which both parties benefit. An example of this is the relationship between a type of acacia tree and a species of stinging ants. The trees produce hollow thorns. The ants make their nests in these thorns and feed on sugars produced by the plant. If any other insect lands on the acacia, the ants quickly surround and kill it. The ants have shelter and a source of food, and the tree is protected from other plant-eating insects.

▉ CHANGING COMMUNITIES

A complex variety of interactions exists within an ecosystem. Many different populations live together and affect each other. In a

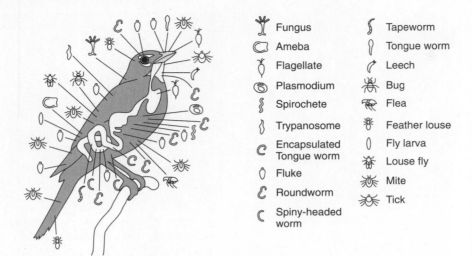

Fungus	Tapeworm
Ameba	Tongue worm
Flagellate	Leech
Plasmodium	Bug
Spirochete	Flea
Trypanosome	Feather louse
Encapsulated Tongue worm	Fly larva
Fluke	Louse fly
Roundworm	Mite
Spiny-headed worm	Tick

Figure 25-5 One bird may be host to many different types of parasites. Many parasites have evolved to be specifically adapted to their host, which is, typically, harmed but not killed by the parasites.

forest, for example, there are populations of grasses, shrubs, trees, fungi, insects, worms, bacteria, birds, reptiles, and mammals that make up the community. Will this community remain the same over time? Probably not. Some populations of organisms may disappear entirely while other populations may move in from somewhere else. Existing populations may increase or decrease in number. The community may even change with the seasons. Despite these kinds of changes, however, the forest—especially a very old forest—may remain essentially the same over many years. If so, it then has become a **stable** community.

But then a sudden, profound change may occur. For example, a fire may destroy much of the forest. Some animals may die, while others are able to escape. What happens now? Did the fire destroy the forest? Definitely not. What scientists observe is the process of ecological **succession**. Usually, succession is a series of slow changes that occur in an area until a stable community is reached. Most naturally occurring successions in an area take longer than a person's lifetime. However, after a sudden disturbance, a stable community quickly begins to go through a series of changes. These changes often follow similar patterns wherever the same kind of disturbance has occurred. (See Figure 25-6.)

Succession also occurs when a new environment appears for the first time, such as when a volcano produces new rock. The gradual succession of communities that appear in these places can tell us a great deal about how communities on Earth evolved during ancient times.

Figure 25-6 This photograph shows that regrowth and succession have taken place in a formerly fire-ravaged forest. Such rapid regrowth after a disaster provides scientists with the opportunity to understand the normally slower process of succession.

Chapter 25 Review

Part A—Multiple Choice

1. A population can best be described as all of the
   1 plants in a particular place
   2 animals in a particular place
   3 different organisms in one place at a particular time
   4 organisms of one species in one place at a particular time

2. An example of a population in a lake is all of the
   1 lake trout
   2 lake trout and brown trout
   3 plants and trout
   4 soil, plants, and fish

3. A community can best be described as all of the
   1 plant species in one particular place
   2 organisms of one species in a particular place
   3 populations that interact in a particular place
   4 animals that interact in a particular place

4. Which ecological term includes everything represented in the illustration below?

   1 ecosystem
   2 population
   3 community
   4 species

5. When population density is low, the
   1 birth rate increases and the death rate drops
   2 death rate increases and the birth rate drops
   3 birth rate and the death rate both increase
   4 birth rate and the death rate both decrease

6. An environment can support only as many organisms as the available energy, minerals, and oxygen will allow. Which term is best defined by this statement?

1 biological feedback
2 homeostatic control
3 carrying capacity
4 biological diversity

7. The carrying capacity of a given environment is *least* dependent upon
   1 recycling of materials
   2 the availability of food and water
   3 the available energy
   4 daily temperature fluctuations

8. When a population experiences zero growth, the
   1 death rate increases faster than the birth rate does
   2 birth rate increases and the death rate drops
   3 death rate increases and the birth rate drops
   4 birth rate and the death rate are about equal

9. Competition between organisms can best be described as an interaction in which the organisms
   1 rely on the same resources
   2 work together to find food
   3 live in the same place but eat different food
   4 eat the same food but live in different places

10. Competition can occur between members of
    1 the same species only
    2 different species only
    3 both the same and different species
    4 two different communities

11. Why is competition important in terms of evolution?
    1 Organisms evolve to be in greater competition with each other.
    2 Organisms evolve so that they can work together.
    3 Only the fittest organisms survive to reproduce.
    4 Competition causes the environment to increase its resources.

12. Although three different bird species inhabit the same type of tree in the same forest, there is very little competition among them. The most likely reason for this is that the birds
    1 are unable to interbreed
    2 have different ecological niches
    3 have a limited supply of food
    4 share food with each other

13. A parasitic relationship differs from a predator-prey relationship in that a

1 host organism is killed right away whereas prey is not
2 prey organism is killed right away whereas a host is not
3 parasite helps its host whereas a predator kills its prey
4 prey organism benefits whereas a parasite's host does not

14. In the symbiotic relationship between cattle egrets and Cape buffalo,

1 both species benefit
2 both species are harmed
3 one species benefits while the other is harmed
4 one species benefits while the other is unaffected

15. Some crocodiles let small birds enter their mouths to pick bits of food from between their teeth. The crocodiles get clean teeth, while the birds get an easy meal. In this type of relationship,

1 both animals benefit
2 both animals are harmed
3 only the crocodiles benefit
4 only the birds benefit

16. The relationship between the crocodiles and birds could be described as a

1 predator-prey relationship
2 parasite-host relationship
3 symbiotic relationship
4 competitive relationship

17. In a certain ecosystem, rattlesnakes are predators of prairie dogs. If the prairie dog population started to increase, how would the ecosystem most likely regain stability?

1 The rattlesnake population would start to decrease.
2 The prairie dog population would increase rapidly.
3 The rattlesnake population would start to increase.
4 The prairie dog population would begin to prey on the rattlesnakes.

18. In the relationship between the hollow-thorned acacia trees and the stinging ants,

1 both the trees and the ants benefit
2 both the trees and the ants are harmed

3 the tree benefits while the ants are harmed
4 the ants benefit while the tree is unaffected

19. Succession in an ecosystem is *usually* a

1 sudden event that changes the ecosystem
2 series of very rapid changes in the area
3 series of slow changes that occur in the area
4 short period of rapid change followed by a stable period

20. A new island formed by volcanic action may eventually become populated by living communities as a result of

1 a decrease in the amount of organic material
2 the lack of abiotic factors in the area
3 decreased levels of carbon dioxide in the area
4 the process of ecological succession

21. Which statement concerning ecosystems is correct?

1 Stable ecosystems that are changed by a natural disaster will slowly recover and may become stable again if left alone for a long time.
2 Climatic change is the principal cause of habitat destruction in ecosystems in the last 50 years.
3 Competition does not influence the number of organisms that live in an ecosystem.
4 Stable ecosystems, once changed by a natural disaster, will never recover and become stable again.

22. Events that occurred in four different ecosystems are listed in the table below. Which ecosystem

Ecosystem	Ecological Events
A	A severe ice storm occurs during the winter, damaging trees and shrubs. No ice storms occur during the next 20 years.
B	A severe drought causes most of the leaves to fall from the trees during a single summer. There are no serious droughts during the next 20 years.
C	An island with a dense shrub population becomes submerged for three years. After the river lowers, the island does not become submerged again for the next 20 years.
D	A fire burns through a large grassy area. Fires do not occur in the area again for the next 20 years.

probably would require the most time for succession to restore it to its previous condition?

1 ecosystem A
2 ecosystem B
3 ecosystem C
4 ecosystem D

Part B—Analysis and Open Ended

23. How would Earth be different if certain factors did not limit population growth?

24. Why is population growth more rapid when population density is low?

Refer to the graph below, which shows the growth curve for a population of Paramecium caudatum, *to answer questions 25 to 27.*

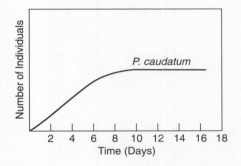

25. Why does the slope of the graph increase from the beginning to the middle?

1 The death rate begins to increase.
2 The growth rate slows after four days.
3 The population grows while it is below carrying capacity.
4 There is intense competition for resources.

26. The level (flat) portion at the top of the graph indicates that the population

1 is growing
2 is shrinking
3 is neither growing nor shrinking
4 no longer exists in that location

27. How does the size of the paramecium population change as it approaches carrying capacity?

28. The following graph shows the growth curves for two different populations of paramecia

species (*P. aurelia and P. caudatum*), grown in the same culture dish for 14 days.

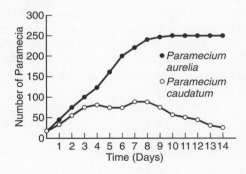

(a) Which ecological concept is best represented by this graph?

1 recycling
2 competition
3 equilibrium
4 decomposition

(b) State *one* possible reason for the decline in the number of *P. caudatum*.

29. Why is competition for food and space usually greatest among members of the same species? How does this relate to the process of evolution?

30. The diagram below illustrates

1 competition between different types of plant life
2 rapid ecological succession after a forest fire
3 gradual succession from bare rock to stable forest
4 evolution of plant life on Earth over two billion years

Base your answers to questions 31 to 33 on the passage below and on your knowledge of biology.

In 1869, the gypsy moth was imported from Europe into Massachusetts. Each gypsy moth caterpillar can eat more than 1 square meter of leaf tissue in its 8-week life, so by 1889 the residents of Boston began to notice many leafless trees.

Every few years, the population of gypsy moths rapidly increases in a season. In the course of two growing seasons, the number of eggs can range from 100 per acre to as many as 1 million per acre. In 1981, about 13 million acres of trees were defoliated (lost their leaves) in the American northeast, and many valuable oak trees died. Between 1979 and 1983, the cost of trying to control these pests totaled 24.2 million dollars. These attempts at control failed.

Rapid growth of a population occurs when there is an abundance of food or when an important environmental factor has been removed. Gypsy moth populations are normally kept in check by phenol chemicals that trees make and release into their leaves. These defensive chemicals stunt caterpillar growth and reduce the number of eggs a female can lay. After several years without caterpillars, the trees stop making these phenols. When this happens, the females eating the phenol-free leaves grow bigger and lay more eggs. Suddenly, a gypsy moth outbreak occurs again, and the cycle is repeated.

When a gypsy moth outbreak occurs, the surrounding ecosystem begins to change as well. Cuckoos, starlings, grackles, mice, and skunks feast on the extra caterpillars, and their numbers increase. Yet all these natural enemies cannot stop the gypsy moth. Trees are stripped of their leaves, weaker trees die at once, and others grow a second set of leaves. If the trees that survive are attacked repeatedly, they also may be weakened beyond recovery.

31. Describe *one* condition that might cause the gypsy moth population to increase rapidly.

32. State *one* reason that a rapid increase in a gypsy moth population may cause some species of herbivores to vanish or be reduced in number.

33. State *one* way some producers protect themselves from gypsy moths.

Refer to the diagram below to answer questions 34 and 35.

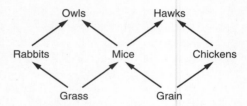

34. Based on the diagram, which of the following statements is true?

1 Rabbits and owls compete for grass.
2 Mice and chickens compete for grain.
3 Rabbits and chickens compete for grass.
4 Chickens and rabbits compete for grain.

35. Owls hunt at night, whereas hawks hunt during the day. This has the effect of

1 reducing competition for mice because the birds occupy separate forests
2 reducing competition for mice because the birds occupy separate niches
3 increasing competition for mice because the birds occupy the same niche
4 reducing competition for rabbits and chickens because the birds eat more mice

Base your answers to questions 36 to 38 on the information and graph below and on your knowledge of biology.

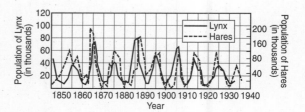

Scientists have hypothesized that the populations of both lynx and snowshoe hares should show cyclical changes—with increases in the predator population size lagging behind increases in the prey population size—if the assumption is made that snowshoe hares are eaten only by lynx.

Does this out-of-phase population cycle of predators and prey actually occur in nature? A classic example of such a cycle was observed by counting all the fur pelts (skins) from northern Canada lynx and snowshoe hares purchased by the Hudson Bay Company between 1845 and 1935. Population cycles of snowshoe hares and their lynx predators, based on the numbers of pelts, are shown in the graph above.

As with any field investigation, many variables could influence the relationship between hare and lynx. One problem is that hare populations have been shown to fluctuate even without lynx present, possibly because the carrying capacity of their environment had been exceeded. To test this hypothesis about population cycles more scientifically, investigators turned to controlled laboratory studies on populations of small predators and their prey.

36. Identify *two* variables other than the size of the lynx population that can affect the size of the hare population.

37. The phrase "carrying capacity" refers to the

1 animals storing extra food for the winter
2 number of organisms a habitat can support
3 transporting of food to organisms in an area
4 maximum possible weight of an individual organism

38. Why would scientists want to have a laboratory study on populations of different predators and their prey?

39. Choose an animal that you find of interest; then list two biotic and two abiotic factors for which it competes with other animals in its environment.

40. Compare and contrast predation and parasitism. Your answer should include:

♦ the definitions of predation, predator, prey

♦ the definitions of parasitism, parasite, host

♦ *one* way predation and parasitism are similar

♦ *one* way predation and parasitism are different

41. Why is evolution important to the relationship between parasites and their hosts?

42. As shown in the following art, the remora has a suckerlike disk on its head by which it attaches to the underside of a shark. The remora feeds on leftovers from the shark's meals, without taking

anything from the shark's body. This is an example of a symbiotic relationship in which

1 both parties benefit by being able to catch more food
2 one party benefits and the other is directly harmed
3 one party benefits and the other is apparently unaffected
4 both parties are harmed by not being able to swim as fast

43. The diagram below shows changes that might occur over time in an area after a forest fire. Which statement is most closely related to the events shown in the diagram?

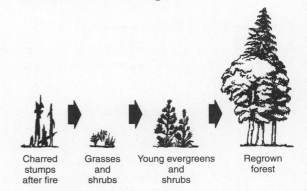

Charred stumps after fire   Grasses and shrubs   Young evergreens and shrubs   Regrown forest

1 The lack of animals in an altered ecosystem speeds natural succession.
2 Abrupt changes in an ecosystem only result from human activities.
3 Stable ecosystems never become established after a natural disaster.
4 An abrupt environmental change can cause long-term changes in an ecosystem.

44. Explain how an event such as a forest fire provides scientists with the opportunity to observe ecological succession.

45. Explain why an ecosystem with a variety of predator species might be more stable over a long period of time than an ecosystem with only one predator species.

Base your answers to questions 46 to 49 on the stages of succession shown below and on your knowledge of biology.

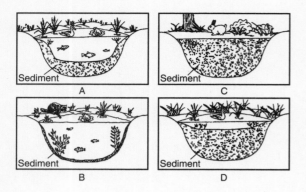

46. What is the correct sequence of these stages?

1 B → A → D → C
2 C → B → A → D
3 A → D → C → B
4 D → A → C → B

47. Which statement helps to explain this type of succession?

1 Species are replaced until an unstable ecosystem is established.
2 Species are replaced until a stable ecosystem is established.
3 Humans replace all species over time and fill all niches.
4 Animals control all changes in the plant species of an area.

48. Which population of organisms would be most harmed by the ecological changes occurring in this succession?

1 trees
2 fish
3 raccoons
4 rabbits

49. Identify *one* factor that could disrupt the final stage of this ecosystem.

Part C—Reading Comprehension

Base your answers to questions 50 to 53 on the information below and on your knowledge of biology. Use one or more complete sentences to answer each question.

In the early 1980s, an Asian elephant was born at the Bronx Zoo in New York City. Its charming antics made it a crowd favorite. It was heralded as the first elephant born in New York in more than 10,000 years. But a year and a half later, it died of unknown causes.

In the 1990s, another Asian elephant was born in America, this time at the National Zoo in Washington, D.C. Like the New York baby, this elephant also died at about the same age. Several years later, scientists learned that the Washington baby elephant died after being infected with a herpes virus. Later, tests of its preserved tissues revealed that the New York baby elephant had also died of a herpes virus infection.

The virus that had killed the baby Asian elephants is commonly found in African elephants. In these animals, the virus produces only a mild skin infection and sores. However, when the virus infects Asian elephants, especially young Asian elephants, it can produce deadly results. The reverse can also happen. It is now thought that a similar virus is found in Asian elephants, but it does not kill them. However, when this virus enters an African elephant, it may cause death. By 2002, 22 baby elephants in the United States had died of the disease.

There is hope for infected baby elephants. One of the earliest symptoms of infection is a purple tongue. If this symptom is noted, treatment with the human antiviral drug famciclovir can be started. This drug can save animals' lives if it is administered early enough. Scientists hope to develop tests to identify elephants that carry the virus and do not become ill. It is thought that these elephants could pass the virus on to other, healthy elephants. In time, scientists hope to develop a vaccine, but that is still far in the future.

50. Why was the birth of the Bronx Zoo elephant given so much attention?

51. In what way were the deaths of the Bronx Zoo and National Zoo elephant babies connected?

52. What do the facts about the herpes virus in elephants show about the relationship of the Asian elephants to the African elephants?

53. Why might a baby Asian elephant that is born in a U.S. zoo today have a better chance of surviving than those that were born in the zoos 20 years ago?

26

Ecosystems

THE BASIC CHARACTERISTICS OF ECOSYSTEMS

An **ecosystem** is made up of living and non-living factors. In other words, biotic factors, such as plants and animals, and abiotic factors, such as water, air, and soil, function together in an ecosystem. For organisms to survive, there must be a source of energy. The flow of energy between organisms and their environment is a basic characteristic of an ecosystem. Organisms are made up of matter. The flow of matter between organisms and their environment is another basic characteristic of an ecosystem.

What is the source of energy for almost all ecosystems on Earth? It is the sun. While energy is constantly reaching Earth from the sun, matter is not. The amount of matter on Earth remains constant. However, matter moves back and forth between organisms and the environment in all ecosystems.

ENERGY FLOW THROUGH ECOSYSTEMS

In most ecosystems, energy arrives as sunlight. Some organisms are able to use this energy directly. Other organisms use it indirectly —they get their energy by eating other organisms. Scientists describe and group all organisms in a system of **trophic**, or feeding, **levels**. (See Figure 26-1.)

On the first level are organisms that use energy, such as sunlight, directly from the environment. These first-level organisms are called **producers**. Green plants and algae are producers, or autotrophs, because they use the process of photosynthesis to make their own food with water, carbon dioxide,

and sunlight. Organisms that feed on producers are in the next trophic level; they are called **consumers**, or heterotrophs. A caterpillar that eats oak leaves is a consumer. It is also a type of **herbivore**, because it feeds on plants. Additional levels exist in which consumers feed on consumers. For example, a small bird may eat a caterpillar. A large hawk may then eat the bird. These animals are called **carnivores**, because they eat other animals. Each of these steps is called a trophic level because it describes the source of the or-

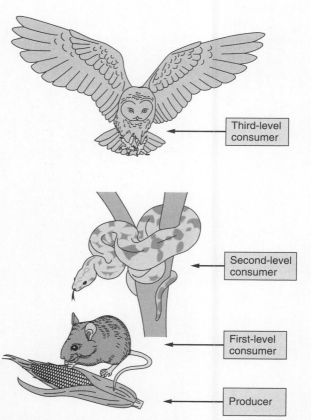

Figure 26-1 The trophic levels describe the flow of energy in an ecosystem, from the producer to the different levels of consumers.

ganisms' food. We can describe the flow of energy in an ecosystem by using these trophic levels.

FOOD CHAINS AND FOOD WEBS

Energy enters an ecosystem at the producer level and is passed along from an organism in one trophic level to an organism in a higher trophic level. This transfer of food energy from one organism to the next is called a **food chain**. From oak leaf to caterpillar to small bird to hawk is a food chain. But in a real ecosystem, a simple food chain like this is never found. Caterpillars are not the only animals that eat oak leaves, small birds are not the only animals that eat caterpillars, and so on. Food chains are actually interconnected in a complex pattern called a **food web**. In a food web, the energy is passed between many different organisms. (See Figure 26-2.)

However, no matter how complex a food web is, energy always moves in one direction—from a lower to a higher trophic level. It does not get recycled. As energy moves through each trophic level, some of it is used and some

of it is lost as heat. The most energy is present at the lowest trophic level (producers); the least is present at the highest trophic level (upper-level consumers). For this reason, additional energy must constantly enter an ecosystem, mainly in the form of sunlight.

Unfortunately, there is a hidden danger in many food chains. If a long-lasting chemical such as DDT enters the environment, it may get passed on from one trophic level to the next. The level of the chemical in each organism increases as it is moved along the food chain. For example, little fish may contain some DDT, the larger fish even more, and finally the fish-eating birds the most. While harmless at very low levels, the chemical may have serious effects at the highest levels. In fact, that is why DDT almost destroyed populations of fish-eating eagles and ospreys. (See Figure 26-3 on page 200.)

THE ENERGY PYRAMID

Ecologists use an **energy pyramid** to describe the flow of energy through an ecosystem. The wide base of the pyramid represents

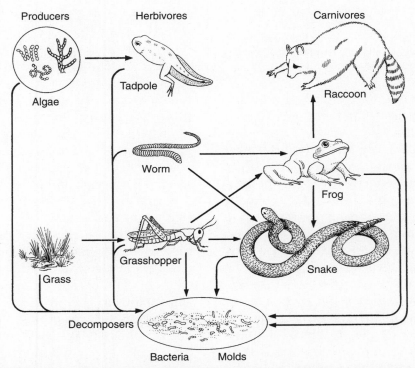

Figure 26-2 Food chains actually interconnect in complex patterns to form a food web, in which the energy passes between many different organisms.

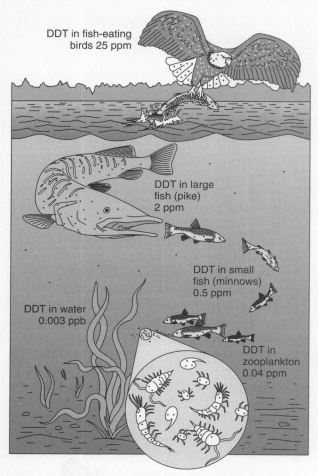

Figure 26-3 A hidden danger may develop in food chains when certain chemicals enter the environment. For example, the level of DDT in each organism increases as the chemical moves up along the food chain. Harmful effects can occur at the highest trophic levels.

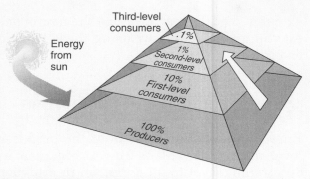

Figure 26-4 This energy pyramid shows that the amount of available energy decreases at each higher trophic level because only 10 percent of all energy gets passed from one level to the next.

the amount of usable energy in the producers, that is, the energy from the sun that is stored in all of the plants. The next step up in the pyramid shows the energy that the first-level consumers get from the producers. This layer is smaller than the energy layer for the producers. Why? Because only about 10 percent of all energy gets passed from one level up to the next. This is true as we move up the pyramid from each level of consumers to the next and then to the top level. (See Figure 26-4.)

An energy pyramid can provide an important lesson in how to feed the ever-increasing human population. Throughout the world, much more food energy is present at the producer level (crops) than at the consumer level (livestock). People may have to make choices based on such questions as: Which type of food is more abundant and available for everyone? Which type of food makes a more efficient use of energy sources?

❌ THE RECYCLING OF MATERIALS IN ECOSYSTEMS

In many parts of the United States, people are now required to recycle consumer wastes such as glass and plastic bottles, newspapers, and metal. Recycling, although a new idea for people, is not a new idea in nature. Natural ecosystems have recycled materials since life began on Earth. In fact, life would not continue without this recycling of matter.

All substances are made up of chemical elements. Of the dozens that occur naturally, only a few elements are found in significant amounts in organisms. These include **carbon**, **hydrogen**, **oxygen**, **nitrogen**, phosphorus, and sulfur. The amount of these elements on Earth today is about the same as when the planet formed. Because they are needed by living things and their supply does not increase, these elements have to be recycled again and again.

How do these elements get recycled in nature? Let's first look at carbon, since all organisms are made of molecules that contain this element. The carbon in organic molecules is obtained from CO_2 in the air. Producers such as grasses and trees take in CO_2 from the air during photosynthesis. They use the carbon from the CO_2 gas to build their carbohydrates (sugars and starches). Consumers obtain carbon from producers and sometimes from other

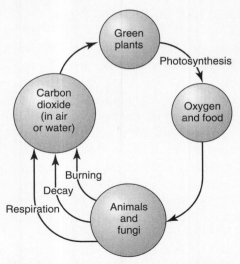

Figure 26-5 Carbon and oxygen are recycled between living things (and the environment) through such processes as photosynthesis, respiration, and decay. Producers, consumers, and decomposers are all part of this natural recycling process.

consumers that serve as food. To complete carbon's recycling, plants and animals return carbon to the atmosphere through respiration when they release CO_2. (See Figure 26-5.)

Recycling of carbon also occurs after a plant or an animal dies. This important part of the recycling process occurs through the actions of **decomposers**, mainly bacteria and fungi. Decomposers are heterotrophs, organisms that are unable to make their own food. They get their food by feeding on the **residue**, or remains, of dead organisms. As they carry out their life processes, decomposers also release CO_2 into the atmosphere. Animals that **scavenge** the remains help recycle the carbon, too.

Oxygen is also recycled between living things. Animals and other organisms need oxygen for respiration—the process that releases the chemical energy stored in food. Land animals obtain oxygen for respiration from the air they breathe. Aquatic animals like fish use the oxygen that is dissolved in the water they live in.

Almost all the oxygen in Earth's atmosphere originally came from the metabolic activities of plants. During photosynthesis, plants and algae give off oxygen as a waste product. Animals breathe in the oxygen given off by plants, just as plants take in the CO_2 released by animals and decomposer organisms. This is natural recycling.

Nitrogen, a gas that is common in the atmosphere, is another element essential for all living things. It is combined and recombined through another complex cycle between biotic and abiotic parts of the environment.

◼ CHANGE AND STABILITY: THE IMPORTANCE OF BIODIVERSITY

The tendency of an ecosystem to resist change and remain the same is known as **stability**. Ecologists have important questions to ask about the stability of ecosystems. Do entire communities in an ecosystem stay the same? What causes a particular community to change? Is the number of species that make up a community critical to its stability? The amount of variety in a community is called species **diversity** or **biodiversity**. A community with only a few species of plants and animals has low biodiversity. A community with many species has great biodiversity. A tropical rain forest community may have the greatest biodiversity of any community on Earth. (See Figure 26-6 on page 202.)

Biodiversity is one major concern of ecologists today. Why? Many known species present on Earth just a few decades ago are already extinct. For the most part, human actions are the cause of these extinctions. As species disappear, biodiversity decreases. Scientists are concerned about the effects of decreased biodiversity on the functioning of ecosystems.

Does an ecosystem need a certain number of species interacting with each other to remain viable? How many species can a community lose without being harmed? For example, if several species of insects in a forest died off, would the forest survive? Would the plants that the insects formerly ate grow too quickly? Would the bird populations suffer with fewer insects to eat? Finally, how much loss of biodiversity can occur before Earth's ecosystems stop functioning properly? This is a serious concern for people and for all species on Earth. Studies that investigate biodiversity and stability in specific communities are being conducted to try to answer important questions such as these.

Figure 26-6 Tropical rain forests, such as this one in Central America, may have the greatest biodiversity of any ecosystem on Earth, containing millions of species of plants and animals.

⬛ HABITAT DESTRUCTION

There is one main reason why biodiversity is decreasing. Many species are disappearing because of habitat loss. Humans are using and changing many places where organisms formerly lived. For example, in the Midwest, many fields contained low-lying areas that remained filled with water all year. Many birds, such as ducks and geese, were able to rest and find food in these ponds during their migrations each spring and fall. However, the farmers could not grow wheat and corn in these wet places. As a result, most of the ponds were filled in. This was a critical loss of habitat for the migrating water birds. The populations of the ducks and geese decreased; and biodiversity was reduced.

This is only one example of the loss of a habitat affecting biodiversity. Habitat and species loss have occurred in many other places, such as on a river when a dam is built. Fish that survive only in moving water die in the still water of a lake that is formed behind a dam. Today, the greatest habitat destruction is occurring in the world's tropical rain forests. It is estimated that most of Earth's

Figure 26-7 Habitat destruction threatens the biodiversity and stability of ecosystems. Part of this rain-forest habitat has been destroyed to make room for a banana plantation.

biodiversity will be lost if the rain forests are destroyed. Sadly, this is happening while scientists are trying to identify and classify the many organisms still being discovered in these forests. In addition, researchers fear that many tropical species, which may contain substances that could prove to be valuable medicines, are being lost forever before even being discovered. (See Figure 26-7.)

Chapter 26 Review

Part A—Multiple Choice

1. A basic trait of ecosystems is the
   1 flow of energy between organisms and the environment
   2 flow of matter between organisms and the environment
   3 flow of water between organisms and the environment
   4 flow of plants between organisms and the environment

2. The source of energy for most ecosystems is
   1 rain   3 flowing water
   2 wind   4 the sun

3. In an ecosystem, which component is *not* recycled?
   1 water   3 energy
   2 oxygen   4 carbon

4. The first trophic level consists of organisms that
   1 use energy to make their own food
   2 eat first-level producers only
   3 eat producers and consumers
   4 add matter to an ecosystem

5. Organisms that eat plants are called both consumers and
   1 producers   3 carnivores
   2 herbivores   4 scavengers

6. What is always transferred in a food chain?
   1 toxins   3 water
   2 energy   4 oxygen

7. Which list indicates a correct flow of energy?
   1 herbivore → sun → carnivore
   2 sun → producer → herbivore
   3 producer → sun → carnivore
   4 carnivore → herbivore → sun

8. One Arctic food chain consists of polar bears, fish, seaweed, and seals. Which sequence demonstrates the correct flow of energy between these organisms?
   1 seals → seaweed → fish → polar bears
   2 seaweed → fish → seals → polar bears
   3 fish → seaweed → polar bears → seals
   4 polar bears → fish → seals → seaweed

9. Which energy transfer is *least* likely to be found in nature?
   1 consumer to consumer
   2 host to parasite
   3 producer to consumer
   4 predator to prey

10. A spider stalks, kills, and then eats an insect. Based on this behavior, which ecological terms describe the spider's roles in a food chain?
   1 producer, carnivore, consumer
   2 carnivore, predator, consumer
   3 predator, herbivore, consumer
   4 scavenger, carnivore, consumer

11. In most habitats, the removal of predators will have the most immediate impact on a population of
   1 producers   3 decomposers
   2 herbivores   4 microbes

12. Which group contains terms that are *all* directly associated with the larger fish in the diagram below?

   1 herbivore, prey, autotroph, host
   2 carnivore, predator, heterotroph, multicellular
   3 predator, scavenger, decomposer, consumer
   4 producer, parasite, fungus, fish

13. In a food web, energy always moves
 1 in a continuous cycle of trophic levels
 2 back and forth between various trophic levels
 3 from lower to higher trophic levels only
 4 from higher to lower trophic levels only

14. A student could best demonstrate knowledge of how energy flows throughout an ecosystem by
 1 labeling a diagram that illustrates ecological succession
 2 drawing a food web using specific organisms living in a pond
 3 conducting an experiment that demonstrates photosynthesis
 4 making a chart to show the role of bacteria in the environment

Base your answers to questions 15 and 16 on the diagram below and on your knowledge of biology.

15. Which organism carries out autotrophic nutrition?
 1 frog   3 plants
 2 snake   4 grasshopper

16. The base of an energy pyramid for this ecosystem would include the
 1 frog   3 plants
 2 snake   4 grasshopper

17. A food web is more stable than a food chain because a food web
 1 transfers all of the producer energy to herbivores
 2 includes alternative pathways for energy flow
 3 reduces the number of niches in the ecosystem
 4 includes more consumers than producers

18. Which trophic level contains the most available food energy?
 1 the producers
 2 first-level consumers

 3 second-level consumers
 4 third-level consumers

19. The hidden danger in many food chains is that
 1 some prey items taste better and thus are eaten too often
 2 harmful chemicals can be passed from one level to another
 3 some foods become poisonous after being eaten too often
 4 the food chains interconnect to form enormous food webs

20. The diagram below represents a pyramid of energy in an ecosystem. Which level in the pyramid would most likely contain members of the plant kingdom?

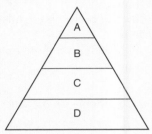

 1 level A
 2 level B
 3 level C
 4 level D

21. In an ecosystem, nutrients can be recycled if they are transferred directly from herbivores to carnivores to
 1 hosts
 2 decomposers
 3 prey
 4 autotrophs

22. Carbon is recycled in nature when
 1 consumers take in carbon dioxide and then release oxygen
 2 producers take in carbon dioxide and consumers release it
 3 decomposers take in carbon dioxide and release oxygen
 4 scavengers and decomposers take in carbon dioxide

23. Which set of statements best illustrates a material cycle in a self-sustaining ecosystem?
 1 In summer, growing plants remove magnesium ions from the soil to make chlorophyll. In autumn, these plants release magnesium

when they die and decompose. In spring, new plants will grow in this same area.

2 DDT is sprayed on a forest ecosystem to control the mosquito population. After a year, the level of DDT is found to be much higher in the tissues taken from a hawk than in the tissues taken from a mouse in this ecosystem.

3 Trees do not live in a desert ecosystem where there is not enough water present in the sandy soil to support their growth. Trees can live in a desert oasis.

4 Plants trap the sun's energy in the chemical bonds of organic molecules. This energy is then used for plant metabolic activities.

24. Which statement best describes what happens to energy and molecules in a stable ecosystem?

1 Both energy and molecules are recycled in the ecosystem.
2 Energy is recycled and molecules are continuously added to the ecosystem.
3 Neither energy nor molecules are recycled in the ecosystem.
4 Molecules are recycled and energy is continuously added to the ecosystem.

25. Decomposers release carbon dioxide as they

1 feed on bacteria and algae
2 carry out photosynthesis
3 build starches and sugars
4 carry out their life processes

26. The organisms that help recycle elements by breaking down organic matter include

1 grass and algae
2 bacteria and algae
3 bacteria and fungi
4 plants and fungi

27. Vultures, which are classified as scavengers, are an important part of an ecosystem because they

1 hunt herbivores, thus limiting their population size in an ecosystem
2 cause the decay of dead organisms, which releases usable energy to living organisms
3 feed on dead animals, which aids in the recycling of environmental materials
4 are the first level in food webs, making energy available to all the other organisms

28. Oxygen is needed for respiration, the process that

1 releases the chemical energy stored in food
2 uses carbon dioxide to produce sugars

3 breaks down the bodies of dead organisms
4 releases oxygen as a waste into the air

29. The oxygen that humans breathe is actually

1 a waste product of respiration
2 a waste product of photosynthesis
3 given off by decomposers
4 produced within the sun

30. The tendency of an ecosystem to stay the same is called

1 diversity
2 resistance
3 stability
4 sterility

31. Which condition would cause an ecosystem to become unstable?

1 Only heterotrophic organisms remain after a change in the region.
2 A variety of nonliving factors are used by the living factors.
3 A slight increase occurs in the number of heterotrophs and autotrophs.
4 The biotic factors and abiotic resources interact more often.

32. Which ecosystem has a better chance of surviving when environmental conditions change over a long period of time?

1 one with a great deal of genetic diversity
2 one with animals and bacteria but no plants
3 one with plants and animals but no bacteria
4 one with little or no genetic diversity

33. Unlike a desert, a tropical rain forest typically has

1 low biodiversity
2 great biodiversity
3 a small variety of organisms
4 a small number of organisms

34. The loss of biodiversity is often related to

1 the search for medical cures
2 too much rain in rain forests
3 a loss of natural habitat
4 evolution not occurring

35. Increased efforts to conserve areas such as rain forests are necessary in order to

1 protect biodiversity
2 exploit finite resources
3 promote extinction of species
4 increase industrialization

Part B—Analysis and Open Ended

36. Briefly describe the two main components of any ecosystem.

37. How do producers differ from consumers?

38. Why are the steps of a food chain called "trophic" levels?

39. Why is a food web a more accurate description than a food chain of interactions in a community?

40. A food web is represented in the diagram below. Which organisms are correctly paired with their roles in this food web?

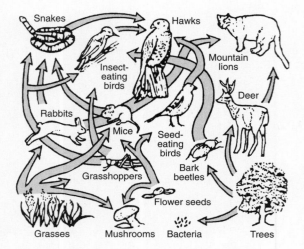

1 producers: mountain lions, snakes; heterotrophs: hawks, mice
2 consumers: all birds, deer; producers: grasses, trees
3 consumers: snakes, grasshoppers; autotrophs: mushrooms, rabbits
4 decomposers: seeds, bacteria; heterotrophs: mice, grasses

41. Explain why the amount of energy in trophic levels can be shown as a pyramid.

42. The diagram at the top of the next column represents a model of a food pyramid. Which statement best describes what happens in this food pyramid?

1 More organisms die at higher levels than at lower levels, decreasing the mass at the top.
2 When organisms die at higher levels, their remains sink to lower levels, increasing the mass at the bottom.

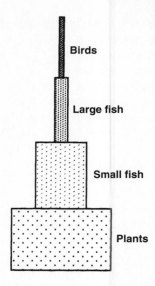

3 Energy is lost to the environment at each level, so less mass can be supported at each higher level.
4 Organisms decay at each level, and thus less mass can be supported at each higher level.

43. According to Figure 26-1 on page 198, a third-level consumer is one that

1 is capable of photosynthesis
2 feeds directly on producers
3 feeds on first-level consumers only
4 feeds on second-level consumers

Refer to the diagram below to answer questions 44 to 47.

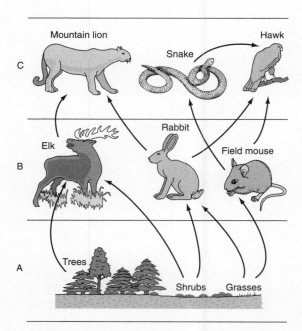

44. Which does the diagram illustrate?

1 a food chain
2 a food web
3 a food cube
4 a food pyramid

45. The organisms shown in level B are classified as both

1 producers and prey
2 consumers and prey
3 scavengers and predators
4 decomposers and prey

46. Which populations would contain the greatest amount of available energy?

1 rabbits and field mice
2 hawks and rabbits
3 trees and shrubs
4 hawks and snakes

47. All of these organisms living together make up a natural

1 population
2 species
3 community
4 consumer

48. Which statement about the producers in the marine food web shown below is correct?

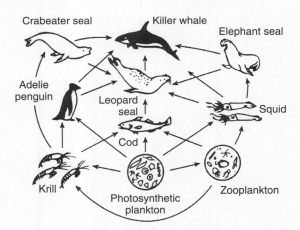

1 An increase in the producers would most likely decrease available energy for the squid.
2 If all the producers in this ecosystem were destroyed, the number of heterotrophs would increase and the ecosystem would reach a new equilibrium.
3 An important producer in this ecosystem is the zooplankton.

4 There is only one group of producers, so they must be numerous enough to supply the energy needed to support the food web.

49. What problem can occur in a food chain when a pollutant enters the ecosystem?

50. According to Figure 26-4 on page 200, as energy moves up each trophic level in an ecosystem, the amount of it that is available becomes

1 10 percent more than it was before
2 10 percent of what it was before
3 50 percent more than it was before
4 50 percent less than it was before

Base your answers to questions 51 to 53 on the passage below, which was written in response to an article about eliminating predators.

In nature, energy flows in only one direction. Transfer of energy must occur in an ecosystem because all life needs energy to live, and only certain organisms can change solar energy into chemical energy.

Producers are eaten by consumers, which are, in turn, eaten by other consumers. Stable ecosystems must contain predators to help control the populations of consumers. Since ecosystems contain many predators, exterminating predators would require a massive effort that would wipe out predatory species from barnacles to blue whales. Without the population control provided by predators, some organisms would soon overpopulate.

51. Draw an energy pyramid that illustrates the sentence, "Producers are eaten by consumers, which are, in turn, eaten by other consumers." Include *three* different, specific organisms in your energy pyramid.

52. Explain the phrase "only certain organisms can change solar energy into chemical energy," which appears in the first paragraph. In your answer be sure to identify:

♦ the type of organisms being described in the statement

♦ the type of nutrition carried out by these organisms

♦ the process being carried out in this type of nutrition

♦ the organelles (in the cells of these organisms) that are involved in this process

53. Explain why an ecosystem with a variety of predator species might be more stable over time than an ecosystem with only one predator species.

Refer to the diagram of a food pyramid below to answer questions 54 and 55.

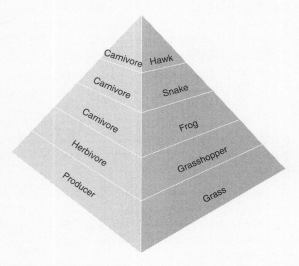

54. The consumer level that would have the largest amount of stored energy is that of the

1 hawk   3 frog
2 snake   4 grasshopper

55. The level that has the smallest amount of stored energy would be that of the

1 top carnivore
2 middle carnivore
3 herbivore
4 producer

Base your answers to questions 56 to 59 on the diagram of a food web below and on your knowledge of biology.

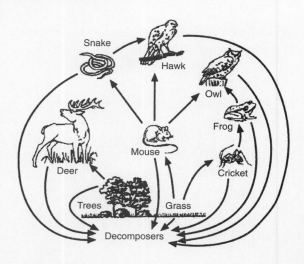

56. If the population of mice were reduced by disease, which change will most likely occur in the food web?

1 The cricket population will increase.
2 The snake population will increase.
3 The grasses will decrease.
4 The deer population will decrease.

57. What is the original source of energy for this food web?

1 chemicals in sugar molecules
2 enzymatic reactions
3 energy from sunlight
4 chemical reactions of bacteria

58. Which organisms are *not* shown in this diagram but are essential to a balanced ecosystem?

1 heterotrophs
2 autotrophs
3 producers
4 decomposers

59. State *one* example of a predator-prey relationship shown in the food web. Indicate which organism is the predator and which is the prey.

60. Use the following terms to complete the categories in the chart below: *decomposers, producers, carnivores,* and *herbivores.*

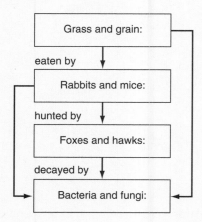

61. How is the amount of matter available on Earth very different from the amount of energy available? How is this related to the recycling of elements in nature?

Refer to the figure at the top of the next column to answer question 62.

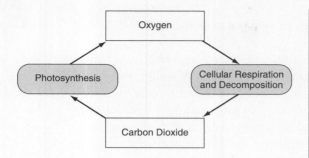

62. Explain how respiration and photosynthesis are involved in the cycling of oxygen and carbon dioxide between living things.

63. Why are ecologists concerned about the number of species in ecosystems?

64. Explain why most ecologists would agree with the following statement: "A forest ecosystem is more stable than a cornfield."

65. How might the biodiversity of an ecosystem be related to its stability?

Base your answer to question 66 on the following passage and on your knowledge of biology.

A tropical rain forest in the country of Belize contains over 100 kinds of trees, in addition to thousands of species of mammals, birds, and insects. Dozens of species living there have not yet been classified and studied. The rain forest could be a commercial source of food as well as a source of medicinal and household products. However, most of this forested area is not accessible because of a lack of roads and, therefore, little commercial use has been made of this region. The building of paved highways into and through this rain forest has been proposed.

66. Discuss some aspects of carrying out this proposal to build paved highways. In your answer be sure to:

♦ state *one* possible impact on biodiversity and *one* reason for this impact

♦ state *one* possible reason for an increase in the number of certain producers as a result of road building

♦ identify *one* type of consumer whose population would most likely increase as a result of an increase in a certain producer population

♦ state *one* possible action the road builders could take to minimize the human impact on the ecology of this region

67. In what ways are humans responsible for the current decrease in biodiversity?

Part C—Reading Comprehension

Base your answers to questions 68 to 71 on the information below and on your knowledge of biology. Use one or more complete sentences to answer each question.

The Everglades is a vast, wide freshwater marsh that covers much of the southern part of Florida. The Everglades begins at the northern edge of Lake Okeechobee, with the overflow of rainwater out of the lake, and extends all the way to the southern tip of the state just before the Florida Keys. The vast majority of the Everglades is covered by a dense growth of saw grass. A very slow, steady flow of water moves through the saw grass from north to south. The Everglades is, therefore, called a "River of Grass."

In 1996, the federal government endorsed the Everglades restoration project. The project will be one of the largest ecological restoration efforts anywhere in the world. Hundreds of millions of dollars will be spent on protecting the fragile Everglades ecosystem. Included in the plan is the removal from sugarcane production of 100,000 acres of farmland in ecologically sensitive areas.

Much of the water that flows through the Everglades has become contaminated by pesticides and fertilizers, which are used to increase crop yields on farms in the area. One of the main goals of the restoration project is to let large areas of land act as natural water filters to remove some of the waterborne contaminants.

Another important part of the Everglades project will restore the natural north-south flow of water. The natural pattern of water flow through the Everglades was disrupted by the canals, pumping stations, and water-control structures that were built to create flood-control and water-supply systems for southern Florida. In fact, these unnatural attempts to control the Everglades' water flow have been harmful to the entire ecosystem. Today, planning is under way to find alternatives that can meet flood-control and water-supply needs while ensuring the long-term health of the Everglades.

68. Why is the Everglades called a "River of Grass?"

69. In two or more complete sentences, describe the types of human activities that have had harmful effects on the Everglades.

70. Describe the steps that are being taken to undo the damage to the Everglades and restore it to its original condition.

71. Based on the information in this essay, what change has occurred in peoples' attitudes toward this vast marsh?

People and the Environment

PEOPLE CHANGING THE ENVIRONMENT

Up until about 10,000 years ago, all humans hunted and gathered their food. Then, people started planting crops and domesticating animals; this marked the beginnings of agriculture. As a result of agriculture, people use the land differently from before. When people cut down and burn trees to make room to plant crops and graze livestock, wild animals are often forced to leave the area—they lose their habitat.

Advances in science and technology have produced even greater changes in the environment. About 200 years ago, developments in science and technology led to the Industrial Revolution, which greatly increased the ways that humans affect the environment.

CHANGES TO THE LAND— ADDING WASTES

All organisms produce wastes as a normal by-product of their life processes. However, since the time of the Industrial Revolution, the amount of wastes produced by humans has increased greatly. Also, since that time, the kinds of wastes have changed. Many of the wastes do not decompose and they often contain harmful chemicals. These waste materials, called *solid wastes*, are often deposited in landfills, areas in which garbage is buried. (See Figure 27-1.)

In a sanitary landfill, attempts are made to limit the effects of the wastes on the environment. Other kinds of landfills are much more harmful to the environment, such as a *toxic* waste dump. The most dangerous of all toxic wastes are radioactive substances. (See Figure 27-2 on page 212.)

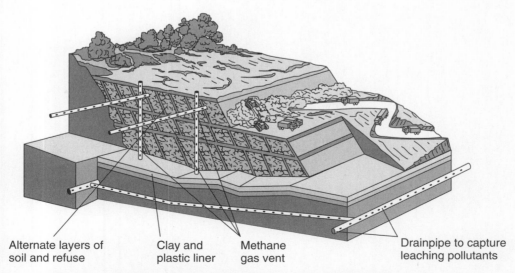

Alternate layers of soil and refuse   Clay and plastic liner   Methane gas vent   Drainpipe to capture leaching pollutants

Figure 27-1 A sanitary landfill is constructed in a way that limits the effects of the waste materials on the surrounding environment.

Figure 27-2 Some landfills contain waste products that are particularly toxic; such sites may pose a health threat to nearby communities.

CHANGES TO THE LAND—LOSING SOIL

Although soil is sometimes called dirt, it is actually a very valuable resource. In fact, this combination of organic and inorganic matter takes hundreds of years to form. Without good nutrient-rich soil, called *topsoil*, we could not grow food. Land ecosystems depend on this resource, too. (See Figure 27-3.)

Soil is now being lost because of human activities. For example, when toxic chemicals enter the ground, the soil becomes unusable. Poor farming practices and overgrazing by livestock can strip an area of all vegetation. The land becomes bare and, if weather patterns change and less rain falls, the land becomes a desert.

As people cut down forests or remove the plants that grow in an area, there is an increase in soil loss, or *erosion*. Both wind and water can cause erosion: rain washes away loose soil, and strong winds blow it away.

CHANGES TO THE WATER

Because flowing water carries away wastes, a stream or river has always seemed the perfect place to dump garbage. Today, as the human population increases in size, many more wastes are placed into streams and rivers. There are simply more wastes introduced than the natural ecosystems can handle. The river or stream, once full of life, begins to lose its ability to support the same species of organisms as before. (See Figure 27-4.)

In addition, the types of wastes deposited have changed. Industries sometimes dump toxic chemicals into rivers. Fish that need clean, well-oxygenated water are replaced by fish that can live in water with lower oxygen levels. If the levels of pollutants keep increas-

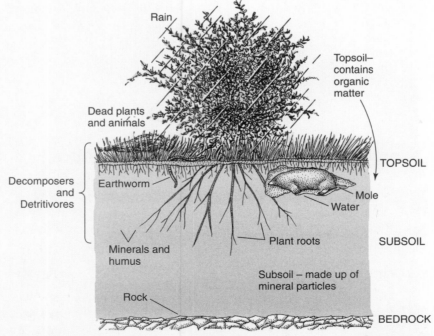

Figure 27-3 Nutrient-rich topsoil takes a long time to form and is of great value to humans and wildlife. Crops cannot grow in soil that lacks adequate nutrients.

Figure 27-4 When a community dumps sewage and industrial wastes into its streams and rivers, the local aquatic ecosystem—and those of communities living downstream—becomes polluted.

ing, these fish will die, too. A major advance in dealing with the problem of water pollution has been the development of sewage treatment plants. In these treatment plants, human organic wastes are treated in large tanks. Wastes in the water are chemically digested by bacteria. The remaining solids, including dead bacteria, then settle to the bottom and are removed. Chlorine is added to the water to kill bacteria. Finally, the purified water is released into a river or stream.

Many cities obtain their water from underground wells. The water from these wells, called *groundwater*, accumulates over time and is stored naturally between layers of rock. There are above-ground reservoirs of fresh water, too. New York City relies on a system that directly transports clean water from upstate reservoirs through underground pipes. Other cities have their water treated first. (See Figure 27-5.)

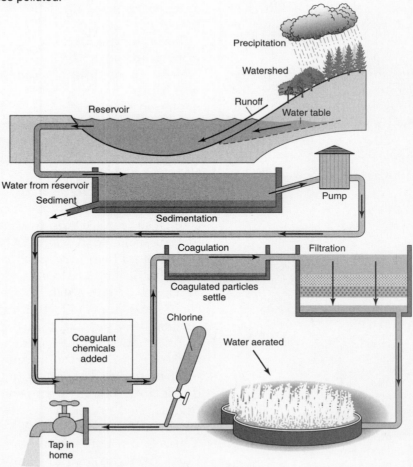

Figure 27-5 Schematic of a city's water treatment process: Fresh water from a reservoir goes through several physical and chemical processes before it is considered clean enough to pipe into people's homes.

CHANGES TO THE AIR

No one owns the air; we all share the air, which forms a continuous blanket over Earth. If the air becomes polluted in one place, that pollution can easily spread to another place. Gases and tiny solid particles are constantly added to the air by human activities. If these substances are not normally found in the air and are harmful, they are called air *pollutants*. The burning of *fossil fuels*—coal, oil, and natural gas—to power cars, heat homes and offices, and produce electricity creates air pollution. In addition, many industries release pollutants into the air from huge smokestacks. (See Figure 27-6.)

Major improvements have been made in the efforts to reduce air pollution. Today, laws require factories to reduce or prevent the release of pollutants from smokestacks. Devices called "scrubbers" are installed, which reduce the emission of harmful compounds. Car engines, too, have built-in pollution control devices that reduce the amount of pollutants added to the air when fuel is burned.

New technologies for producing energy also have been developed. Solar collectors and photovoltaic cells can provide us with heat or electricity without polluting the air. The Clean Air Acts of 1970 and 1977 began many of these changes. As a result, the air is now cleaner than it was just a few decades ago.

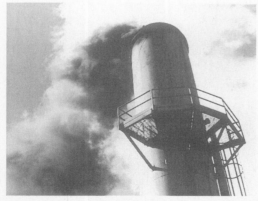

Figure 27-6 Factories that produce enormous quantities of manufactured goods are typical of our industrial society. Unfortunately, these factories may also release some air and water pollutants.

GLOBAL AIR POLLUTION PROBLEMS

Acid rain is a form of air pollution that produces far-ranging effects. Sulfur dioxide and nitrogen oxides are produced when fossil fuels are burned. Winds carry these gases high into the atmosphere and over long distances. They combine with water droplets in the air, which fall back to the ground as acid rain. Many forests and lakes in North America and Europe have been severely damaged by acid rain. (See Figure 27-7.)

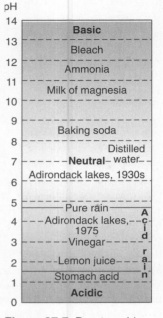

Figure 27-7 Due to acid rain, lakes in New York's Adirondack Mountains have become more acidic, causing harm to wildlife. This pH chart shows how much the lakes' acidity increased in 40 years.

Perhaps even more important are the effects of **global warming** and ozone depletion. Carbon dioxide (CO_2) in the **atmosphere** helps keep Earth warm by trapping heat. This is called the "greenhouse effect." But the amount of CO_2 has been increasing in the atmosphere due to the burning of fossil fuels and deforestation, in particular, the destruction of countless trees in rain forests, which used to absorb CO_2. With more CO_2 in the atmosphere, more heat is trapped. Many people are concerned that, as a result, Earth's climate is getting warmer. Such a change in climate could have major effects on habitats and organisms everywhere. (See Figure 27-8.)

Scientists are also concerned about the effects of certain air pollutants that harm the **ozone shield**. This effect is known as ozone **depletion**. The layer of ozone gas that surrounds Earth high in the atmosphere blocks out harmful ultraviolet (UV) radiation. The UV rays are part of the energy that reaches Earth from the sun, and they can damage the DNA in our cells, causing skin cancer.

Chlorofluorocarbons (CFCs), found in air

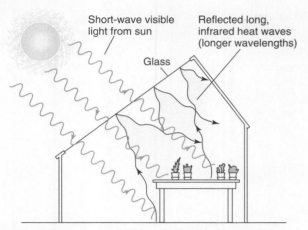

Figure 27-8 The greenhouse effect: Carbon dioxide in the air traps infrared energy, warming the atmosphere. This is similar to the way that the glass roof of a greenhouse traps heat, keeping the plants warm.

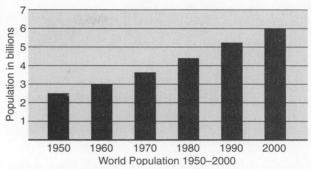

Figure 27-9 Industrial and scientific advances in the last 50 years have caused the human population to more than double in size—an increase that may put us past Earth's carrying capacity and cause environmental problems.

conditioners and refrigerators, are suspected of causing the most ozone depletion. In 1987, an agreement was signed by many countries to protect the ozone layer by limiting or banning the use of these chemicals. Progress has been made since the agreement was signed.

▪▪ HUMAN POPULATION GROWTH

The most serious problem that now affects all life on Earth is the rate at which the human population is increasing. Over most of its history, the human population increased slowly. However, the rise of agriculture caused a rapid increase as people settled down with a more secure source of food. More recently, the Industrial Revolution, combined with scientific advances in farming and medicine, caused an explosion in human population size. (See Figure 27-9.)

What is Earth's carrying capacity for humans? It is now known that the growth rate for the human population peaked in 1990 and is now declining. However, the population is still growing and it is believed it will reach a peak of about 9 billion around 2050. Others think that the population is already past Earth's carrying capacity and that serious environmental problems have already begun.

An exploding human population may lead to the extinction of vast numbers of species. Many organisms are already endangered due to the loss of habitat and other human factors. Industrialization (which causes more air, water, and land pollution), acid rain, global warming, and ozone shield depletion are worldwide concerns. Earth—as one large, complex, intact ecosystem—is threatened by an ever-increasing human population. (See Figure 27-10.)

In 1994, participants from 160 countries at an international conference agreed that

Figure 27-10 The exploding human population leads to overcrowding, pressure on limited resources, loss of wildlife habitat, increased pollution, and other problems that may threaten Earth's health as well as our own.

Earth's population cannot continue to grow at its current rate; yet they disagreed about how to lower the growth rate. What is clear is that our population, like that of any other organism, cannot increase forever. Either we will find a way to control human population size or nature will do it for us. Polluted soil, air, and water; lack of food and space; and widespread disease may ultimately limit human population size. However, individual choices and government planning could also limit it. People can still try to make the right decisions.

Chapter 27 Review

Part A—Multiple Choice

1. Over time, human populations have

  1 produced fewer and fewer wastes
  2 increased the amount of wastes produced
  3 prevented harmful chemicals from being produced
  4 decreased the amount of waste in landfills

2. Which animal has had the greatest negative impact on Earth's ecosystems?

  1 gypsy moth
  2 human
  3 zebra mussel
  4 shark

3. Which phrase would be appropriate for area A in the chart below?

Technological Device	Positive Impact	Negative Impact
Nuclear power plant	Provides efficient, inexpensive energy	A

  1 produces radioactive wastes
  2 provides light from radioactive substances
  3 results in greater biodiversity
  4 reduces dependence on fossil fuels

4. The most dangerous of all toxic wastes are

  1 solid wastes
  2 nutrient-rich soils
  3 radioactive substances
  4 plastic garbage

5. When humans cut down forests in an area,

  1 soil is lost through erosion
  2 the soil becomes richer
  3 new soil forms quickly
  4 flooding is prevented

6. When too many wastes are dumped into a river, the wastes

  1 eventually disappear through dilution
  2 gradually cause harm to the river ecosystem
  3 are carried away, where they can cause no harm
  4 are broken down immediately

7. Which is *not* a cause of increased water pollution?

  1 dumping sewage into streams and rivers
  2 addition of chlorine to treated water
  3 pouring industrial wastes into rivers
  4 eroded soil washing off land into streams

8. Dumping raw sewage into a river will lead to a reduction in the dissolved oxygen in the water. This condition will most likely cause

  1 an increase in all fish populations
  2 an increase in the depth of the water
  3 a decrease in most fish populations
  4 a decrease in water temperature

9. In a sewage treatment plant, bacteria are

  1 added to the water before it is released into a river
  2 killed by chlorine at the beginning of the process
  3 used to chemically digest wastes in the water first
  4 left in the purified water because they are harmless

10. "Natural ecosystems provide an array of basic processes that affect humans." Which statement does *not* support this quotation?

  1 Bacteria of decay help recycle materials.
  2 Treated sewage is less damaging to the environment than untreated sewage.
  3 Trees add to the amount of atmospheric oxygen.

4 Lichens and mosses on rocks help to break down the rocks, forming soil.

11. A negative impact of technology is an increase in the
    1 development of new products
    2 availability of different foods
    3 wastes released into the environment
    4 societal awareness of the environment

12. Which type of waste will decompose most quickly?
    1 foam cup
    2 plastic bag
    3 glass bottle
    4 banana peel

13. Which practice will best protect the soil?
    1 removing excess trees from it
    2 planting vegetation on it
    3 allowing cattle to feed on the land
    4 adding lots of chemicals to it

14. An increase in the use of fossil fuels is an indication of which type of society?
    1 hunter-gatherer
    2 agricultural
    3 industrial
    4 horticultural

15. Acid rain forms when
    1 carbon dioxide traps heat near Earth
    2 ozone is depleted from the atmosphere
    3 gases from fossil fuels combine with water droplets in the air
    4 chlorine is added to waste water

16. Methods used by people to reduce the emission of pollutants from smokestacks are an attempt to
    1 lessen the amount of insecticides in the environment
    2 lessen the formation and harmful impact of acid rain
    3 eliminate diversity in natural habitats
    4 use non-chemical controls on pest species

17. Changes in the chemical composition of the atmosphere that may produce acid rain are most closely associated with
    1 insects that excrete acids
    2 factory smokestack emissions
    3 runoff from acidic soils
    4 flocks of migrating birds

18. How is the Industrial Revolution related to the greenhouse effect?
    1 It caused the start of the greenhouse effect on Earth.
    2 It marked the end of the greenhouse effect on Earth.
    3 It caused a decrease in the amount of CO_2 released into the atmosphere.
    4 It caused an increase in the amount of CO_2 released into the atmosphere.

19. People can have a large negative impact on ecosystems when they
    1 conserve natural resources
    2 modify the environment
    3 restrict the use of chemicals
    4 pass laws to protect habitats

20. By causing atmospheric changes through activities such as polluting and careless tree harvesting, humans have
    1 caused the destruction of habitats
    2 established equilibrium in ecosystems
    3 affected global stability in a positive way
    4 replaced nonrenewable resources

21. The effect of CO_2 and other greenhouse gases on the atmosphere can best be likened to that of a
    1 blanket
    2 balloon
    3 pitcher of water
    4 crowd of people

22. Deforestation will most directly result in an increase in
    1 atmospheric carbon dioxide
    2 wildlife populations
    3 atmospheric ozone
    4 renewable resources

23. Which human activity would have the most direct impact on the oxygen–carbon dioxide cycle?
    1 reducing the rate of ecological succession
    2 destroying large forested areas
    3 decreasing the use of water
    4 banning the use of leaded gasoline

24. Chlorofluorocarbons are harmful to the environment because they
    1 kill fish in lakes
    2 form acid rain
    3 cause ozone depletion
    4 increase the greenhouse effect

25. The ozone layer of Earth's atmosphere helps to filter ultraviolet radiation. As the ozone layer is depleted, more ultraviolet radiation reaches Earth's surface. This increase in ultraviolet radiation may be harmful because it can directly cause

    1 photosynthesis to stop in all marine plants
    2 mutations in the DNA of organisms
    3 abnormal migration patterns in waterfowl
    4 sterility in most species of mammals and birds

26. Which factor is often responsible for the other three?

    1 increase in levels of toxins in fresh water
    2 increased poverty and malnutrition
    3 increase in human population
    4 increased depletion of finite resources

Part B—Analysis and Open Ended

27. Briefly explain how humans change the land through agriculture. Your answer should include the impact on the following:

    ◆ topsoil
    ◆ forests
    ◆ wildlife

28. How has industrialization changed the types of wastes produced by humans?

29. Why might building a landfill near an aquatic ecosystem cause harm to it?

Base your answers to questions 30 and 31 on the graph below, which shows pollution from nitrogen-containing compounds (nitrates) in a brook flowing through a forested area and a deforested area between 1965 and 1968.

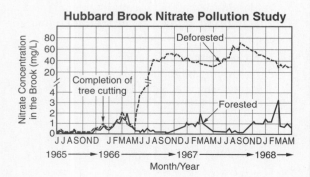

Hubbard Brook Nitrate Pollution Study

30. State how nitrate pollution in the brook changed after it flowed through the deforested area.

31. Explain how deforestation contributed to this change.

32. Explain why toxic waste dumps are most harmful to the environment.

33. Why is good topsoil considered so valuable? List three human activities that cause loss of soil.

Refer to the following graphs to answer questions 34 and 35.

Oxygen content and fish population in a lake

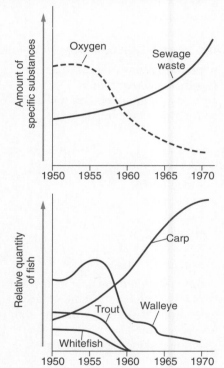

34. According to the graphs, an increase in sewage waste in a lake over time would be associated with

    1 an increase in dissolved oxygen and an increase in most fish populations
    2 a decrease in dissolved oxygen and an increase in most fish populations
    3 an increase in dissolved oxygen and a decrease in most fish populations
    4 a decrease in dissolved oxygen and a decrease in most fish populations

35. According to the graphs, the fish species that adapted most successfully to the change in oxygen content over time was the

1 trout, because it can live in highly oxygenated water
2 carp, because it can live in poorly oxygenated water
3 walleye, because it can live in highly oxygenated water
4 whitefish, because it can live in poorly oxygenated water

36. Explain how a new power plant built on the banks of the Rocky River could have an environmental impact on the Rocky River ecosystem downstream from the plant. Your explanation must include the effects of the power plant on:

♦ water temperature

♦ dissolved oxygen

♦ local fish species

37. Describe the importance of sewage treatment to both people and wildlife.

38. Both car exhaust and factory emissions add pollutants to the air. For each case, tell *how* it adds to air pollution and *what* is being done to reduce the problem.

39. The map below shows the movement of some air pollution across part of the United States. Which statement is a correct inference that can be drawn from this information?

Movement of Air Polllution

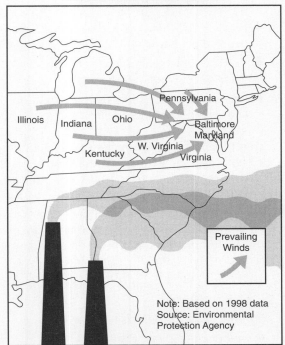

Note: Based on 1998 data
Source: Environmental Protection Agency

1 Illinois produces more air pollution than the other states shown.
2 The air pollution problem in Baltimore is increased by the addition of pollution from other areas.
3 There are no air pollution problems in the southern states.
4 The air pollution problems in Virginia clear up quickly as the air moves toward the sea.

Refer to the illustration below to answer questions 40 and 41.

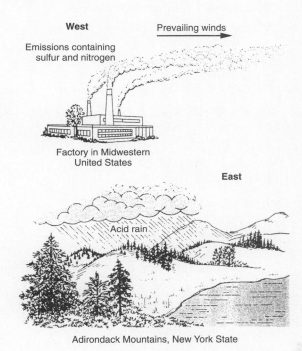

40. According to the illustration, acid rain is both an air pollutant and a water pollutant. Explain why this is true.

41. For each type of pollution (air and water), give *one* example of the kind of habitat the acid rain affects. By what means does it reach these different ecosystems?

Base your answers to questions 42 and 43 on the information below and on the diagram at the top of the next page.

Acid rain can have a pH between 1.5 and 5.0. The effect of acid rain on the environment depends on the pH of the rain and the characteristics of the environment. It appears that acid rain has a negative effect on plants. The following scale shows the pH of normal rain.

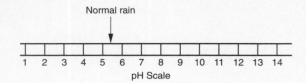

42. Provide the information that should be included in a research plan to test the effect of pH on the early growth of bean plants in the laboratory. In your answer be sure to:

   ♦ state a hypothesis
   ♦ identify the independent variable
   ♦ state *two* factors that should be kept constant

43. Construct a data table in which you could organize the research results.

Refer to the following graph to answer questions 44 and 45.

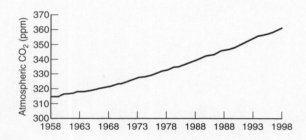

44. According to the graph, from the late 1950s to the late 1980s, the amount of CO_2 in Earth's atmosphere has been

   1 steadily decreasing
   2 steadily increasing
   3 staying about the same
   4 going up and down

45. Changes in the amount of CO_2 in Earth's atmosphere have been correlated with steadily increasing average global temperatures over the past 50 years. Based on this statement and the data in the graph, you could reason that

   1 as the amount of CO_2 in the air increases, the average temperature decreases
   2 as the amount of CO_2 in the air increases, the average temperature increases
   3 as the amount of CO_2 in the air decreases, the average temperature stabilizes
   4 as the amount of CO_2 in the air decreases, the average temperature increases

46. Some scientists are urging that immediate action be taken to stop activities that contribute to global warming. Discuss the effects of global warming on the environment and describe some human activities that may contribute to it. Your answer *must* include:

   ♦ an explanation of what is meant by the term *global warming*
   ♦ *one* human activity that is thought to be a major contributor to global warming
   ♦ an explanation of *how* the human activity may contribute to the problem
   ♦ *one* negative effect of global warming if it continues for many years

47. What substance is thought to cause the depletion of Earth's ozone shield? What has been done to solve this problem?

48. In the early 1980s, scientists discovered "holes" in the ozone shield that surrounds Earth. State one negative effect (that is, threat to health) that this environmental change could have on humans.

49. Choose *one* ecological problem to discuss from the following list: global warming; destruction of the ozone shield; acid rain; increased nitrogen and phosphorous in lakes; loss of biodiversity. In your answer be sure to state:

   ♦ the type of ecological problem you have chosen
   ♦ *one* human action that may have caused the problem
   ♦ *one* way in which the problem may negatively affect humans
   ♦ *one* way in which the problem may negatively affect the ecosystem
   ♦ *one* positive action that people can or did take to reduce the problem

50. Describe *two* specific methods that have been recently used by people to reduce the amount of chemicals being added to the environment.

51. All living organisms are dependent on a stable environment. Describe how humans have made the environment *less* stable for other organisms by:

   ♦ changing the chemical composition of air, soil, and water
   ♦ reducing the biodiversity of an area
   ♦ introducing advanced technologies

Base your answers to questions 52 and 53 on the information below and on your knowledge of biology.

Amphibians have long been considered an indicator of the health of life on Earth. Scientists are concerned because amphibian populations have been declining worldwide since the 1980s. In fact, in the past decade, twenty species of amphibians have become extinct and many others are endangered.

Scientists have linked this decline in amphibians to global climatic changes. Warmer weather during the last three decades has resulted in the destruction of many of the eggs produced by the Western toad. Warmer weather has also led to a decrease in rain and snow in the Cascade Mountain Range in Oregon, reducing the water level in lakes and ponds that serve as the reproductive sites for the Western toad. As a result, the eggs are exposed to more ultraviolet light. This makes the eggs more susceptible to a water mold that kills the embryos by the hundreds of thousands.

52. The term that is commonly used to describe the worldwide climatic changes mentioned in the passage is
    1 global warming
    2 deforestation
    3 mineral depletion
    4 industrialization

53. State *two* ways in which the decline in amphibian populations could disrupt the stability of the ecosystems they inhabit.

Refer to Figure 27-9 on page 215 to answer the following question.

54. By the year 2000, the worldwide human population had reached
    1 three times the size it was in 1950
    2 three times the size it was in 1960
    3 two times the size it was in 1970
    4 two times the size it was in 1960

Part C—Reading Comprehension

Base your answers to questions 55 to 57 on the information below and on your knowledge of biology. Use one or more complete sentences to answer each question.

Dr. David Vaughan is a British scientist who has been studying glaciers for a long time. As a glaciologist, he is very interested in the ice that covers and surrounds the great landmass at the South Pole. If the climate change that is occurring on Earth causes global warming, then the ice of Antarctica will start to melt. The melting ice would, eventually, raise sea levels around the world and the results would be disastrous. There is enough ice in just the western part of Antarctica to cause a rise of five meters in sea levels. This would flood many coastlines where millions of people live. However, knowing what is happening to the ice of Antarctica is very difficult.

Much attention has been given in recent years to ice shelves, floating masses of ice that surround much of Antarctica. While their melting will not directly affect sea levels—that ice is already in the sea—the loss of the ice shelves would make it much easier for the huge masses of land ice to melt. So, Dr. Vaughan and other glaciologists have been closely monitoring a series of collapses of ice shelves that began in 1995. In January 1995, a 770-square-mile section of an ice shelf along the Antarctic Peninsula broke apart suddenly. This area of ice was 35 times larger than all of Manhattan. Another even larger ice shelf that was at least 400 years old broke apart in 1998. And, in just 35 days beginning on January 31, 2002, the largest collapse to be seen in 30 years occurred. About this ice shelf, Dr. Vaughan said, "We knew what was left would collapse, but the speed of it is staggering." The area of ice that disappeared was 220 meters thick and contained 720 billion tons of ice!

These ice-shelf collapses are not entirely unexpected. The temperatures in the area of the Antarctic Peninsula, a long sliver of land pointing toward South America from Antarctica, have been rising steadily since the 1940s. The average temperature is now 2.5 degrees Celsius higher than it was in 1945. This is the fastest rate of warming seen any place on Earth. With the higher temperatures comes the melting of ice. The really big question remains: Is this change occurring only in this one place on Earth, or is this an early warning sign of global warming everywhere?

Scientists cannot agree on the answer to this big question. However, they are determined to study ice around the world—especially in the Antarctic—even more closely to get an answer. But how can this study be done, knowing how difficult it is to get to Antarctica? Go into outer space! Which is exactly what has been done. A satellite, called *ICESat* (*I*ce *C*loud & Land *E*levation *Sat*ellite), launched in 2003, is now orbiting Earth to track precise changes in ice sheets around the world.

55. Why is the possible melting of Antarctic ice of such concern?

56. How is the melting of ice shelves related to a possible rise in sea levels?

57. How is the use of outer space helping in the study of the Antarctic?

28

Saving the Biosphere

WHAT NEEDS TO BE SAVED?

In 1970, the first Earth Day marked the beginning of the modern environmental movement. Many environmental organizations were founded at this time, and the government passed several environmental protection laws.

Can we protect the environment not only for people but for all species? The total area of land, water, and air on Earth's surface where life is found—and that needs protection—is called the **biosphere**. (See Figure 28-1.) Saving the biosphere means paying attention to local, regional, and global problems. What must we do now to protect the environment for organisms that will be alive after us?

Figure 28-1 The biosphere is the total area of Earth's land, air, and water in which life is found. Earth is the ultimate ecosystem; although environmental problems may start out as local ones, they can become regional and, eventually, have a global impact.

A CHANGE IN ATTITUDE

Sometimes the most important, and difficult, changes concern accepted attitudes in our society. For example, what if we thought our lifestyle should not harm the environment and other living species? Would we be willing to make the necessary changes to accomplish this?

An industrialized society mainly views Earth as a source of valuable resources for its use. In contrast, ecology teaches us that humans are just one of many interdependent species that also need resources to live. In order for our species to survive, we must make sure that these important relationships within ecosystems also survive.

THINK GLOBALLY, ACT LOCALLY

The future health of the environment will depend on people's attitudes and behaviors. It has been suggested that people should learn to appreciate the "hidden costs" of many consumer goods. In other words, the environment pays a price for the products used by people in an industrial society.

Environmentalists have encouraged people to live by the "3 Rs": reduce, reuse, recycle. To *reduce* consumption, you would use less of a product or resource; for example, fewer paper towels can be used to clean up a spill. You can also *reuse* a product; for example, paper or plastic grocery bags that you bring food home in can be taken back the next week and used again. Finally, many used products can be made into other products; for example, in many cities, you now have to *recycle* plastic, glass, metal, and paper. (See Figure 28-2 on page 224.) These materials are used again in

Figure 28-2 Recycling of plastic, glass, and metal containers is required in some cities. As shown here, students can help in the recycling effort by sorting and recycling metal soda cans.

other products, such as benches made up of a "wooden" building material that is actually a form of recycled plastic. Such recycling helps to conserve our natural resources. (See Figure 28-3.)

Great care must also be taken to avoid the environmental damage that can occur when a species from a distant place is introduced into a new environment. Without any predators or natural controls, the new alien species can reproduce without limit, upsetting the stability of the ecosystem.

Much is also being learned about how the harmful effects resulting from the chemical control of pests in the environment can be avoided by the use of biological control. For example, native predators of insect pests can be used as a means of natural or biological control.

▓ RENEWABLE VERSUS NONRENEWABLE RESOURCES

Air, water, and sunlight are some of the important resources that are basic to life on Earth. Modern industrialized society requires other resources, too, such as coal, oil, and metal ores.

Resources can be considered renewable or nonrenewable. A *renewable resource* can be replaced within a generation. Enough of the resource is being made (by natural processes) to replace what is being used. For example, the wood used to build houses can be replaced if enough trees are replanted. The sun's **solar energy** and the wind can be considered renewable resources. (See Figure 28-4.)

A *nonrenewable resource* cannot be replaced within our lifetime; it exists in limited amounts and takes a very long time to form. This includes such energy sources as coal, oil, and natural gas. In addition, metals such as gold, silver, iron, copper, and aluminum and nonmetals such as sand, gravel, and limestone are nonrenewable resources.

One way to protect the biosphere is to use renewable energy sources. For example, electricity can be made in a dam from the power of falling water, rather than from the burning of coal. Wind power can turn the blades on

Figure 28-3 The teenagers in this photograph are helping to conserve natural resources by recycling newspapers.

Figure 28-4 The solar panels on this school's roof capture solar energy—a renewable resource—which is used to provide 70 percent of the building's heat and hot water.

Figure 28-5 Windfarms, such as this one in California, harness wind power—another renewable resource—to generate electricity.

giant windmills to generate electricity. (See Figure 28-5.) Sunlight can be used to heat water and buildings, and to produce electricity. (See Figure 28-6.) Finally, hot water from deep beneath Earth's surface—geothermal energy—can be used to heat buildings and to make electricity. All of these are renewable energy sources.

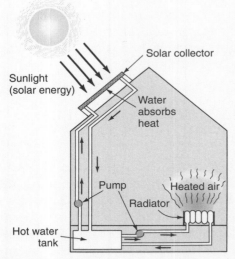

Figure 28-6 The diagram illustrates how energy from sunlight can be collected by a solar panel and used to heat a building.

❖ SUSTAINABLE DEVELOPMENT

If the economy is not growing, people think that something is wrong. Yet, unlimited growth is not possible. We cannot use more and more of Earth's resources indefinitely; and we cannot add more and more pollution to our environment. We need to find a way to live that is sustainable—a way that does not ruin Earth's ability to support life in the future. Improving the way we live without harming the environment is called *sustainable development*. By making these changes now, we may secure a healthier future for our planet.

❖ SUSTAINABLE DEVELOPMENT OF FORESTS

Some of the greatest damage to the environment is done by cutting down trees. The most economical way to harvest trees is to cut down an entire forest, a method called *clear-cutting*. (See Figure 28-7.) Such **deforestation** causes animals to lose habitat; and the rain—no longer absorbed by tree roots—flows right into streams, carrying topsoil with it. Due to soil erosion, the land cannot support plant growth; and the freshwater habitat for fish is disturbed, too.

Sustainable development in forestry means replacing every tree that is cut with a seedling and making sure that the seedling survives. A forest—with trees of all sizes and different ages, continually regrowing—would provide a healthy habitat for other woodland species and would prevent soil erosion.

Figure 28-7 The wood from forests has many uses: lumber, paper pulp, and fuel. Although cutting all the trees in one area is economical for loggers, the environment pays a high price in terms of loss of habitats.

ENVIRONMENTAL PROTECTION IN A DEVELOPED COUNTRY

Like other industrialized, developed countries, the United States uses up more than its share of energy and resources. Also, since the environment has been seriously affected, environmental awareness has increased since 1970.

One important result of this increased awareness is the Environmental Protection Agency (EPA), which is responsible for safeguarding the environment for future generations.

Another response to environmental issues was passage of the Endangered Species Act, which regulates a wide range of activities that affect threatened or endangered plant and animal species. (See Figure 28-8.)

Figure 28-8 The spotted owl of the Pacific Northwest is an endangered species. By protecting the owl, the Endangered Species Act also protects the old-growth forests it lives in.

ENVIRONMENTAL PROTECTION IN A DEVELOPING COUNTRY

People's lives in developed countries are very different from those in developing countries. In industrialized nations, the average standard of living is high and most people can expect to live for 70 or more years. But in many developing countries, most people are poor and have a much shorter life span. So, saving

Figure 28-9 In Africa, near Kenya's Ewasu River, local guides take tourists on camel safaris. This is one way in which "parks for people" programs can help local people earn a living from their natural environment without doing it harm.

the biosphere means different things in rich nations and poor nations. Environmental leaders are now learning about these differences.

Environmentalists realize that it is very difficult to set aside parks for endangered animals if doing so stops people from getting enough food and housing. One solution that has worked—called "parks for people"—directly involves local people in protecting their environment. For example, in Kenya, villagers work as guides for tourists who come to see the wildlife. (See Figure 28-9.)

SAVING THE BIOSPHERE: A WORLDWIDE EFFORT

In 1992, representatives from 178 countries attended the largest environmental meeting ever held, known as the Earth Summit. The main theme of the meeting was sustainable development. Work has continued since then, but differences among the countries interfere with more progress. While local and regional efforts to save the environment are important, the global effort matters most. The protection of Earth's biosphere for the future requires the efforts of all people.

Chapter 28 Review

Part A—Multiple Choice

1. The biosphere is the total area where life exists on or in Earth's

   1 land only
   2 water only
   3 land and water
   4 land, water, and air

2. A major reason that humans have negatively affected the environment in the past is that they

   1 often lacked an understanding of how their activities affect the environment
   2 attempted to control their population growth
   3 passed laws to protect certain wetlands
   4 discontinued the use of certain chemicals used to control insects

3. Recycling of materials such as glass, metal, and plastic helps to

   1 keep the cost of groceries low
   2 conserve our natural resources
   3 prevent natural resources
   4 build more wooden houses

4. Which human activity would be *least* likely to disrupt the stability of an ecosystem?

   1 disposing of wastes in the ocean
   2 increasing the human population
   3 using more fossil fuels
   4 recycling bottles and cans

5. An industrialized society mainly views Earth as a

   1 home for wildlife
   2 hazardous place to live
   3 source of natural resources
   4 barren landscape

6. Coal and wood are found in nature. They are both examples of

   1 enzymes   3 metals
   2 resources   4 proteins

7. An example of a renewable resource is

   1 natural gas   3 coal
   2 silver   4 wood

8. The definition of a renewable resource is that it

   1 can be replaced by nature within a generation

   2 cannot be replaced by nature within a generation
   3 is manufactured by humans
   4 is not too expensive to use

9. A nonrenewable resource is one that

   1 is replaced by nature as fast as it is used
   2 exists in a limited supply that can run out
   3 is recycled naturally by Earth's systems
   4 does not pollute ecosystems when it is used

10. Suppose that the average life span of most people is about 70 years and that there is only a 70-year supply of fossil fuel left on Earth. Then imagine that you are a member of a government panel that is deciding on how to handle the fuel situation. The best possible decision you could suggest would be to

    1 use up all the fuel in the present generation and not worry about the future
    2 find some alternative energy sources so that the fossil fuel lasts longer
    3 destroy the remaining fossil fuel so that no nations will fight over it
    4 have people return to a farming society so that they do not need the fuel

11. Which practice would most likely deplete a nonrenewable natural resource?

    1 harvesting pine trees on a tree farm
    2 restricting water usage during a period of water shortage
    3 burning coal to generate electricity in a power plant
    4 building a dam and a power plant to use water to generate electricity

12. All of the following are nonrenewable energy sources *except*

    1 coal   3 falling water
    2 gas   4 natural gas

13. Which statement is true about endangered species and nonrenewable resources?

    1 They are both living things that need to be protected.
    2 They are both nonliving things that need to be protected.
    3 They both need to be protected so they do not disappear.
    4 They both can be renewed quickly if they do disappear.

14. The goal of sustainable development is to
    1 achieve unlimited economic growth at any cost
    2 make sure the economy expands at a steady rate
    3 improve the way we live without harming the environment
    4 expand our lifestyle even if we run out of natural resources

15. Which method would you *not* use to solve environmental problems?
    1 promote global awareness
    2 cooperation among people
    3 "parks for people" programs
    4 increasing the population

16. Which action would best illustrate people's concern for the biosphere?
    1 passing game laws that limit the number of animals that may be hunted
    2 increasing the use of pesticides that may drain off farms into river systems
    3 allowing air to be polluted by only those factories that use new technology
    4 removing resources from nature at a faster rate than they are being replaced

17. One way to help provide a suitable environment for the future is to urge individuals to
    1 apply ecological principles when making decisions that have an impact
    2 agree that population controls have no impact on environmental matters
    3 control all aspects of natural environments
    4 work toward increasing global warming

Part B—Analysis and Open Ended

18. What is meant by "think globally, act locally" in terms of environmental protection?

19. Define the "3 Rs" proposed by environmentalists. Give an example of each one.

20. Why is it important for industries to be involved in recycling programs?

21. Compare and contrast renewable and nonrenewable resources. Your answer should include:
    ♦ a definition of *renewable* and of *nonrenewable*
    ♦ *one* example of a renewable resource
    ♦ *two* examples of a renewable *energy* source
    ♦ *two* examples of a nonrenewable resource

22. Recycling can extend the use of nonrenewable resources but *cannot* restore them. Humans can restore renewable resources to reduce some negative effects of increased consumption. Identify *one* resource that is renewable, and describe *one* specific way people can restore this resource if it is being depleted.

23. Use the following terms to complete the concept map below, which lists a variety of natural resources: *flowing water; copper; wind power; wood (charcoal); oil; gold; trees (lumber); limestone; geothermal energy; gravel; natural gas; coal; sunlight; sand; silver.*

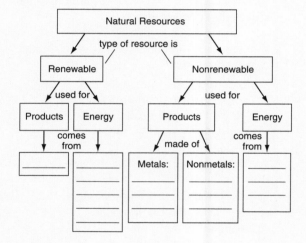

24. Explain what is meant by the term "sustainable development." Give *one* example of sustainable use of a natural resource.

25. Refer to Figure 28-3 on page 224 to answer the question. How might the activity these young people are involved in actually help to conserve forests and wildlife?

26. In what way does the protection of habitats help enforce the Endangered Species Act?

27. Refer to Figure 28-7 on page 225 to answer the question. How does the method of "clear-cutting" shown here damage habitats both on land and in the water?

28. Why are people in developing nations sometimes more directly affected by conservation

programs than are people in developed (industrialized) nations?

29. How do "parks for people" programs help local people and their environment at the same time?

30. Give at least one reason why *all* nations—developed and developing—should be involved in global efforts to save the biosphere.

31. List *five* changes that could be made in your own home that could help protect the environment.

32. List *five* changes that could be made in your community that could help protect the environment.

Part C—Reading Comprehension

Base your answers to questions 33 to 36 on the information below and on your knowledge of biology. Use one or more complete sentences to answer each question.

A decade after the *Exxon Valdez* oil tanker spilled millions of gallons of crude (oil) off Prince William Sound in Alaska, most of the fish and wildlife species that were injured have not fully recovered.

Only two out of the 28 species, the river otter and the bald eagle, listed as being injured from the 1989 spill are considered to be recovered said a new report, which was released by a coalition of federal and Alaska agencies working to help restore the oil spill region.

Eight species are considered to have made little or no progress toward recovery since the spill, including killer whales, harbor seals, and common loons (a type of bird). Several other species, including sea otters and Pacific herring, have made significant progress toward recovery, but are still not at levels seen before the accident. More than 10.8 million gallons of crude oil spilled into the water when the tanker *Exxon Valdez* ran aground 25 miles south of Valdez on March 24, 1989.

The spill killed an estimated 250,000 seabirds, 2800 sea otters, 300 harbor seals, 250 bald eagles, and up to 22 killer whales. Billions of salmon and herring eggs, as well as tidal plants and animals, were also smothered in oil. (*Source:* Reuters)

33. The oil spilled by the *Exxon Valdez* tanker is an example of a
1 nonrenewable resource and is a source of energy
2 renewable resource and is a source of ATP
3 nonrenewable resource and synthesizes ATP
4 renewable resource and is a fossil fuel

34. Which two species appear to have suffered the least damage from the oil spill over time?

35. The impact that the oil spill had on the environment is still occurring. Give information from the reading passage that supports this statement.

36. Which autotrophic organisms were negatively affected by the oil spill?

Base your answers to questions 37 to 41 on the information on the following page and on your knowledge of biology. Use one or more complete sentences to answer each question.

The location of a former fuel storage depot and packaging operation in the industrial port of Toronto, Canada, was the proposed site of a sports arena and entertainment complex. The problem was that the soil in this area was contaminated with gasoline, diesel fuel, home heating oil, and grease from the operation of the previous facility. Unless these substances were removed, the project could not proceed.

The traditional method of cleaning up such sites is the "dig-and-dump" method, in which the contaminated soil is removed, deposited in landfills, and replaced with clean soil. This dig-and-dump method is messy and costly, and it adds to landfills that are already overloaded. A technique known as *bioremediation*, which was used to help in the cleanup of the *Exxon Valdez* oil spill in Alaska, offered a relatively inexpensive way of dealing with this pollution problem.

The bioremediation cleanup process cost $1.4 million, one-third the cost of the dig-and-dump method, and involved encasing 85,000 tons of soil in a plastic "biocell" the size of a football field. This plastic-encased soil already contained naturally occurring bacteria that would have cleaned up the area after 50 years or more with the amounts of oxygen and nutrients normally found in the soil. But air, water, and fertilizer were piped into the biocell, stimulating the bacteria to reproduce rapidly and speed up the process. The cleanup by this technique was begun in August and completed in November of the same year. The bacteria attack parts of the contaminating molecules by breaking the carbon-to-carbon bonds that hold them together. This helps to change these molecules in the soil into carbon dioxide and water.

Although this method is effective for cleaning up some forms of pollution, bioremediation is not effective for inorganic materials, such as lead or other heavy metals, since these wastes are already in a base state that cannot be degraded any further.

37. The use of bioremediation by humans is an example of

1 interfering with nature so that natural processes cannot take place
2 using a completely unnatural method to solve a problem
3 solving a problem by speeding up natural processes
4 being unaware of and not using natural processes

38. The bacteria convert the contaminants into

1 carbon dioxide and water
2 toxic substances
3 proteins and fats
4 diesel fuel and grease

39. State an ecological problem associated with the use of the dig-and-dump method.

40. Explain why the cleanup of the proposed sports and entertainment site took only four months.

41. Bioremediation is *not* an effective method for breaking down

1 grease and heating oil
2 gasoline for vehicles
3 fuel for diesel engines
4 heavy metals such as lead

MST Learning Standards 4 and 1

The Regents Living Environment Examination is based on the New York State Learning Standards for Mathematics, Science, and Technology (MST). In particular, the examination will test student performance on the commencement (high school) level of **MST Learning Standard 4: The Living Environment and MST Learning Standard 1: Scientific Inquiry.**

◻ MST LEARNING STANDARD 4: THE LIVING ENVIRONMENT

There are seven key ideas for MST Learning Standard 4: The Living Environment. The material in the six themes of this review book covers all the required knowledge of these seven key ideas, which are as follows:

1. Living things are both similar to and different from each other and nonliving things.

2. Organisms inherit genetic information in a variety of ways that result in continuity of structure and function between parents and offspring.

3. Individual organisms and species change over time.

4. The continuity of life is sustained through reproduction and development.

5. Organisms maintain a dynamic equilibrium that sustains life.

6. Plants and animals depend on each other and their physical environment.

7. Human decisions and activities have had a profound impact on the physical and living environment.

◻ MST LEARNING STANDARD 1: SCIENTIFIC INQUIRY

There are three key ideas for MST Learning Standard 1: Scientific Inquiry, which are as follows:

1. The central purpose of scientific inquiry is to develop explanations of natural phenomena in a continuing, creative process.

2. Beyond the use of reasoning and consensus, scientific inquiry involves the testing of proposed explanations involving the use of conventional techniques and procedures and usually requiring considerable ingenuity.

3. The observations made while testing proposed explanations, when analyzed using conventional and invented methods, provide new insights into phenomena.

What students should know and be able to do in relation to each of these key ideas is described below. This content may be tested on the Regents Living Environment Examination.

Introduction to Scientific Inquiry

Science is a body of knowledge about the world. It is also a thinking process that has been designed by people in order to learn about how the world works. Based on observations and evidence collected from experimentation, and using what is already known, scientific explanations are developed about the world. These explanations are always subject to change as new observations and evidence are presented. Scientific methods exist to constantly test and re-test what we know in relation to existing explanations. In this way, scientific knowledge advances toward a more complete understanding of the world around us.

Key Idea 1: *The central purpose of scientific inquiry is to develop explanations of natural phenomena in a continuing, creative process.*

Natural phenomena, events, and occurrences are understood by people on the basis of existing explanations. To think about these explanations, it is necessary to be able to visualize, that is, to create mental pictures and to develop mathematical models. To develop scientific explanations, one uses evidence that can be observed as well as what people already know about the world. It is also important to learn about the history of science and the particular individuals who have contributed to scientific understanding. While science provides knowledge about the world, it also challenges people to develop the values to use this knowledge ethically and effectively.

To develop scientific ideas, one needs to think, do library research, and discuss one's ideas with others, including experts. Scientific inquiry involves asking questions and locating information from a variety of sources. It also involves making wise judgments about how reliable and relevant the information is.

At times there may be more than one explanation for the same phenomenon. A science student needs to work to resolve the differences. This is done through the use of evidence from experiments and direct observation. Also, an explanation that results in predictions that turn out to be accurate is more likely to be the correct explanation. All scientific explanations may be changed if new evidence suggests the change. This leads to a continually better understanding of how things work in the world.

It is necessary to develop explanations about small things and large things, that is, at different levels of scale. Also, one must consider phenomena from different points of view and degrees of complexity. In order to do this, experts from different subject areas often need to be consulted.

Key Idea 2: *Beyond the use of reasoning and consensus, scientific inquiry involves the testing of proposed explanations involving the use of conventional techniques and procedures and usually requiring considerable ingenuity.*

To test an explanation that has been put forward, a scientist must design ways to make observations related to the explanation. The design of the research must make use of library investigation in order to review scientific literature and also through discussions with other scientists. The research plan requires that one understand the big concepts being investigated. The plan should also include a variety of techniques, proper equipment, and safety procedures.

In creating a research plan, scientists use hypotheses to test the proposed explanations. A hypothesis is a prediction of what should be observed under specific conditions if the explanation is true. Hypotheses are very useful in science for helping one to determine what data should be collected and how the data should be interpreted. The research plan to test a hypothesis must include procedures to make sure that there is a fair interpretation of the data. This is called avoiding bias. These procedures include repeated trials, large sample size, and objective data-collection techniques.

Once a research plan is created, it must be carried out. This means doing the actual experiment, obtaining and putting together the necessary equipment, and recording one's observations.

Key Idea 3: *The observations made while testing proposed explanations, when analyzed using conventional and invented methods, provide new insights into phenomena.*

Observations made during scientific research usually need to be analyzed to see what they mean. The data may be organized and represented in a variety of ways in order to do this; for example, diagrams, tables, charts, graphs, equations, and matrices. When the data are interpreted, the result may be the statement of a new hypothesis. Another result may be the conclusion that a general understanding or explanation of a natural phenomenon is, in fact, correct.

The mathematical processes of statistical

analysis are used to determine if the results obtained might have been simply due to chance. Statistics also allows one to conclude the degree to which predicted results based upon the hypothesis match the actual results. From this matching comes the conclusion as to whether the proposed explanation is, or is not, supported by the data.

The analysis of the data, followed by public discussion, can lead to a revision of the explanation, the development of new hypotheses, and the design of new research plans.

When claims are made based on the collected evidence, the claims should be questioned if the design of the experiment was at fault; for example, if there were small sample sizes, incomplete or misleading use of data, or the lack of controlled conditions. Also, great care should be taken to not confuse fact with opinion.

When all research and data analysis is concluded, a written report is prepared for the public to study. The report includes a literature review, the research, the results and suggestions for further research. One purpose of making the results public is to allow the research to be repeated. Science assumes that through the collection of similar evidence, different individuals will come to the same explanations of nature. Peer review—the study of research reports by fellow scientists—is important as a check on the quality of the research. It also results, at times, in the suggestion of alternative explanations for the same observations.

Required Laboratory Activities for the Regents Exam, Part D

The *Regents Examination in Living Environment* includes multiple-choice and open-ended questions based on a series of required laboratory activities completed during the school year. Part D of the Living Environment examination (given in January, June, and August each year) will test at least three of the four laboratory activities that are required for that year. There are now seven different laboratory activities scheduled for implementation and testing through 2007. (*Note:* Over time, new labs will be introduced to replace the current lab activities. Lab #4: Adaptations for Reproductive Success in Flowering Plants may be implemented in the near future to replace Lab #3: The Beaks of Finches.) The first four labs are described below.

Lab #1: Relationships and Biodiversity

Lab #2: Making Connections

Lab #3: The Beaks of Finches

Lab #5: Diffusion Through a Membrane

While completing the laboratory activities, you will record your results and answers to questions in the Student Laboratory Packets. You are to keep these sheets for review before taking the Regents examination. You will also transfer your answers to separate Student Answer Packets, which will be used and kept by the school as evidence of your completion of the laboratory requirement for the Living Environment Regents exam. All directions to the teachers and printed materials for the students have been prepared and distributed by the New York State Education Department.

Required Laboratory Activity #1: Relationships and Biodiversity

This activity is a simulation that consists of six tests done in the lab as well as a seventh task, a reading assignment. Your goal is to collect and analyze data on several different plant species in order to determine which of the species is most closely related to a valuable but endangered species. The evidence, both structural and molecular, is used to develop a hypothesis about the evolutionary relationships between the plant species. The final task of the activity, the reading passage, focuses on the importance of preserving biodiversity.

This laboratory activity is most closely correlated with topics covered in Theme I—Evolution, Theme V—Genetics and Molecular Biology, and Theme VI—Interaction and Interdependence. The activity could be used during the teaching of these themes or as a performance assessment of laboratory skills. The lab requires that you have an understanding of DNA and protein synthesis.

Required Laboratory Activity #2: Making Connections

The purpose of this activity is to help you learn how to design and use a controlled experiment in order to draw a conclusion. In particular, you are to determine which of two conflicting claims is supported by your experimentation. The laboratory activity consists of two parts. In Part A, you will practice two simple techniques—taking a person's pulse and measuring muscle fatigue by squeezing a clothespin. Part B is the main portion of the activity. You will design your own investigation after reviewing guidelines for conducting a controlled experiment. You will use your experiment to determine which claim is supported, namely that a person can squeeze a clothespin more times by exercising first or more times by *not* exercising first. Results are put in writing and some students make oral presentations of their reports to the class for peer review.

This laboratory activity is most closely correlated with topics of human physiology covered in "Chapter 8—Getting Food to Cells: Nutrition" and "Chapter 9—Matter on the Move: Gas Exchange and Transport" in Theme II—Energy, Matter, and Organization. However, a minimum of content knowledge is required for the activity. The main focus of the activity is to understand the concepts involved in experimental design, and this activity could be used to introduce this topic.

Required Laboratory Activity # 3:
The Beaks of Finches

The purpose of this activity is to use a simulation to study how structural differences affect the survival rate of members within a species. The lab is based on the observations of the many finch species on the Galápagos Islands that Charles Darwin used in support of the process of natural selection. You will work in pairs to represent a finch. Each pair is randomly assigned a grasping tool—such as forceps, tongs, pliers, or tweezers—that represents a type of beak to be used to pick up seeds of different sizes. The efficiency of the tool-beaks at picking up small seeds such as lentils determines whether the "finches"

survive and stay in the same "environment" with these seeds or "migrate" in search of food to a different environment with larger seeds such as lima beans. The survivors in the activity now compete with others to continue to explore the efficiency of their "beaks."

This laboratory activity is most closely correlated with topics covered in "Chapter 1, The Process of Evolution" within Theme I—Evolution. You will need to be familiar with the concepts of adaptation, variation, and natural selection.

Required Laboratory Activity #5:
Diffusion Through a Membrane

In this laboratory activity, you will study the process of diffusion by using a model "cell" to test selective permeability of the cell membrane. The "cell" is made of dialysis tubing or a plastic bag that contains a glucose and starch solution, which is immersed in water for a period of time. The water in the beaker is then tested for the presence of glucose and starch. In the second part of the laboratory activity, you will use a microscope to observe the effects of salt water and distilled water on red onion cells. You will see the effects of the diffusion (osmosis) of water out of the cells when they are surrounded by salt water.

This laboratory activity is most closely correlated with sections on cell processes found in "Chapter 6, Chemical Activity in the Cell" within Theme II—Energy, Matter, and Organization, as well as topics in "Chapter 10, The Need for Homeostasis" and "Chapter 13, Excretion and Water Balance" within Theme III—Maintaining a Dynamic Equilibrium.

Living Environment Part D—
Sample Lab Questions

Beginning with the June 2004 administration, the Regents Examination in Living Environment will include a new section, Part D. The questions on Part D will consist of a combination of multiple-choice and open-ended questions related to at least three of the four required living environment laboratory activities and will comprise approximately 15% of the examination.

These sample questions are provided to help teachers and students become familiar with the format of questions for this part of the examination. They provide examples of ways the required laboratory experiences may be assessed. A rating guide is also included.

Sample Items Related to Lab Activity #1: *Relationships and Biodiversity*

1   In the *Relationships and Biodiversity* laboratory activity, students were instructed to use a clean dropper to place each of four different samples of plant extracts on the chromatography paper. A student used the same dropper for each sample without cleaning it between each use. State one way this student's final chromatogram would be different from a chromatogram that resulted from using the correct procedure.   [1]

2   State one reason that safety goggles were required during the indicator test for enzyme M.   [1]

Base your answers to questions 3 through 6 on the information and data table below and on your knowledge of biology.

A student was told that three different plant species are very closely related. She was provided with a short segment of the same portion of the DNA molecule that coded for enzyme *X* from each of the three species.

Information Regarding Enzyme X

DNA sequence from plant species *A*	CAC	GTG	GAC
Amino acid sequence for enzyme *X* coded for by that DNA	Val	His	Leu
DNA sequence from plant species *B*	CAT	GTG	CAA
Sequence of bases in mRNA produced by that DNA	_____	_____	_____
Amino acid sequence for enzyme *X* coded for by the DNA	Val	His	Val
DNA sequence from plant species *C*	CAG	GTA	CAG
Sequence of bases in mRNA produced by that DNA	GUC	CAU	GUC
Amino acid sequence for enzyme *X* coded for by the DNA	_____	_____	_____

3   The correct sequence of mRNA bases for plant species *B* is

    (1) GUA CAC GUU
    (2) GTA CAC GTT
    (3) CAU GUG CAA
    (4) TCG TGT ACC

4   Use the mRNA Codon Chart on the next page to determine the amino acid sequence for enzyme *X* in plant species *C* and record the sequence in the appropriate place in the data table.   [1]

5   Is it possible to determine whether species *B* or species *C* is more closely related to species *A* by comparing the amino acid sequences that would result from the three given DNA sequences? Support your answer.   [1]

6   Determine whether species *B* or species *C* appears more closely related to species *A*. Support your answer with data from the data table. (*Base your answer only on the DNA sequences provided* for enzyme *X* in these three plant species.)   [1]

Universal Genetic Code Chart

Messenger RNA codons and the amino acids they code for.

		SECOND BASE			
	U	**C**	**A**	**G**	

FIRST BASE		U	C	A	G	THIRD BASE
U	UUU } PHE UUC UUA } LEU UUG	UCU UCC UCA } SER UCG	UAU } TYR UAC UAA } STOP UAG	UGU } CYS UGC UGA } STOP UGG } TRP	U C A G
C	CUU CUC CUA } LEU CUG	CCU CCC CCA } PRO CCG	CAU } HIS CAC CAA } GLN CAG	CGU CGC CHA } ARG CGG	U C A G
A	AUU AUC } ILE AUA AUG } MET or START	ACU ACC ACA } THR ACG	AAU } ASN AAC AAA } LYS AAG	AGU } SER AGC AGA } ARG AGG	U C A G
G	GUU GUC GUA } VAL GUG	GCU GCC GCA } ALA GCG	GAU } ASP GAC GAA } GLU GAG	GGU GGC GGA } GLY GGG	U C A G

Sample Items Related to Lab Activity #2: *Making Connections*

Base your answers to questions 7 through 9 on the information and data table below and on your knowledge of biology.

In the Making Connections laboratory activity, a group of students obtained the following data:

Student Tested	Pulse Rate at Rest	Pulse Rate After Exercising
1	70	97
2	75	106
3	84	120
4	60	91
5	78	122

7   Explain how this change in pulse rate is associated with homeostasis in muscle cells.   [1]

8   Identify the system of the human body whose functioning is represented by this data.   [1]

9   Identify *one* other system of the human body whose functioning would be expected to be altered as a direct result of the exercise. Describe how this system would most likely be altered.   [1]

Base your answers to questions 10 and 11 on the information below and on your knowledge of biology.

A biology class performed an investigation to determine the influence of exercise on pulse rate. During the investigation, one group of twelve students, Group *A*, counted how many times they could squeeze a clothespin in a 1-minute period, then exercised for 4 minutes, and repeated the clothespin squeeze for an additional 1 minute. Another group of twelve students, Group *B*, also counted how many times they could squeeze a clothespin in a 1-minute period, but then they rested for 4 minutes, and repeated the clothespin squeeze for an additional 1 minute. The data table below shows the average results obtained by the students.

Effect of Exercise on Number of Clothespin Squeezes

Groups of Student	Average Number of Clothespin Squeezes During First Minute	Average Number of Clothespin Squeezes During Second Minute
Group A (exercise)	75	79
Group B (rest)	74	68

10 State *two* specific examples from the description of the investigation and the data table that support this investigation being a well-designed experiment. [2]

11 The chart below shows relative blood flow through various organs during exercise and at rest.

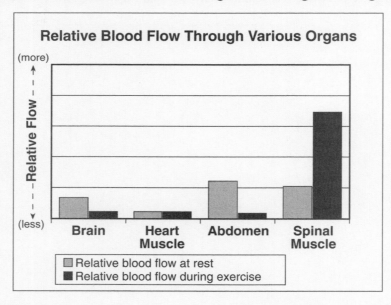

Using information from both the data table and the chart, explain how muscle fatigue and blood circulation could account for the results the students obtained. [2]

Sample Items Related to Lab Activity #5: *Diffusion Through a Membrane*

Base your answers to questions 12 and 13 on the diagrams below and on your knowledge of biology.

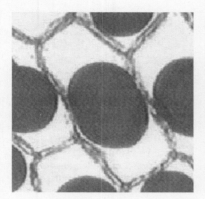 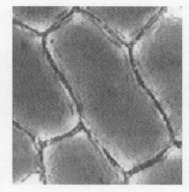

Diagram 1: red onion cells   Diagram 2: red onion cells

12 Describe how to prepare a wet-mount slide of red onion cells with the cell membrane shrinking away from the cell wall, as shown in diagram 1. The following materials are available: microscope slide, pipettes, cover slips, paper towels, water, salt solution, and red onion sections. [3]

13 List the laboratory procedures to follow that would cause the cells in diagram 1 to resemble the cells in diagram 2.   [2]

_____

14 A student places an artificial cell, similar to the one used in the laboratory activity *Diffusion Through a Membrane*, in a beaker containing water. The artificial cell contains starch and sugar. A starch indicator is added to the water in the beaker. Explain how the student will know if the starch is able to diffuse out of the artificial cell.   [1]

GLOSSARY

abiotic describes the nonliving parts of an organism's environment

acidity describes a low pH level due to dissolved acids, such as in acid rain

active transport movement of substances across a membrane from an area of lower concentration to an area of higher concentration; requires energy

adaptations special characteristics that make an organism well suited for a particular environment

AIDS (acquired immunodeficiency syndrome) an immunodeficiency disease, caused by HIV in humans

algae plantlike organisms, often single-celled, that carry out photosynthesis

alleles the two different versions of a gene for a particular trait

allergic reactions conditions caused by an overreaction of the immune system

amino acids organic compounds that are the building blocks of proteins

antibiotic a chemical that kills specific microorganisms; frequently used to combat infectious diseases

antibodies molecules that individuals produce as a defense against foreign objects in the body; antibodies bind to specific antigens

antigen a protein on a foreign object that stimulates the immune system to produce antibodies

artificial selection see *selective breeding*

asexually describes reproduction that requires only one parent to pass on genetic information, e.g., budding

atmosphere the blanket of gases that covers Earth; usually called "air"

atoms the smallest units of an element that can combine with other elements

ATP (adenosine triphosphate) the substance used by cells as an immediate source of chemical energy for the cell

autotrophic describes a self-feeding organism that obtains its energy from inorganic sources, e.g., plants (producers)

bacteria single-celled organisms that have no nuclear membrane to surround and contain their DNA molecule

behavior every action that an animal takes, either learned or instinctive; usually to aid survival

biodiversity the variety of different species in an ecosystem or in the world

biome a very large area characterized by a certain climate and types of plants and animals

biotechnology describes new procedures and devices that utilize discoveries in biology

biotic describes the living parts of an organism's environment

budding a form of asexual reproduction in which the offspring grows out of the side of the parent

cancer a disease that results from uncontrolled cell division, which damages normal tissues

carbon one of the six most important chemical elements for living things; carbon atoms form the backbone of nearly all organic compounds

carbon dioxide the inorganic molecule from which plants get carbon for photosynthesis; waste product of cellular respiration; a greenhouse gas

carnivores animals that obtain their energy by eating other animals; see also *consumers* and *heterotrophic*

carrying capacity the size of a population that an ecosystem can support

catalysts substances that increase the rate of a chemical reaction, but are not changed during the reaction

cell the smallest living unit of an organism; all organisms are made up of at least one cell

cell membrane a selectively permeable plasma membrane that separates and regulates substances that pass between the inside and outside of a cell

cellular respiration the process that uses oxygen to create ATP for energy use

chloroplast the organelle in a plant that contains the pigment chlorophyll and carries out photosynthesis

chromosomes structures composed of DNA that contain the genetic material

circulation the movement of blood throughout the body of an animal

cloning the production of identical individuals from the cell of another individual

community populations of different species that interact within an area

competition the struggle between organisms for limited resources such as food and space

consumers organisms that obtain their energy by feeding on other organisms; heterotrophic life-forms

coordination the means by which body systems work together to maintain homeostasis; a property of living things

cytoplasm the watery fluid that fills a cell, surrounding its organelles

decomposers heterotrophic organisms that obtain their energy by feeding on decaying organisms

deforestation the cutting down and clearing away of forests; clear-cutting

deplete to use up natural resources that cannot be replaced within our lifetimes

depletion (ozone) a reduction in the amount of ozone in Earth's ozone layer

development the changes in an organism that occur from fertilization until death

deviations changes in the body's normal functions that are detected by control mechanisms, which maintain a balanced internal environment

differentiation the creation of specialized cells from less specialized parent cells through controlled gene expression

diffusion the movement of molecules from an area of higher concentration to an area of lower concentration

digestion the process of breaking down food particles into molecules small enough to be absorbed by cells

diversity the variety of different traits in a species or different species in an ecosystem

DNA (deoxyribonucleic acid) the hereditary material of all organisms, which contains the instructions for all cellular activities

dynamic equilibrium in the body, a state of homeostasis in which conditions fluctuate yet always stay within certain limits

ecology the study of the interactions of living things with their environment

ecosystem an area that contains all living and nonliving parts that interact

egg the female gamete that supplies half the genetic information to the zygote

embryo an organism in an early stage of development before it is hatched, born, or germinated

energy pyramid describes the flow of energy through an ecosystem; most energy is at the base (producers), and decreases at each higher level (consumers)

environment the physical surroundings of an organism, with which it interacts

enzymes proteins that act as catalysts for a biological reaction

equilibrium in ecosystems, an overall stability in spite of cyclic changes

estrogen in females, along with progesterone, a major sex hormone that affects secondary sex characteristics and reproduction

evolution the change in organisms over time due to natural selection acting on genetic variations that enable them to adapt to changing environments

excretion the removal of metabolic wastes from the body

expression the use of genetic information in a gene to produce a particular characteristic, which can be modified by interactions with the environment

extinction the death of all living members of a species

feedback mechanisms systems that reverse an original response that was triggered by a stimulus

fertilization in sexual reproduction, process by which an egg cell and a sperm cell unite to form a zygote

fetus a developing embryo after the first three months of development

food chain the direct transfer of energy from one organism to the next

food web the complex, interconnecting food chains in a community

fossil the traces or remains of a dead organism, preserved by natural processes

fungus (*plural,* fungi) heterotrophic organisms that obtain their energy by feeding on decaying organisms, e.g., yeast and mushrooms

gametes the male and female sex cells that combine to form a zygote during fertilization

genes the segments of DNA that contain the genetic information for a given trait or protein

gene expression see *expression*

genetic engineering recombinant DNA technology, i.e., the insertion of genes from one organism into the genetic material of another; see also *biotechnology*

genetic variation the differences among offspring in their genetic makeup

geologic time Earth's history divided into vast units of time by which scientists mark important changes in Earth's climate, surface, and life-forms

global warming an increase in the average atmospheric temperature of Earth due to more heat-trapping CO_2 in the air, which causes the "greenhouse effect"

glucose a simple sugar that has six carbon atoms bonded together; a subunit of complex carbohydrates

habitat the place in which an organism lives; a specific environment that has an interacting community of organisms

herbivores animals that obtain their energy by eating plants; see also *consumers* and *heterotrophic*

hereditary describes the genetic information that is passed from parents to offspring

heterotrophic describes an organism that obtains its energy by feeding on other living things, e.g., animals (consumers)

homeostasis in the body, the maintenance of a constant internal environment

hormones chemical messengers that bind with receptor proteins to affect gene activity, resulting in long-lasting changes in the body

host the organism that a parasite uses for food and shelter by living in or on it

hydrogen one of the six most important chemical elements for living things

immune system recognizes and attacks specific invaders, such as bacteria, to protect the body against infection and disease

immunity the ability to resist or prevent infection by a particular microbe

inheritance the process by which traits are passed from one generation to the next

inorganic in cells, substances that allow chemical reactions to take place; in ecosystems, substances that are cycled between living things and the environment

insulin substance secreted by the pancreas that maintains normal blood sugar levels

internal development occurs when the embryo develops within the female's body

internal fertilization occurs when the sperm fertilizes the egg cell within the female's body

kingdoms the major groupings into which scientists categorize all living things

level of organization a scale for looking at the structure of a system, e.g., from atoms to cells to tissues to organs to organisms to populations to ecosystems

lipids the group of organic compounds that includes fats and oils

malfunction occurs when an organ or body system stops functioning properly, which may lead to disease or death

meiosis the division of one parent cell into four daughter cells; reduces the number of chromosomes to one-half the normal number

membrane see *cell membrane*

metabolic describes the chemical reactions (building up and breaking down) that take place in an organism

microbes microscopic organisms that may cause disease when they invade another organism's body; microorganisms, e.g., bacteria and viruses

mitochondria the organelles at which the cell's energy is released

mitosis the division of one cell's nucleus into two identical daughter cell nuclei

molecules the smallest unit of a compound, made up of atoms

movement the flow of materials between the cell and its environment; a property of living things, i.e., locomotion

multicellular describes organisms that are made up of more than one cell

mutation an error in the linear sequence (gene) of a DNA molecule

natural selection the process by which organisms having the most adaptive traits for an environment are more likely to survive and reproduce

nerve cells in animals, the cells that transmit nerve impulses to other nerve cells and to other types of cells

niche an organism's role in, or interaction with, its ecosystem

nitrogen one of the six most important chemical elements for living things

nucleotides the building blocks, or subunits, of DNA; they include four types of nitrogen bases, which occur in two pairs

nucleus the dense region of a (eukaryotic) cell that contains the genetic material

nutrients important molecules in food, such as lipids, proteins, and vitamins

nutrition the life process by which organisms take in and utilize nutrients

organ describes a level of organization in living things, i.e., a structure made up of similar tissues that work together to perform the same task, e.g., the liver

organelles structures within a cell that perform a particular task, e.g., the vacuole

organic relating to compounds that contain carbon and hydrogen (in living things)

organisms living things; life-forms

organ system a group of organs that works together to perform a major task, e.g., the digestive system

ovaries the female reproductive organs that produce the mature egg cells

oxygen one of the six most important chemical elements for living things; released as a result of photosynthesis; essential to cellular (aerobic) respiration

ozone depletion see *depletion (ozone)*

ozone shield the layer of ozone gas that surrounds Earth high in the atmosphere and blocks out harmful ultraviolet (UV) radiation

pancreas gland that secretes pancreatic juice (containing enzymes that aid digestion), and insulin (maintains normal blood sugar levels)

parasites organisms that live in or on another organism (a host), causing it harm

passive transport movement of substances across a membrane; requires no use of energy

pathogens microscopic organisms that cause diseases, such as certain bacteria and viruses; see also *microbes*

pesticides chemicals used to kill agricultural pests, mainly insects, some of which have evolved resistance to the chemicals

pH a measurement (on a scale of 0 to 14) of how acidic or basic a solution is

photosynthesis the process that, in the presence of light energy, produces chemical energy (glucose) and water

placenta the organ that forms in the uterus of mammals to nourish a developing embryo and remove its waste products

population all the individuals of the same species that live in the same area

predator an organism that feeds on another living organism (the prey); a consumer

predator-prey an interaction in which the prey is usually killed right away

pregnancy in animals, the condition of having a developing embryo within the body

prey an organism that is eaten by another organism (the predator)

producers organisms on the first trophic level, which obtain their energy from inorganic sources, e.g., by photosynthesis; autotrophic life-forms

progesterone in females, along with estrogen, a major sex hormone; see *estrogen*

proteins a group of organic compounds that are made up of chains of amino acids

radiation a form of energy that can cause genetic mutations in sex cells and body cells

receptors molecules that play an important role in the interactions between cells, e.g., molecules that bind with hormones

recombination the formation of new combinations of genetic material due to crossing-over during meiosis or due to genetic engineering

recombining during meiosis, the process that causes an increase in genetic variability due to the exchange of material between chromosomes

replicate the process by which DNA makes a copy of itself during cell division and protein synthesis

reproduction the production of offspring (i.e., passing on of hereditary information), either by sexual or asexual means

residue the remains of dead organisms, which are recycled in ecosystems by the activities of bacteria and fungi

response an organism's reaction to a stimulus; can be inborn or learned

respiration in the lungs, the process of exchanging gases; in cells, the process that releases the chemical energy stored in food; see also *cellular respiration*

ribosomes the organelles at which protein synthesis occurs, and which contain RNA

scavenge to gather the remains of a kill, rather than to hunt living animals

selective breeding the process by which humans encourage the development of specific traits by breeding the plants or animals that have those traits

sex cells the male and female gametes; they have one-half the normal chromosome number as a result of meiosis

sexually describes reproduction that requires two parents to pass on genetic information

simple sugars single sugars that have six carbon atoms, e.g., glucose

solar energy radiant energy from the sun that is a renewable resource

species a group of related organisms that can breed and produce fertile offspring

sperm the male gamete that supplies half the genetic information to the zygote

stability the ability of an ecosystem to continue and to remain healthy; usually, the greater the species diversity, the more stable the ecosystem

starches complex carbohydrates made up of many glucose molecules; used for energy storage in plants

stimulus (*plural*, stimuli) any event, change, or condition in the environment that causes an organism to make a response (i.e., to react)

subunits the four types of nucleotide bases that make up the DNA molecule

succession the gradual replacement of one ecological community by another until reaching a point of stability

symbiosis a close relationship between two or more different organisms that live together, which is often but not always beneficial

synthesis the building of compounds that are essential to life, e.g., protein synthesis

systems describes a level of organization in living things, i.e., groups of organs that work together to perform the same task; see also *organ system*

template in DNA replication, the original molecule that is used to make a copy

territory the area in which an animal lives, and which it usually defends

testes the pair of male reproductive organs that produces the sperm cells

testosterone in males, the main sex hormone that influences secondary sex characteristics and reproduction

theory of evolution see *evolution*

tissues describes a level of organization in living things, i.e., groups of similar cells that work together to perform the same function

toxins chemicals that can harm a developing fetus if taken in by the mother during pregnancy; also, chemicals that may get passed from one trophic level to the next (and increase in each organism) as they move up the food chain

trophic level a feeding level on a food chain or in a food web

uterus in mammals, the reproductive organ that holds the developing embryo

vaccinations injections that prepare the immune system to better fight a specific disease in the future

vacuole an organelle that stores materials, including wastes, for the cell

variability see *genetic variation*

viruses particles of genetic material that can only replicate within a host cell, where they usually cause harm

white blood cells several types of cells that work to protect the body from disease-causing microbes and foreign substances

zygote the fertilized egg cell that is formed when the nuclei of two gametes (a male and a female) fuse

INDEX

PHOTO CREDITS

Photographs are provided courtesy of the following:

National Aeronautics and Space Administration (NASA): 223

N.Y. Public Library/Science, Industry, and Business Library: 166

N.Y.S. Department of Environmental Conservation: 214

Photo Researchers, Inc.: 1, Biophoto Associates; 4, Mary Eleanor Browning; 10, Tom McHugh; 26 (left), Robert C. Hermes; 26 (right), Tim Davis; 27, John Reader; 28, Tim Davis; 33, Ricardo Arias; 48, Van D. Bucher; 56, Michael Austin; 89 (left), A. Cosmos Blank; 89 (right), Stephen Dalton; 101, Meckes/Ottawa; 102, Biophoto Associates; 104, Francis Leroy, Biocosmos/Science Photo Library (artwork based on SEM); 112, Omikron; 119, Biophoto Associates; 136 (four images), Dr. Yorgos Nikas; 137, Biophoto Associates; 144, A. Barrington Brown; 161, Mary Eleanor Browning; 168 (left), Biophoto Associates; 168 (right), Doug Martin; 172, Laguna Design (computer artwork); 179, Tom McHugh; 180 (left), Ylla; 180 (right), Fletcher & Baylis; 82, Richard Parker; 187, NASA; 189 (left), Eric Husking; 189 (right), Tom McHugh; 190, M. P. Kahl; 202 (top), Carl Frank; 202 (bottom), Karl Weidmann; 212, Michael Hayman; 213, Michael P. Gadomski; 215, Rafael Macia; 224 (top), Jeff Isaac Greenberg; 224 (bottom left), Christa Armstrong Rapho; 224 (bottom right), Tom McHugh; 225 (top), Georg Gerster; 225 (bottom), Simon Fraser/Science Photo Library; 226 (left), Doug Plummer; 226 (right), William & Marcia Levy.

Visuals Unlimited, Inc., © SIU: 174

LIVING ENVIRONMENT
JUNE 2006

Part A

Answer all questions in this part. [30]

Directions (1–30): For *each* statement or question, write on your separate answer sheet the *number* of the word or expression that, of those given, best completes the statement or answers the question.

1 The diagram below represents levels of organization in living things.

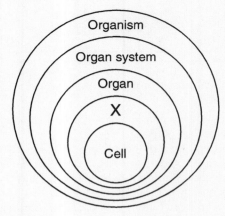

Which term would best represent *X*?

(1) human   (3) stomach
(2) tissue   (4) organelle

2 The evolutionary pathways of ten different species are represented in the diagram below.

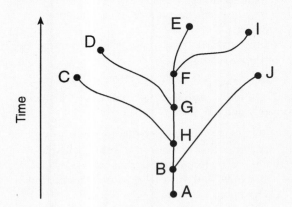

Which two species are the most closely related?

(1) *C* and *D*   (3) *G* and *J*
(2) *E* and *I*   (4) *A* and *F*

3 Which row in the chart below best describes the active transport of molecule *X* through a cell membrane?

Row	Movement of Molecule X	ATP
(1)	high concentration → low concentration	used
(2)	high concentration → low concentration	not used
(3)	low concentration → high concentration	used
(4)	low concentration → high concentration	not used

4 Hereditary information is stored inside the

(1) ribosomes, which have chromosomes that contain many genes
(2) ribosomes, which have genes that contain many chromosomes
(3) nucleus, which has chromosomes that contain many genes
(4) nucleus, which has genes that contain many chromosomes

5 A human liver cell is very different in structure and function from a nerve cell in the same person. This is best explained by the fact that

(1) different genes function in each type of cell
(2) liver cells can reproduce while the nerve cells cannot
(3) liver cells contain fewer chromosomes than nerve cells
(4) different DNA is present in each type of cell

6 Most of the starch stored in the cells of a potato is composed of molecules that originally entered these cells as

(1) enzymes
(2) simple sugars
(3) amino acids
(4) minerals

7 Hereditary traits are transmitted from generation to generation by means of

(1) specific sequences of bases in DNA in reproductive cells
(2) proteins in body cells
(3) carbohydrates in body cells
(4) specific starches making up DNA in reproductive cells

8 Which process can produce new inheritable characteristics within a multicellular species?

(1) cloning of the zygote
(2) mitosis in muscle cells
(3) gene alterations in gametes
(4) differentiation in nerve cells

9 Which two processes result in variations that commonly influence the evolution of sexually reproducing species?

(1) mutation and genetic recombination
(2) mitosis and natural selection
(3) extinction and gene replacement
(4) environmental selection and selective breeding

10 The illustration below shows an insect resting on some green leaves.

The size, shape, and green color of this insect are adaptations that would most likely help the insect to

(1) compete successfully with all birds
(2) make its own food
(3) hide from predators
(4) avoid toxic waste materials

11 A food web is represented below.

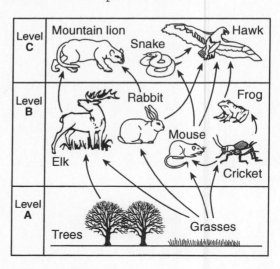

Which statement best describes energy in this food web?

(1) The energy content of level *B* depends on the energy content of level *C*.
(2) The energy content of level *A* depends on energy provided from an abiotic source.
(3) The energy content of level *C* is greater than the energy content of level *A*.
(4) The energy content of level *B* is transferred to level *A*.

12 Which statement concerning proteins is *not* correct?

(1) Proteins are long, usually folded, chains.
(2) The shape of a protein molecule determines its function.
(3) Proteins can be broken down and used for energy.
(4) Proteins are bonded together, resulting in simple sugars.

13 All chemical breakdown processes in cells directly involve

(1) reactions that are controlled by catalysts
(2) enzymes that are stored in mitochondria
(3) the production of catalysts in vacuoles
(4) enzymes that have the same genetic base sequence

14 Steps in a reproductive process used to produce a sheep with certain traits are listed below.

Step 1 — The nucleus was removed from an unfertilized egg taken from sheep A.

Step 2 — The nucleus of a body cell taken from sheep B was then inserted into this unfertilized egg from sheep A.

Step 3 — The resulting cell was then implanted into the uterus of sheep C.

Step 4 — Sheep C gave birth to sheep D.

Which sheep would be most genetically similar to sheep D?

(1) sheep A, only
(2) sheep B, only
(3) both sheep A and B
(4) both sheep A and C

15 Which diagram best illustrates an event in sexual reproduction that would most directly lead to the formation of a human embryo?

(1)

(2)

(3)

(4)

16 Offspring that result from meiosis and fertilization each have

(1) twice as many chromosomes as their parents
(2) one-half as many chromosomes as their parents
(3) gene combinations different from those of either parent
(4) gene combinations identical to those of each parent

17 Which developmental process is represented by the diagram below?

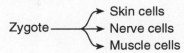

(1) fertilization
(2) differentiation
(3) evolution
(4) mutation

18 The diagram below represents human reproductive systems.

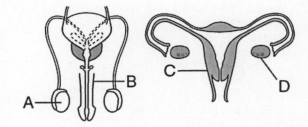

Which statement best describes part of the human reproductive process?

(1) Testosterone produced in A is transferred to D, where it influences embryonic development.
(2) Testosterone produced in D influences formation of sperm within B.
(3) Estrogen and progesterone influence the activity of C.
(4) Progesterone stimulates the division of the egg within C.

19 Which order of metabolic processes converts nutrients consumed by an organism into cell parts?

(1) digestion → absorption → circulation → diffusion → synthesis
(2) absorption → circulation → digestion → diffusion → synthesis
(3) digestion → synthesis → diffusion → circulation → absorption
(4) synthesis → absorption → digestion → diffusion → circulation

20 The diagram below represents a cell organelle involved in the transfer of energy from organic compounds.

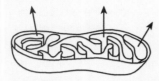

The arrows in the diagram could represent the release of

(1) ATP from a chloroplast carrying out photosynthesis
(2) oxygen from a mitochondrion carrying out photosynthesis
(3) glucose from a chloroplast carrying out respiration
(4) carbon dioxide from a mitochondrion carrying out respiration

21 Which process illustrates a feedback mechanism in plants?

(1) Chloroplasts take in more nitrogen, which increases the rate of photosynthesis.
(2) Chloroplasts release more oxygen in response to a decreased rate of photosynthesis.
(3) Guard cells change the size of leaf openings, regulating the exchange of gases.
(4) Guard cells release oxygen from the leaf at night.

22 Which human activity would have the most positive effect on the environment of an area?

(1) using fire to eliminate most plants in the area
(2) clearing the area to eliminate weed species
(3) protecting native flowers and grasses in the area
(4) introducing a foreign plant species to the area

23 What impact do the amounts of available energy, water, and oxygen have on an ecosystem?

(1) They act as limiting factors.
(2) They are used as nutrients.
(3) They recycle the residue of dead organisms.
(4) They control environmental temperature.

24 Many years ago, a volcanic eruption killed many plants and animals on an island. Today the island looks much as it did before the eruption. Which statement is the best possible explanation for this?

(1) Altered ecosystems regain stability through the evolution of new plant species.
(2) Destroyed environments can recover as a result of the process of ecological succession.
(3) Geographic barriers prevent the migration of animals to island habitats.
(4) Destroyed ecosystems always return to their original state.

25 The growth of a population is shown in the graph below.

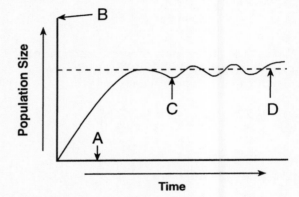

Which letter indicates the carrying capacity of the environment for this population?

(1) A   (3) C
(2) B   (4) D

26 When habitats are destroyed, there are usually fewer niches for animals and plants. This action would most likely *not* lead to a change in the amount of

(1) biodiversity
(2) competition
(3) interaction between species
(4) solar radiation reaching the area

27 Which set of terms best identifies the letters in the diagram below?

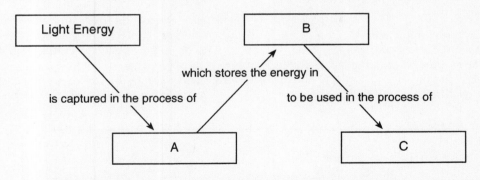

	A	B	C
(1)	photosynthesis	inorganic molecules	decomposition
(2)	respiration	organic molecules	digestion
(3)	photosynthesis	organic molecules	respiration
(4)	respiration	inorganic molecules	photosynthesis

28 The diagram below represents some energy transfers in an ecosystem.

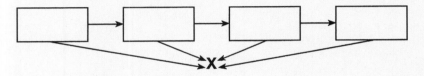

Which type of organism is most likely represented by letter *X*?

(1) decomposer
(2) autotroph
(3) producer
(4) herbivore

29 Some farmers currently grow genetically engineered crops. An argument *against* the use of this technology is that

(1) it increases crop production
(2) it produces insect-resistant plants
(3) its long-term effects on humans are still being investigated
(4) it always results in crops that do not taste good

30 The removal of nearly all the predators from an ecosystem would most likely result in

(1) an increase in the number of carnivore species
(2) a decrease in new predators migrating into the ecosystem
(3) a decrease in the size of decomposers
(4) an increase in the number of herbivores

Part B–1

Answer all questions in this part. [13]

Directions (31–43): For *each* statement or question, write on the separate answer sheet the *number* of the word or expression that, of those given, best completes the statement or answers the question.

31 The graph below shows the effect of moisture on the number of trees per acre of five tree species.

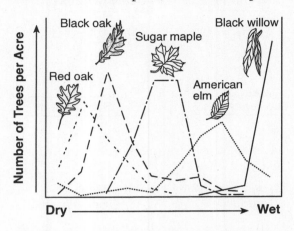

Which observation best represents information shown in the graph?

(1) All five species grow in the same habitat.
(2) The American elm grows in the widest range of moisture conditions.
(3) Red oaks can grow in wetter conditions than black willows.
(4) Sugar maples can grow anywhere black oaks can grow.

32 A science researcher is reviewing another scientist's experiment and conclusion. The reviewer would most likely consider the experiment *invalid* if

(1) the sample size produced a great deal of data
(2) other individuals are able to duplicate the results
(3) it contains conclusions not explained by the evidence given
(4) the hypothesis was not supported by the data obtained

33 The graph below shows how the human population has grown over the last several thousand years.

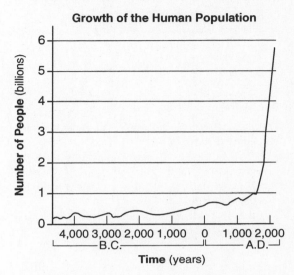

Which statement is a valid inference that can be made if the human population continues to grow at a rate similar to the rate shown between 1000 A.D. and 2000 A.D.?

(1) Future ecosystems will be stressed and many animal habitats may be destroyed.
(2) Global warming will decrease as a result of a lower demand for fossil fuels.
(3) One hundred years after all resources are used up, the human population will level off.
(4) All environmental problems can be solved without a reduction in the growth rate of the human population.

34 Cellular communication is illustrated in the diagram below.

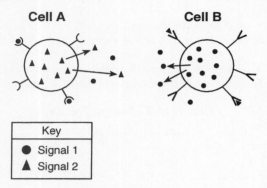

Cell A   **Cell B**

Key
● Signal 1
▲ Signal 2

Information can be sent from

(1) cell *A* to cell *B* because cell *B* is able to recognize signal 1
(2) cell *A* to cell *B* because cell *A* is able to recognize signal 2
(3) cell *B* to cell *A* because cell *A* is able to recognize signal 1
(4) cell *B* to cell *A* because cell *B* is able to recognize signal 2

35 The diagram below represents single-celled organism *A* dividing by mitosis to form cells *B* and *C*.

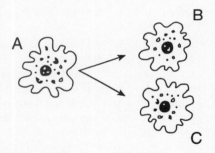

B

A

C

Cells *A*, *B*, and *C* all produced protein *X*. What can best be inferred from this observation?

(1) Protein *X* is found in all organisms.
(2) The gene for protein *X* is found in single-celled organisms, only.
(3) Cells *A*, *B*, and *C* ingested food containing the gene to produce protein *X*.
(4) The gene to produce protein *X* was passed from cell *A* to cells *B* and *C*.

Base your answers to questions 36 and 37 on the information in the diagram below and on your knowledge of biology.

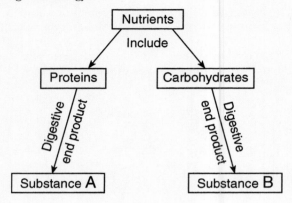

36 In an autotrophic organism, substance *B* functions as a

(1) source of energy   (3) vitamin
(2) hormone   (4) biotic resource

37 In a heterotrophic organism, substance *A* could be used directly for

(1) photosynthesis
(2) synthesis of enzymes
(3) a building block of starch
(4) a genetic code

38 The dichotomous key shown below can be used to identify birds *W, X, Y,* and *Z*.

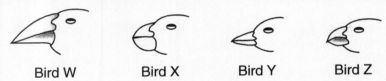

Bird W   Bird X   Bird Y   Bird Z

Dichotomous Key to Representative Birds

1. a. The beak is relatively long and slender...............................*Certhidea*
   b. The beak is relatively stout and heavy...................................go to 2
2. a. The bottom surface of the lower beak is flat and straight ......*Geospiza*
   b. The bottom surface of the lower beak is curved ....go to 3
3. a. The lower edge of the upper beak has a distinct bend ....*Camarhynchus*
   b. The lower edge of the upper beak is mostly flat ....*Platyspiza*

Bird *X* is most likely

(1) *Certhidea*
(2) *Geospiza*
(3) *Camarhynchus*
(4) *Platyspiza*

39 An experimental setup is shown in the diagram below.

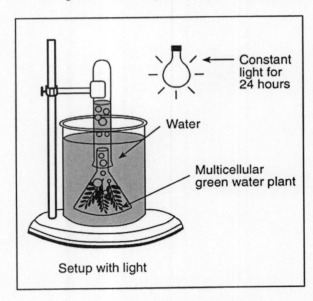
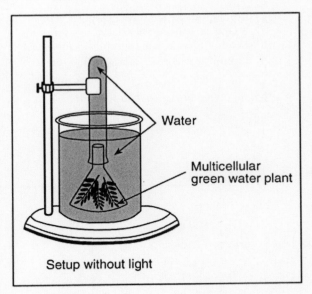

Setup with light   Setup without light

Which hypothesis would most likely be tested using this setup?

(1) Green water plants release a gas in the presence of light.
(2) Roots of water plants absorb minerals in the absence of light.
(3) Green plants need light for cell division.
(4) Plants grow best in the absence of light.

Base your answers to questions 40 through 42 on the passage below which describes an ecosystem in New York State and on your knowledge of biology.

The Pine Bush ecosystem near Albany, New York, is one of the last known habitats of the nearly extinct Karner Blue butterfly. The butterfly's larvae feed on the wild green plant, lupine. The larvae are in turn consumed by predatory wasps. The four groups below represent other organisms living in this ecosystem.

Group A	Group B	Group C	Group D
algae mosses ferns pine trees oak trees	rabbits tent caterpillars moths	hawks moles hognosed snakes toads	soil bacteria molds mushrooms

40 The Karner Blue larvae belong in which group?

(1) A
(2) B
(3) C
(4) D

41 Which food chain best represents information in the passage?

(1) lupine → Karner Blue larvae → wasps
(2) wasps → Karner Blue larvae → lupine
(3) Karner Blue larvae → lupine → wasps
(4) lupine → wasps → Karner Blue larvae

42 Which group contains decomposers?

(1) A
(2) B
(3) C
(4) D

43 A graph of the population growth of two different species is shown below.

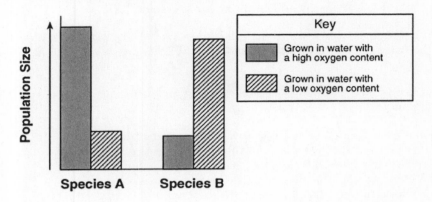

Which conclusion can be drawn from information in the graph?

(1) Oxygen concentration affects population sizes of different species in the same manner.
(2) Species *A* requires a high oxygen concentration for maximum population growth.
(3) Species *B* requires a high oxygen concentration to stimulate population growth.
(4) Low oxygen concentration does not limit the population size of either species observed.

Part B–2

Answer all questions in this part. [12]

Directions (44–55): For those questions that are followed by four choices, circle the *number* of the choice that best completes the statement or answers the question. For all other questions in this part, follow the directions given in the question.

Base your answers to questions 44 through 48 on the passage and data table below and on your knowledge of biology.

The amount of oxygen gas dissolved in water is important to the organisms that live in a river. The amount of dissolved oxygen varies with changes in both physical factors and biological processes. The temperature of the water is one physical factor affecting dissolved oxygen levels as shown in the data table below. The amount of dissolved oxygen is expressed in parts per million (ppm).

Dissolved Oxygen Levels at Various Temperatures

Water Temperature (°C)	Level of Dissolved Oxygen (ppm)
1	14
10	11
15	10
20	9
25	8
30	7

Directions (44–45): Using the information given, construct a line graph on the grid on page 13, following the directions below.

44 Mark an appropriate scale on each labeled axis. [1]

45 Plot the data for dissolved oxygen on the grid. Surround each point with a small circle and connect the points. [1]

Example:

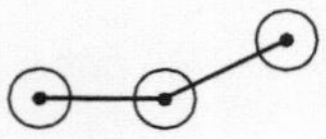

Dissolved Oxygen Levels at Various Temperatures

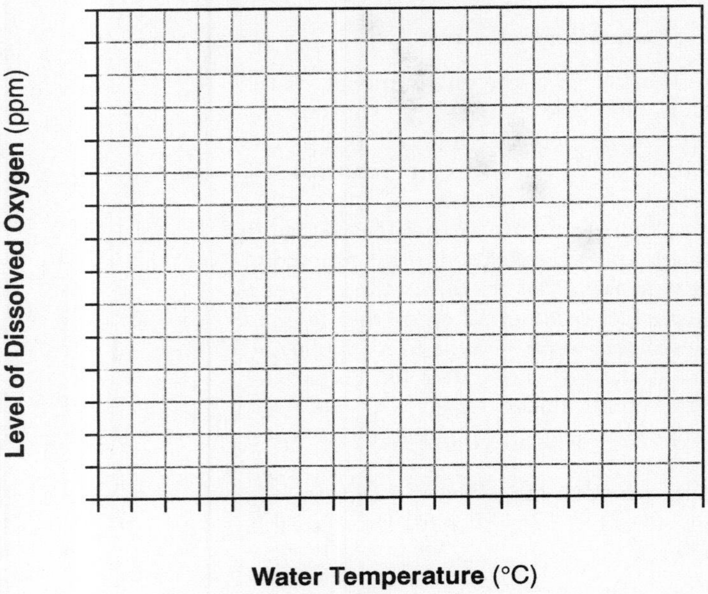

Level of Dissolved Oxygen (ppm)

Water Temperature (°C)

44 ☐

45 ☐

46 If the trend continues as shown in the data, what would the dissolved oxygen level most likely be if the temperature of the water was 35°C? [1]

_____ **ppm**

46 ☐

47 State the relationship between the level of dissolved oxygen and water temperature. [1]

_____

_____

_____

_____

47 ☐

48 Identify *one* physical or biological process taking place within the river, other than temperature change, that would affect the level of dissolved oxygen and state whether this process would increase or decrease the level of dissolved oxygen. [1]

_____

_____

_____

48 ☐

Base your answers to questions 49 through 51 on the passage below and on your knowledge of biology.

For Teacher Use Only

In Search of a Low-Allergy Peanut

Many people are allergic to substances in the environment. Of the many foods that contain allergens (allergy-inducing substances), peanuts cause some of the most severe reactions. Mildly allergic people may only get hives. Highly allergic people can go into a form of shock. Some people die each year from reactions to peanuts.

A group of scientists is attempting to produce peanuts that lack the allergy-inducing proteins by using traditional selective breeding methods. They are searching for varieties of peanuts that are free of the allergens. By crossing those varieties with popular commercial types, they hope to produce peanuts that will be less likely to cause allergic reactions and still taste good. So far, they have found one variety that has 80 percent less of one of three complex proteins linked to allergic reactions. Removing all three of these allergens may be impossible, but even removing one could help.

Other researchers are attempting to alter the genes that code for the three major allergens in peanuts. All of this research is seen as a possible long-term solution to peanut allergies.

49 Allergic reactions usually occur when the immune system produces

(1) antibiotics against usually harmless antigens

(2) antigens against usually harmless antibodies

(3) antibodies against usually harmless antigens

(4) enzymes against usually harmless antibodies

49 ☐

50 How does altering the DNA of a peanut affect the proteins in peanuts that cause allergic reactions?

(1) The altered DNA is used to synthesize changed forms of these proteins.

(2) The altered DNA leaves the nucleus and becomes part of the allergy-producing protein.

(3) The altered DNA is the code for the antibodies against the allergens.

(4) The altered DNA is used as an enzyme to break down the allergens in peanuts.

50 ☐

51 Explain how selective breeding is being used to try to produce commercial peanuts that will *not* cause allergic reactions in people. [1]

_____

_____

_____

_____

For Teacher Use Only

51 ☐

Base your answers to questions 52 through 55 on the diagram below and on your knowledge of biology. The arrows in the diagram represent biological processes.

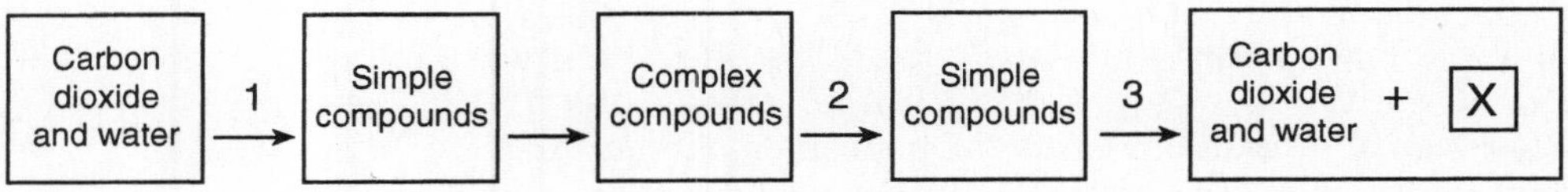

| Carbon dioxide and water | 1 → | Simple compounds | → | Complex compounds | 2 → | Simple compounds | 3 → | Carbon dioxide and water | + | X |

52 Identify *one* type of organism that carries out process 1. [1]

_____

52 ☐

53 Explain why process 2 is essential in humans. [1]

_____

_____

_____

53 ☐

54 Identify process 3. [1]

_____

54 ☐

55 Identify what letter *X* represents. [1]

_____

55 ☐

Part C

Answer all questions in this part.   [17]

Directions (56–62): Record your answers in the spaces provided in this examination booklet.

56 Growers of fruit trees have always had problems with insects. Insects can cause visible damage to fruits, making them less appealing to consumers. As a result of this damage, much of the fruit cannot be sold. Insecticides have been useful for controlling these insects, but, in recent years, some insecticides have been much less effective. In some cases, insecticides do nothing to stop the insect attacks.

Provide a biological explanation for this loss of effectiveness of the insecticides. In your answer, be sure to:

- identify the original event that resulted in the evolution of insecticide resistance in some insects   [1]
- explain why the percentage of resistant insects in the population has increased   [1]
- describe *one* alternative form of insect control, other than using a different insecticide, that fruit growers could use to protect their crops from insect attack   [1]

For Teacher Use Only

_____

_____

_____

_____

_____

_____

_____

_____

_____

_____

56

57 The concentration of salt in water affects the hatching of brine shrimp eggs. Brine shrimp eggs will develop and hatch at room temperature in glass containers of salt solution. Describe a controlled experiment using three experimental groups that could be used to determine the best concentration of salt solution in which to hatch brine shrimp eggs. Your answer must include at least:

- a description of how the control group and each of the three experimental groups will be different [1]
- *two* conditions that must be kept constant in the control group and the experimental groups [2]
- data that should be collected [1]
- *one* example of experimental results that would indicate the best concentration of salt solution in which to hatch brine shrimp eggs [1]

57

Base your answers to questions 58 and 59 on the statement and diagram below and on your knowledge of biology.

Women are advised to avoid consuming alcoholic beverages during pregnancy.

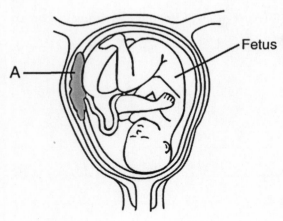

A

Fetus

58 Identify the structure labeled *A* and explain how the functioning of structure *A* is essential for the normal development of the fetus. [2]

Structure *A*: _____

_____

_____

_____

58 □

59 Explain why consumption of alcoholic beverages by a pregnant woman is likely to be more harmful to her fetus than to herself. [1]

_____

_____

_____

59 □

Base your answers to questions 60 and 61 on the statement below and on your knowledge of biology.

Some internal environmental factors may interfere with the ability of an enzyme to function efficiently.

60 Identify *two* internal environmental factors that directly influence the rate of enzyme action. [2]

_____

_____

60 ☐

61 Explain why changing the shape of an enzyme could affect the ability of the enzyme to function. [1]

_____

_____

_____

61 ☐

62 Deforestation is viewed as a problem in the world today. Describe a cause and an effect of deforestation and a way to lessen this effect. In your answer, be sure to:

• state *one* reason deforestation is occurring [1]
• state *one* environmental problem that results from widespread deforestation [1]
• state *one* way to lessen the effects of deforestation, other than planting trees [1]

_____

_____

_____

_____

_____

_____

_____

_____

62 ☐

Part D

Answer all questions in this part.   [13]

Directions (63–74): For those questions that are followed by four choices, circle the *number* of the choice that best completes the statement or answers the question. For all other questions in this part, follow the directions given in the question.

Base your answers to questions 63 through 65 on the Universal Genetic Code Chart on page 21 and on your knowledge of biology. Some DNA, RNA, and amino acid information from four similar sequences of four plant species is shown in the chart below.

For Teacher Use Only

63 Using the information given, fill in the missing mRNA base sequence for species *B* in the chart below.   [1]

64 Using the Universal Genetic Code Chart on page 21, fill in the missing amino acid sequence for species *C* in the chart below.   [1]

Species A	DNA base sequence	CCG	TGC	ATA	CAG	GTA
	mRNA base sequence	GGC	ACG	UAU	GUC	CAU
	Amino acid sequence	**GLY**	**THR**	**TYR**	**VAL**	**HIS**
Species B	DNA base sequence	TGC	TGC	ATA	CAG	GTA
	mRNA base sequence	___	___	___	___	___
	Amino acid sequence	**THR**	**THR**	**TYR**	**VAL**	**HIS**
Species C	DNA base sequence	CCG	TGC	ATA	CAG	GTT
	mRNA base sequence	GGC	ACG	UAU	GUC	CAA
	Amino acid sequence	___	___	___	___	___
Species D	DNA base sequence	CCT	TGT	ATG	CAC	GTC
	mRNA base sequence	GGA	ACA	UAC	GUG	CAG
	Amino acid sequence	**GLY**	**THR**	**TYR**	**VAL**	**GLN**

63 ☐

64 ☐

65 According to these amino acid sequences, which *two* plant species are the most closely related? Support your answer. [1]

Species _____ and _____

_____

_____

_____

65 ☐

Universal Genetic Code Chart
Messenger RNA Codons and the Amino Acids They Code For

		SECOND BASE				
		U	C	A	G	
FIRST BASE	**U**	UUU } PHE UUC UUA } LEU UUG	UCU UCC } SER UCA UCG	UAU } TYR UAC UAA } STOP UAG	UGU } CYS UGC UGA] STOP UGG] TRP	U C A G
	C	CUU CUC } LEU CUA CUG	CCU CCC } PRO CCA CCG	CAU } HIS CAC CAA } GLN CAG	CGU CGC } ARG CGA CGG	U C A G
	A	AUU AUC } ILE AUA AUG } MET or START	ACU ACC } THR ACA ACG	AAU } ASN AAC AAA } LYS AAG	AGU } SER AGC AGA } ARG AGG	U C A G
	G	GUU GUC } VAL GUA GUG	GCU GCC } ALA GCA GCG	GAU } ASP GAC GAA } GLU GAG	GGU GGC } GLY GGA GGG	U C A G

THIRD BASE

66 A student was comparing preserved specimens of three plant species, X, Y, and Z, in a classroom. Which statement is an example of an observation the student could have made and *not* an inference?

(1) The leaves produced by plant X are 4 cm across and 8 cm in length.

(2) Plant Y has large purple flowers that open at night.

(3) Plant X produces many seeds that are highly attractive to finches.

(4) The flowers of plant Z are poisonous to household pets.

66 □

Base your answers to questions 67 and 68 on the information below and on your knowledge of biology.

A student squeezes and releases a clothespin as often as possible for 2 minutes and then takes his pulse for 20 seconds. After a 2-minute rest, he repeats the procedure. This pattern is repeated one more time. The student's 20-second pulse counts were 23, 26, and 21.

67 Complete the "Pulse/Min" column in the data table below for all three trials as well as the average pulse rate per minute. [1]

Pulse Rate After Activity

Trial	20-Second Pulse Counts	Pulse/Min
1	23	
2	26	
3	21	
Average		

67 □

68 What additional data should the student have collected in order to determine the effect of squeezing a clothespin on his pulse rate? [1]

_____

_____

_____

68 □

Base your answers to questions 69 through 71 on the passage below and on your knowledge of biology.

> When Charles Darwin traveled to the Galapagos Islands, he observed 14 distinct varieties of finches on the islands. Darwin also observed that each finch variety ate a different type of food and lived in a slightly different habitat from the other finches. Darwin concluded that the finches all shared a common ancestor but had developed different beak structures.

69 The 14 varieties of finches are most likely the result of

   (1) absence of biodiversity

   (2) biological evolution

   (3) asexual reproduction

   (4) lack of competition

69 ☐

70 The second sentence best describes

   (1) an ecosystem

   (2) a food web

   (3) a niche

   (4) a predator/prey relationship

70 ☐

71 The different beak structures mentioned in the last sentence were most likely influenced by

   (1) selection for favorable variations

   (2) environmental conditions identical to those of the common ancestor

   (3) abnormal mitotic cell division

   (4) characteristics that are acquired during the bird's lifetime

71 ☐

72 The diagram below represents a laboratory setup used by a student during an investigation of diffusion.

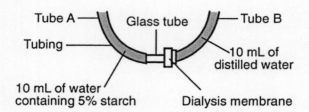

Which statement best explains why the liquid in tube A will rise over a period of time?

(1) The starch concentrations are equal on both sides of the membrane.

(2) The water will pass from a region of lower starch concentration to one of higher starch concentration.

(3) Water and starch volumes are the same in both tubes A and B.

(4) The fluids in both tubes A and B will change from a higher temperature to a lower temperature.

73 A red onion cell has undergone a change, as represented in the diagram below.

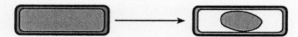

This change is most likely due to the cell being placed in

(1) distilled water

(2) light

(3) salt water

(4) darkness

For Teacher Use Only

74 A laboratory setup for a demonstration is represented in the diagram below.

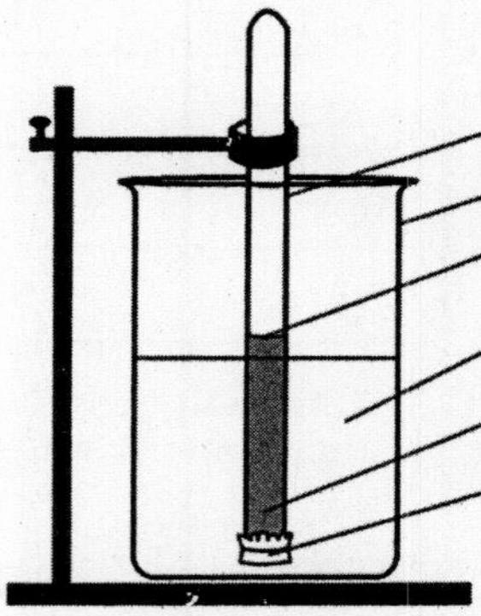
Test tube
Beaker
Meniscus
Water
Starch-water mixture
Dialysis membrane

Describe how an indicator can be used to determine if starch diffuses through the membrane into the beaker. In your answer, be sure to include:

- the procedure used [1]
- how to interpret the results [1]

74

LIVING ENVIRONMENT
JUNE 2006

Part	Maximum Score	Student's Score
A	30	
B–1	13	
B–2	12	
C	17	
D	13	
Total Raw Score (maximum Raw Score: 85)		
Final Score (from conversion chart)		

Raters' Initials

Rater 1 . . . .   Rater 2 . . . .

ANSWER SHEET

Student . . . .

Teacher . . . .

School . . . . Grade . . . .

Record your answers to Part A and Part B–1 on this answer sheet.

Part A

1	11	21 . . . .
2	12	22 . . . .
3	13	23 . . . .
4	14	24 . . . .
5	15	25 . . . .
6	16	26 . . . .
7	17	27 . . . .
8	18	28 . . . .
9	19	29 . . . .
10	20	30 . . . .

Part A Score

Part B–1

31	38 . . . .
32	39 . . . .
33	40 . . . .
34	41 . . . .
35	42 . . . .
36	43 . . . .
37	

Part B–1 Score

LIVING ENVIRONMENT
AUGUST 2006

Directions (1–30): For *each* statement or question, write on your separate answer sheet the *number* of the word or expression that best completes the statement or answers the question.

1 The levels of organization for structure and function in the human body from least complex to most complex are

(1) systems → organs → tissues → cells
(2) cells → organs → tissues → systems
(3) tissues → systems → cells → organs
(4) cells → tissues → organs → systems

2 Genes are inherited, but their expressions can be modified by the environment. This statement explains why

(1) some animals have dark fur only when the temperature is within a certain range
(2) offspring produced by means of sexual reproduction look exactly like their parents
(3) identical twins who grow up in different homes have the same characteristics
(4) animals can be cloned, but plants cannot

3 Meat tenderizer contains an enzyme that interacts with meat. If meat is coated with tenderizer and then placed in a refrigerator for a short time, how would the enzyme be affected?

(1) It would be broken down.
(2) Its activity would slow down.
(3) Its shape would change.
(4) It would no longer act as an enzyme.

4 Which row in the chart below contains correct information concerning synthesis?

Row	Building Blocks	Substance Synthesized Using the Building Blocks
(1)	glucose molecules	DNA
(2)	simple sugars	protein
(3)	amino acids	enzyme
(4)	molecular bases	starch

5 Molecule *X* moves across a cell membrane by diffusion. Which row in the chart below best indicates the relationship between the relative concentrations of molecule *X* and the use of ATP for diffusion?

Row	Movement of Molecule X	Use of ATP
(1)	high concentration → low concentration	used
(2)	high concentration → low concentration	not used
(3)	low concentration → high concentration	used
(4)	low concentration → high concentration	not used

6 Which statement best compares a multicellular organism to a single-celled organism?

(1) A multicellular organism has organ systems that interact to carry out life functions, while a single-celled organism carries out life functions without using organ systems.
(2) A single-celled organism carries out fewer life functions than each cell of a multicellular organism.
(3) A multicellular organism always obtains energy through a process that is different from that used by a single-celled organism.
(4) The cell of a single-celled organism is always much larger than an individual cell of a multicellular organism.

7 Which statement indicates that different parts of the genetic information are used in different kinds of cells, even in the same organism?

(1) The cells produced by a zygote usually have different genes.
(2) As an embryo develops, various tissues and organs are produced.
(3) Replicated chromosomes separate during gamete formation.
(4) Offspring have a combination of genes from both parents.

8 Three structures are represented in the diagram below.

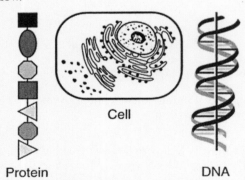

Protein   Cell   DNA

What is the relationship between these three structures?

(1) DNA is made up of proteins that are synthesized in the cell.

(2) Protein is composed of DNA that is stored in the cell.

(3) DNA controls the production of protein in the cell.

(4) The cell is composed only of DNA and protein.

9 In a group of mushrooms exposed to a poisonous chemical, only a few of the mushrooms survived. The best explanation for the resistance of the surviving mushrooms is that the resistance

(1) was transmitted to the mushrooms from the poisonous chemical

(2) resulted from the presence of mutations in the mushrooms

(3) was transferred through the food web to the mushrooms

(4) developed in response to the poisonous chemical

10 Which statement correctly describes the genetic makeup of the sperm cells produced by a human male?

(1) Each cell has pairs of chromosomes and the cells are usually genetically identical.

(2) Each cell has pairs of chromosomes and the cells are usually genetically different.

(3) Each cell has half the normal number of chromosomes and the cells are usually genetically identical.

(4) Each cell has half the normal number of chromosomes and the cells are usually genetically different.

11 In an environment that undergoes frequent change, species that reproduce sexually may have an advantage over species that reproduce asexually because the sexually reproducing species produce

(1) more offspring in each generation

(2) identical offspring

(3) offspring with more variety

(4) new species of offspring in each generation

12 Mutations that occur in skin or lung cells have little effect on the evolution of a species because mutations in these cells

(1) usually lead to the death of the organism

(2) cannot be passed on to offspring

(3) are usually beneficial to the organism

(4) lead to more serious mutations in offspring

13 The teeth of carnivores are pointed and are good for puncturing and ripping flesh. The teeth of herbivores are flat and are good for grinding and chewing. Which statement best explains these observations?

(1) Herbivores have evolved from carnivores.

(2) Carnivores have evolved from herbivores.

(3) The two types of teeth most likely evolved as a result of natural selection.

(4) The two types of teeth most likely evolved as a result of the needs of an organism.

14 What would most likely happen if most of the bacteria and fungi were removed from an ecosystem?

(1) Nutrients resulting from decomposition would be reduced.

(2) Energy provided for autotrophic nutrition would be reduced.

(3) The rate of mutations in plants would increase.

(4) Soil fertility would increase.

15 A certain bacterial colony originated from the division of a single bacterial cell. Each cell in this colony will most likely

(1) express adaptations unlike those of the other cells

(2) replicate different numbers of genes

(3) have a resistance to different antibiotics

(4) synthesize the same proteins and enzymes

16 Removal of one ovary from a human female would most likely

  (1) affect the production of eggs
  (2) make fertilization impossible
  (3) make carrying a fetus impossible
  (4) decrease her ability to provide essential nutrients to an embryo

17 Which substance usually passes in the greatest amount through the placenta from the blood of the fetus to the blood of the mother?

  (1) oxygen
  (2) carbon dioxide
  (3) amino acids
  (4) glucose

18 An enzyme known as rubisco enables plants to use large amounts of carbon dioxide. This enzyme is most likely active in the

  (1) nucleus
  (2) vacuoles
  (3) mitochondria
  (4) chloroplasts

19 Starch molecules present in a maple tree are made from materials that originally entered the tree from the external environment as

  (1) enzymes
  (2) simple sugars
  (3) amino acids
  (4) inorganic compounds

20 Which change in a sample of pond water could indicate that heterotrophic microbes were active?

  (1) increase in ozone level
  (2) increase in glucose level
  (3) decrease in oxygen level
  (4) decrease in carbon dioxide level

21 Some human white blood cells help destroy pathogenic bacteria by

  (1) causing mutations in the bacteria
  (2) engulfing and digesting the bacteria
  (3) producing toxins that compete with bacterial toxins
  (4) inserting part of their DNA into the bacterial cells

22 Four students each drew an illustration to show the flow of energy in a field ecosystem. Which illustration is *most* accurate?

(1)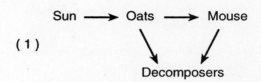

(2)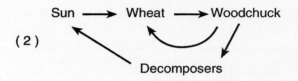

(3)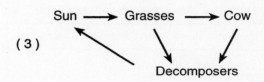

(4)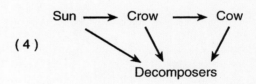

23 As succession proceeds from a shrub community to a forest community, the shrub community modifies its environment, eventually making it

  (1) more favorable for itself and less favorable for the forest community
  (2) more favorable for itself and more favorable for the forest community
  (3) less favorable for itself and more favorable for the forest community
  (4) less favorable for itself and less favorable for the forest community

24 The diagram below represents a series of events in the development of a bird.

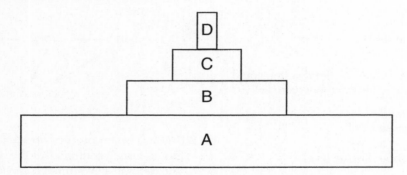

Zygote

Which series of terms best represents the sequence of processes shown?

(1) meiosis → growth → differentiation
(2) meiosis → differentiation → growth

(3) mitosis → meiosis → differentiation
(4) mitosis → differentiation → growth

25 Bacteria that are removed from the human intestine are genetically engineered to feed on organic pollutants in the environment and convert them into harmless inorganic compounds. Which row in the table below best represents the most likely negative and positive effects of this technology on the ecosystem?

Row	Negative Effect	Positive Effect
(1)	Inorganic compounds interfere with cycles in the environment.	Human bacteria are added to the environment.
(2)	Engineered bacteria may out-compete native bacteria.	The organic pollutants are removed.
(3)	Only some of the pollutants are removed.	Bacteria will make more organic pollutants.
(4)	The bacteria will cause diseases in humans.	The inorganic compounds are buried in the soil.

26 An energy pyramid is represented below.

D

C

B

A

How much energy would be available to the organisms in level *C*?

(1) all of the energy in level *A*, plus the energy in level *B*
(2) all of the energy in level *A*, minus the energy in level *B*
(3) a percentage of the energy contained in level *B*
(4) a percentage of the energy synthesized in level *B* and level *D*

27 Which graph illustrates changes that indicate a state of dynamic equilibrium in a mosquito population?

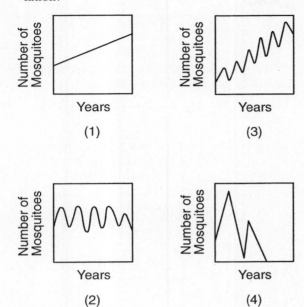

28 Which condition is necessary for enzymes and hormones to function properly in the human body?

(1) These chemicals must have a specific shape.
(2) These chemicals must be able to replicate.
(3) Body temperature must be above 40°C.
(4) Body pH must be above 10.

29 Four environmental factors are listed below.

    A. energy
    B. water
    C. oxygen
    D. minerals

Which factors limit environmental carrying capacity in a land ecosystem?

(1) A, only
(2) B, C, and D, only
(3) A, C, and D, only
(4) A, B, C, and D

30 Which human activity would have the *least* negative impact on the quality of the environment?

(1) adding animal wastes to rivers
(2) cutting down tropical rain forests for plywood
(3) using species-specific sex attractants to trap and kill insect pests
(4) releasing chemicals into the groundwater

Part B–1

Answer all questions in this part. [10]

Directions (31–40): For *each* statement or question, write on the separate answer sheet the *number* of the word or expression that best completes the statement or answers the question.

Base your answers to questions 31 through 34 on the diagram below and on your knowledge of biology. The diagram represents a single-celled organism, such as an ameba, undergoing the changes shown.

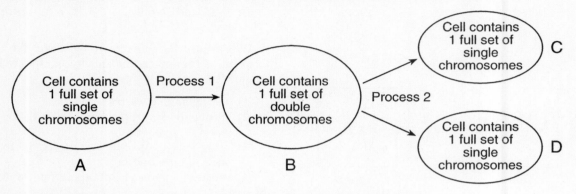

31 As a result of these processes, the single-celled organism accomplishes

    (1) gamete production   (3) sexual reproduction
    (2) energy production   (4) asexual reproduction

32 Process 1 is known as

    (1) replication   (3) differentiation
    (2) meiosis   (4) digestion

33 Process 1 and process 2 are directly involved in

    (1) meiotic cell division   (3) fertilization
    (2) mitotic cell division   (4) recombination

34 The genetic content of *C* is usually identical to the genetic content of

    (1) *B* but not *D*   (3) *D* but not *A*
    (2) both *B* and *D*   (4) both *A* and *D*

Base your answers to questions 35 through 37 on the diagram below that shows some evolutionary pathways. Each letter represents a different species.

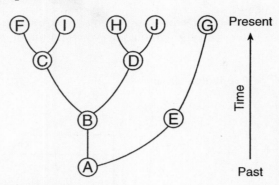

35 Which two organisms are most closely related?

(1) *F* and *I*   (3) *A* and *G*
(2) *F* and *H*   (4) *G* and *J*

36 The most recent ancestor of organisms *D* and *F* is

(1) *A*   (3) *C*
(2) *B*   (4) *I*

37 If *A* represents a simple multicellular heterotrophic organism, *B* would most likely represent

(1) a single-celled photosynthetic organism
(2) an autotrophic mammal
(3) a complex multicellular virus
(4) another type of simple multicellular heterotroph

38 A scientist studied iguanas inhabiting a chain of small ocean islands. He discovered two species that live in different habitats and display different behaviors. His observations are listed in the table below.

Observations of Two Species of Iguanas

Species A	Species B
spends most of its time in the ocean	spends most of its time on land
is rarely found more than 10 meters from shore	is found many meters inland from shore
eats algae	eats cactus and other land plants

Which statement best describes these two species of iguanas?

(1) Both species evolved through the process of ecological succession.
(2) Each species occupies a different niche.
(3) The two species can interbreed.
(4) Species *A* is a scavenger and species *B* is a carnivore.

Base your answers to questions 39 and 40 on the information and graph below and on your knowledge of biology.

A population of paramecia (single-celled aquatic organisms) was grown in a 200-mL beaker of water containing some smaller single-celled organisms. Population growth of the organisms for 28 hours is shown in the graph below.

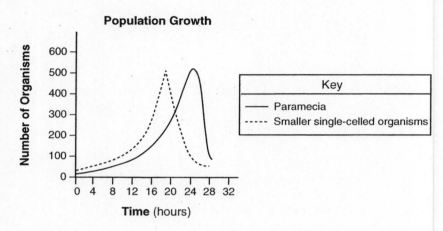

Population Growth

39 Which factor most likely accounts for the change in the paramecium population from 8 to 20 hours?

(1) an increase in the nitrogen content of water
(2) an increase in wastes produced
(3) an increase in available food
(4) an increase in water pH

40 One likely explanation for the change in the paramecium population from 26 hours to 28 hours is that the

(1) carrying capacity of the beaker was exceeded
(2) rate of reproduction increased
(3) time allowed for growth was not sufficient
(4) oxygen level was too high

Part B–2

Answer all questions in this part.   [15]

Directions (41–55): For those questions that are followed by four choices, circle the *number* of the choice that best completes the statement or answers the question. For all other questions in this part, follow the directions given in the questions and record your answers in the spaces provided.

Base your answers to questions 41 through 45 on the passage below and on your knowledge of biology.

Better Rice

The production of new types of food crops will help raise the quantity of food grown by farmers. Research papers released by the National Academy of Sciences announced the development of two new superior varieties of rice—one produced by selective breeding and the other by biotechnology.

One variety of rice, called Nerica (New Rice for Africa), is already helping farmers in Africa. Nerica combines the hardiness and weed resistance of rare African rice varieties with the productivity and faster maturity of common Asian varieties.

Another variety, called Stress-Tolerant Rice, was produced by inserting a pair of bacterial genes into rice plants for the production of trehalose (a sugar). Trehalose helps plants maintain healthy cell membranes, proteins, and enzymes during environmental stress. The resulting plants survive drought, low temperatures, salty soils, and other stresses better than standard rice varieties.

41 Why is the production of new varieties of food crops necessary?

(1) Essential food crops are rapidly becoming extinct.

(2) Technology for producing fresh water for agriculture has improved.

(3) Burning fossil fuels has decreased agricultural areas.

(4) World population continues to increase.

41 ☐

42 Which substance from bacteria was most likely inserted into rice plants in the development of the trehalose-producing rice?

(1) sugar

(2) enzymes

(3) DNA

(4) trehalose

42 ☐

43 Nerica was most likely produced by

    (1) crossing a variety of African rice with a variety of Asian rice

    (2) cloning genes for hardiness and weed resistance from Asian rice

    (3) using Asian rice to compete with rare African varieties

    (4) inserting genes for productivity and faster maturity into Asian rice

43 □

44 Which strain of rice was produced as a result of genetic engineering? Support your answer. [1]

_____

_____

_____

44 □

45 State *one* reason that further testing must be done before rice plants that produce trehalose are approved for human consumption. [1]

_____

_____

_____

45 □

Base your answers to questions 46 through 49 on the information and data table below and on your knowledge of biology.

A number of bean seeds planted at the same time produced plants that were later divided into two groups, *A* and *B*. Each plant in group *A* was treated with the same concentration of gibberellic acid (a plant hormone). The plants in group *B* were not treated with gibberellic acid. All other growth conditions were kept constant. The height of each plant was measured on 5 consecutive days, and the average height of each group was recorded in the data table below.

Data Table

	Average Plant Height (cm)				
	Day 1	**Day 2**	**Day 3**	**Day 4**	**Day 5**
Group A	5	7	10	13	15
Group B	5	6	6.5	7	7.5

Directions (46–48): Using the information in the data table, construct a line graph on the grid on the next page, following the directions below.

46 Mark an appropriate scale on the axis labeled "Average Plant Height (cm)." [1]

47 Plot the data for the average height of the plants in group *A*. Surround each point with a small circle and connect the points. [1]

Example:

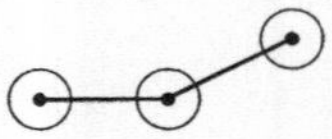

48 Plot the data for the average height of the plants in group *B*. Surround each point with a small triangle and connect the points. [1]

Example:

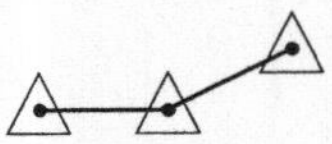

For Teacher Use Only

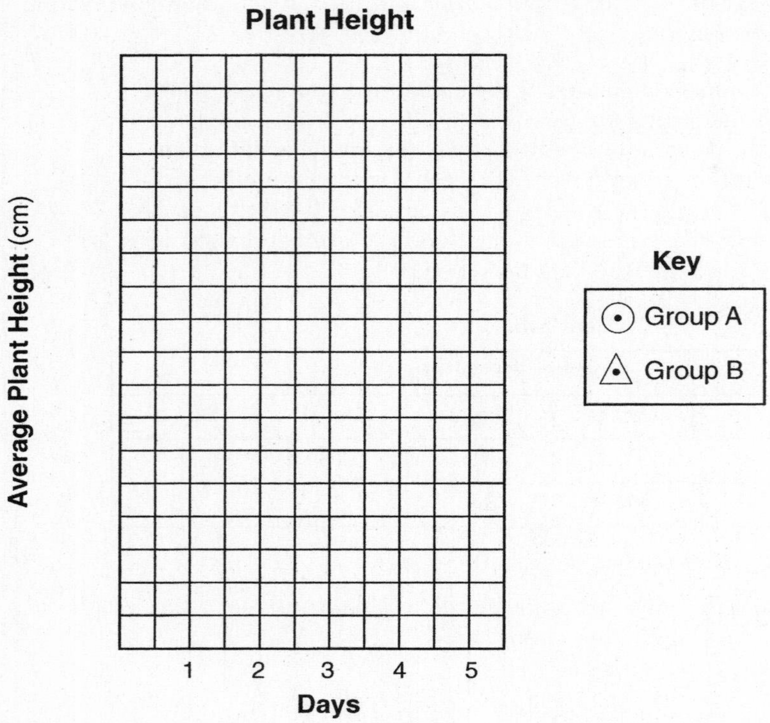
Plant Height
Average Plant Height (cm)
1
2
3
4
5
Days
Key
Group A
Group B

46

47

48

49 State a valid conclusion that can be drawn concerning the effect of gibberellic acid on bean plant growth. [1]

____

____

____

49

Base your answers to questions 50 through 55 on the data table below and on your knowledge of biology. The table contains information about glucose production in a species of plant that lives in the water of a salt marsh.

Temperature (°C)	Glucose Production (mg/hr)
10	5
20	10
30	15
40	5

50 Which terms describe temperature in this investigation?

(1) abiotic factor and independent variable

(2) abiotic factor and dependent variable

(3) biotic factor and independent variable

(4) biotic factor and dependent variable

51 What evidence from the data table shows that a salt-marsh plant is sensitive to its environment? [1]

_____

_____

52 At which temperature would the plants most likely use the greatest amount of carbon dioxide?

(1) 10°C

(2) 20°C

(3) 30°C

(4) 40°C

50 ☐

51 ☐

52 ☐

53 How much oxygen will plants that live in water at 10°C most likely produce?

(1) twice the amount of oxygen produced at 20°C

(2) the same amount of oxygen produced at 40°C

(3) the most oxygen produced at any temperature

(4) more oxygen than is produced at 30°C

53 ☐

54 State *one* possible reason for the change in glucose production when the temperature was increased from 30°C to 40°C. [1]

_____

_____

54 ☐

55 Which level of the energy pyramid below would contain the plant species of this salt marsh?

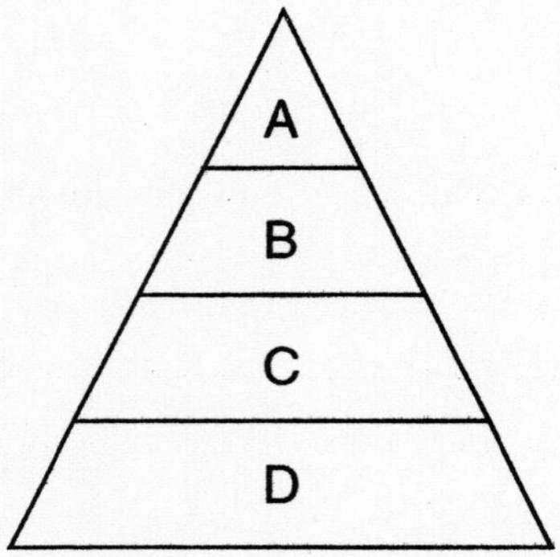

(1) *A*

(2) *B*

(3) *C*

(4) *D*

55 ☐

Part C

Answer all questions in this part. [17]

Directions (56–59): Record your answers in the spaces provided in this examination booklet.

56 A scientist wants to determine the best conditions for hatching brine shrimp eggs. In a laboratory, brine shrimp hatch at room temperature in glass containers of salt water. The concentration of salt in the water is known to affect how many brine shrimp eggs will hatch.

Design an experiment to determine which of three saltwater concentrations (2%, 4%, or 6%) is best for hatching brine shrimp eggs. In your experimental design, be sure to:

- state how many containers to use in the experiment, and describe what would be added to each container in addition to the eggs   [1]
- state *two* factors that must be kept constant in all the containers   [1]
- state what data must be collected during this experiment   [1]
- state *one* way to organize the data so that they will be easy to analyze   [1]
- describe a result that would indicate the best salt solution for hatching brine shrimp eggs [1]

For Teacher Use Only

56

57 Not all diseases are caused by pathogenic organisms. Other factors, such as inheritance, poor nutrition, and toxic substances, may also cause disease.

Describe a disease or disorder that can occur as a result of one of these other factors. Your answer must include at least:

- the name of the disease   [1]
- *one* specific factor that causes this disease   [1]
- *one* major effect of this disease on the body, other than death   [1]
- *one* way this disease can be prevented, treated, or cured   [1]

_____

_____

_____

_____

_____

_____

_____

_____

_____

_____

_____

_____

_____

_____

_____

_____

_____

_____

_____

_____

57

58 Describe how *two* of the cell structures listed below interact to help maintain a balanced internal environment in a cell.

       mitochondrion
       ribosome
       cell membrane
       nucleus
       vacuole

In your answer be sure to:

- select *two* of these structures, write their names, and state *one* function of each [2]
- describe how each structure you selected contributes to the functioning of the other [2]

58

59 Currently, Americans rely heavily on the burning of fossil fuels as sources of energy. As a result of increased demand for energy sources, there is a continuing effort to find alternatives to burning fossil fuels.

Discuss fossil fuels and alternative energy sources. In your answer be sure to:

- state *one disadvantage* of burning fossil fuels for energy [1]
- identify *one* energy source that is an alternative to using fossil fuels [1]
- state *one* advantage of using this alternative energy source [1]
- state *one disadvantage* of using this alternative energy source [1]

59

Part D

Answer all questions in this part. [13]

Directions (60–69): For those questions that are followed by four choices, circle the *number* of the choice that best completes the statement or answers the question. For all other questions in this part, follow the directions given in the questions and record your answers in the spaces provided.

Base your answers to questions 60 and 61 on the information and diagram below and on your knowledge of biology. The diagram illustrates an investigation carried out in a laboratory activity on diffusion. The beaker and the artificial cell also contain water.

For Teacher Use Only

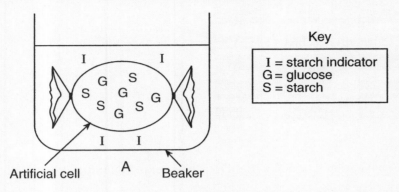

60 Predict what would happen over time by showing the location of molecules *I*, *G*, and *S* in diagram *B* below. [3]

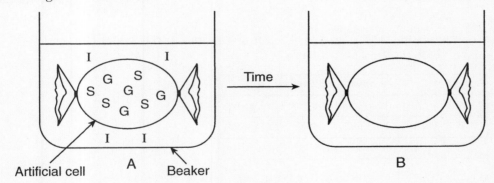

60 ☐

61 State what is observed when there is a positive test for starch using the starch indicator. [1]

_____

_____

_____

61 ☐

Base your answers to questions 62 through 64 on the information and diagram below and on your knowledge of biology.

The DNA of three different species of birds was analyzed to help determine if there is an evolutionary relationship between these species. The diagram shows the results of this analysis.

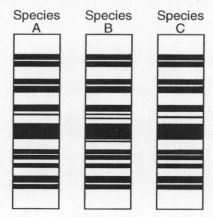

Species A   Species B   Species C

62 Identify the technique normally used to separate the DNA fragments to produce the patterns shown in the diagram.   [1]

_____

62 ☐

63 The chart below contains amino acid sequences for part of a protein that is found in the feathers on each of these three species of birds.

Species	Amino Acid Sequence
A	Arg-Leu-Glu-Gly-His-His-Pro-Lys-Arg
B	Arg-Gly-Glu-Gly-His-His-Pro-Lys-Arg
C	Arg-Leu-Glu-Gly-His-His-Pro-Lys-Arg

State *one* way this data supports the inference that these three bird species may be closely related.   [1]

_____

_____

63 ☐

64 State *one* type of additional information that could be used to determine if these three species are closely related.   [1]

_____

_____

64 ☐

Base your answers to questions 65 through 67 on the information and graph below and on your knowledge of biology.

Pulse-rate data were collected from some students during their lunch time for the lab activity, *Making Connections*. The data are represented in the histogram below.

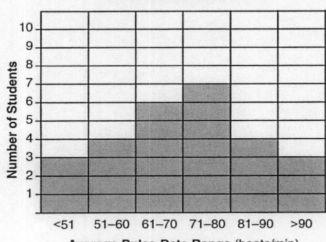

Student Pulse-Rate Data

Number of Students

Average Pulse-Rate Range (beats/min)

65 The histogram includes data from a total of how many students?

(1) 6

(2) 7

(3) 10

(4) 27

65 ☐

66 Describe *one* way in which a pulse rate below 45 would disrupt homeostasis in an individual whose average resting pulse rate falls in the range of 71–80. [1]

_____

_____

66 ☐

67 State *one* way the data would most likely be different if the pulse rates were collected immediately after exercising instead of during lunch. [1]

_____

_____

67 ☐

Base your answers to questions 68 and 69 on the finch diversity chart below, which contains information concerning the finches found on the Galapagos Islands.

Finch Diversity

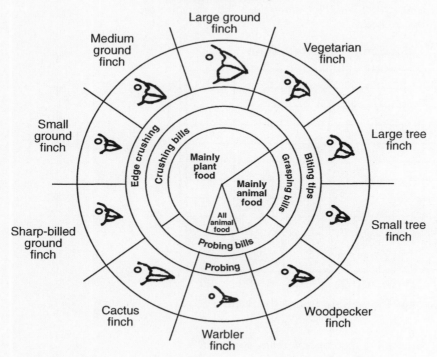

68 Identify *one* bird that would most likely compete for food with the large tree finch. Support your answer. [1]

_____

_____

69 Identify *one* trait, other than beak characteristics, that would contribute to the survival of a finch species and state *one* way this trait contributes to the success of this species. [2]

_____

_____

_____

LIVING ENVIRONMENT
AUGUST 2006

ANSWER SHEET

Student . . . .

Teacher . . . .

School . . . . Grade . . . .

Part	Maximum Score	Student's Score
A	30	
B–1	10	
B–2	15	
C	17	
D	13	

Total Raw Score (maximum Raw Score: 85)

Final Score (from conversion chart)

Raters' Initials

Rater 1 . . . . Rater 2 . . . .

Record your answers to Part A and Part B–1 on this answer sheet.

Part A

1	11	21 . . . .
2	12	22 . . . .
3	13	23 . . . .
4	14	24 . . . .
5	15	25 . . . .
6	16	26 . . . .
7	17	27 . . . .
8	18	28 . . . .
9	19	29 . . . .
10	20	30 . . . .

Part A Score

Part B–1

31	36 . . . .
32	37 . . . .
33	38 . . . .
34	39 . . . .
35	40 . . . .

Part B–1 Score

LIVING ENVIRONMENT
JANUARY 2007

Part A

Answer all questions in this part. [30]

Directions (1–30): For *each* statement or question, write on your separate answer sheet the *number* of the word or expression that, of those given, best completes the statement or answers the question.

1 When brown tree snakes were accidentally introduced onto the island of Guam, they had no natural predators. These snakes sought out and ate many of the eggs of insect-eating birds. What probably occurred following the introduction of the brown tree snakes?

(1) The bird population increased.
(2) The insect population increased.
(3) The bird population began to seek a new food source.
(4) The insect population began to seek a new food source.

2 What will most likely happen to wastes containing nitrogen produced as a result of the breakdown of amino acids within liver cells of a mammal?

(1) They will be digested by enzymes in the stomach.
(2) They will be removed by the excretory system.
(3) They will be destroyed by specialized blood cells.
(4) They will be absorbed by mitochondria in nearby cells.

3 Which sequence represents the correct order of organization in complex organisms?

(1) tissues → organs → systems → cells
(2) organs → tissues → systems → cells
(3) systems → organs → cells → tissues
(4) cells → tissues → organs → systems

4 Which organelle is correctly paired with its specific function?

(1) cell membrane—storage of hereditary information
(2) chloroplast—transport of materials
(3) ribosome—synthesis of proteins
(4) vacuole—production of ATP

5 Homeostasis in unicellular organisms depends on the proper functioning of

(1) organelles   (3) guard cells
(2) insulin   (4) antibodies

6 Which statement best explains the change shown in the diagram below?

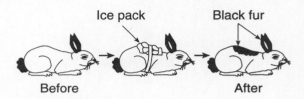

(1) Gene expression in an organism can be modified by interactions with the environment.
(2) Certain rabbits produce mutations that affect genes in specific areas of the body.
(3) Sorting and recombination of genes can be influenced by very cold temperatures.
(4) Molecular arrangement in existing proteins can be altered by environmental factors.

7 After a rabbit population reaches the carrying capacity of its habitat, the population of rabbits will most likely

(1) decrease, only
(2) increase, only
(3) alternately increase and decrease
(4) remain unchanged

8 Variation in the offspring of sexually reproducing organisms is the direct result of

(1) sorting and recombining of genes
(2) replication and cloning
(3) the need to adapt and maintain homeostasis
(4) overproduction of offspring and competition

9 An error in genetic information present in a body cell of a mammal would most likely produce

(1) rapid evolution of the organism in which the cell is found

(2) a mutation that will affect the synthesis of a certain protein in the cell

(3) an adaptation that will be passed on to other types of cells

(4) increased variation in the type of organelles present in the cell

10 Which process is illustrated in the diagram below?

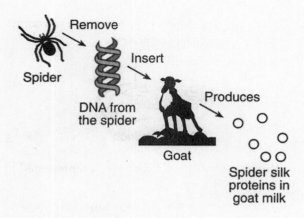

(1) chromatography
(2) direct harvesting
(3) meiosis
(4) genetic engineering

11 Which statement is most closely related to the modern theory of evolution?

(1) Characteristics that are acquired during life are passed to offspring by sexual reproduction.

(2) Evolution is the result of mutations and recombination, only.

(3) Organisms best adapted to a changed environment are more likely to reproduce and pass their genes to offspring.

(4) Asexual reproduction increases the survival of species.

12 In 1993, there were only 30 panthers in Florida. They were all closely related and many had reproductive problems. To avoid extinction and restore health to the population, biologists introduced 8 female panthers from Texas. Today, there are more than 80 panthers in Florida and most individuals have healthy reproductive systems. The success of this program was most likely due to the fact that the introduced females

(1) produced more reproductive cells than the male panthers in Texas

(2) solved the reproductive problems of the species by asexual methods

(3) increased the genetic variability of the panther population in Florida

(4) mated only with panthers from Texas

13 The *least* genetic variation will probably be found in the offspring of organisms that reproduce using

(1) mitosis to produce a larger population
(2) meiosis to produce gametes
(3) fusion of eggs and sperm to produce zygotes
(4) internal fertilization to produce an embryo

14 Woolly mammoths became extinct thousands of years ago, while other species of mammals that existed at that time still exist today. These other species of mammals most likely exist today because, unlike the mammoths, they

(1) produced offspring that all had identical inheritable characteristics

(2) did not face a struggle for survival

(3) learned to migrate to new environments

(4) had certain inheritable traits that enabled them to survive

15 Marine sponges contain a biological catalyst that blocks a certain step in the separation of chromosomes. Which cellular process would be directly affected by this catalyst?

(1) mitosis   (3) respiration
(2) diffusion   (4) photosynthesis

16 A tree produces only seedless oranges. A small branch cut from this tree produces roots after it is planted in soil. When mature, this new tree will most likely produce

(1) oranges with seeds, only
(2) oranges without seeds, only
(3) a majority of oranges with seeds and only a few oranges without seeds
(4) oranges and other kinds of fruit

17 The diagram below represents a human reproductive system.

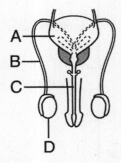

Meiosis occurs within structure

(1) A   (3) C
(2) B   (4) D

18 Which statement about embryonic organ development in humans is accurate?

(1) It is affected primarily by the eating habits and general health of the father.
(2) It may be affected by the diet and general health of the mother.
(3) It will not be affected by any medication taken by the mother in the second month of pregnancy.
(4) It is not affected by conditions outside the embryo.

19 Experiments revealed the following information about a certain molecule:

— It can be broken down into amino acids.
— It can break down proteins into amino acids.
— It is found in high concentrations in the small intestine of humans.

This molecule is most likely

(1) an enzyme
(2) an inorganic compound
(3) a hormone
(4) an antigen

20 The diagram below represents a structure involved in cellular respiration.

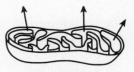

Mitochondrion

The release of which substance is represented by the arrows?

(1) glucose   (3) carbon dioxide
(2) oxygen   (4) DNA

21 Scientists have genetically altered a common virus so that it can destroy the most lethal type of brain tumor without harming the healthy tissue nearby. This technology is used for all of the following *except*

(1) treating the disease
(2) curing the disease
(3) controlling the disease
(4) diagnosing the disease

22 Many species of plants interact with harmless underground fungi. The fungi enable the plants to absorb certain essential minerals and the plants provide the fungi with carbohydrates and other nutrients. This describes an interaction between a

(1) parasite and its host
(2) predator and its prey
(3) scavenger and a decomposer
(4) producer and a consumer

23 In an ocean, the growth and survival of seaweed, small fish, and sharks depends on abiotic factors such as

(1) sunlight, temperature, and minerals
(2) sunlight, pH, and type of seaweed
(3) number of decomposers, carbon dioxide, and nitrogen
(4) number of herbivores, carbon, and food

24 A basketball player develops speed and power as a result of practice. This athletic ability will *not* be passed on to her offspring because

(1) muscle cells do not carry genetic information
(2) mutations that occur in body cells are not inherited
(3) gametes do not carry complete sets of genetic information
(4) base sequences in DNA are not affected by this activity

25 Carbon dioxide containing carbon-14 is introduced into a balanced aquarium ecosystem. After several weeks, carbon-14 will most likely be present in

(1) the plants, only
(2) the animals, only
(3) both the plants and animals
(4) neither the plants nor animals

26 Which situation is a result of human activities?

(1) decay of leaves in a forest adds to soil fertility
(2) acid rain in an area kills fish in a lake
(3) ecological succession following volcanic activity reestablishes an ecosystem
(4) natural selection on an island changes gene frequencies

27 Which human activity will most likely have a *negative* effect on global stability?

(1) decreasing water pollution levels
(2) increasing recycling programs
(3) decreasing habitat destruction
(4) increasing world population growth

28 Which process helps reduce global warming?

(1) decay
(2) industrialization
(3) photosynthesis
(4) burning

29 Which phrase belongs in box *X* of the flowchart below?

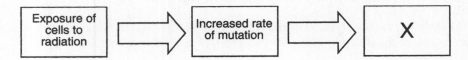

(1) Increased chance of cancer
(2) Increase in the production of functional gametes
(3) Decrease in genetic variability of offspring
(4) Decreased number of altered genes

30 The data in the table below indicate the presence of specific reproductive hormones in blood samples taken from three individuals. An *X* in the hormone column indicates a positive lab test for the appropriate levels necessary for normal reproductive functioning in that individual.

Data Table

Individuals	Hormones Present		
	Testosterone	Progesterone	Estrogen
1		X	X
2			X
3	X		

Which processes could occur in individual 3?

(1) production of sperm, only
(2) production of sperm and production of eggs
(3) production of eggs and embryonic development
(4) production of eggs, only

Part B–1

Answer all questions in this part. [10]

Directions (31–40): For *each* statement or question, write on the separate answer sheet the *number* of the word or expression that, of those given, best completes the statement or answers the question.

31 While viewing a specimen under high power of a compound light microscope, a student noticed that the specimen was out of focus. Which part of the microscope should the student turn to obtain a clearer image under high power?

(1) eyepiece   (3) fine adjustment
(2) coarse adjustment (4) nosepiece

32 The diagram below shows the relative concentration of molecules inside and outside of a cell.

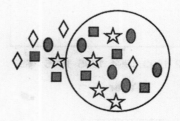 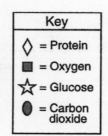

Key	
◇	= Protein
■	= Oxygen
☆	= Glucose
⬭	= Carbon dioxide

Which statement best describes the general direction of diffusion across the membrane of this cell?

(1) Glucose would diffuse into the cell.
(2) Protein would diffuse out of the cell.
(3) Carbon dioxide would diffuse out of the cell.
(4) Oxygen would diffuse into the cell.

33 Which statement most accurately describes scientific inquiry?

(1) It ignores information from other sources.
(2) It does not allow scientists to judge the reliability of their sources.
(3) It should never involve ethical decisions about the application of scientific knowledge.
(4) It may lead to explanations that combine data with what people already know about their surroundings.

34 The diagram below represents a pyramid of energy that includes both producers and consumers.

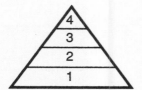

The greatest amount of available energy is found at level

(1) 1   (3) 3
(2) 2   (4) 4

35 How much water should be removed from the graduated cylinder shown below to leave 5 milliliters of water in the cylinder?

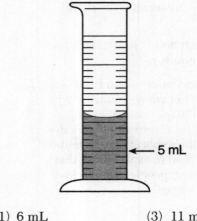

← 5 mL

(1) 6 mL
(2) 7 mL
(3) 11 mL
(4) 12 mL

36 The diagram below represents a food web.

A Meadow Environment

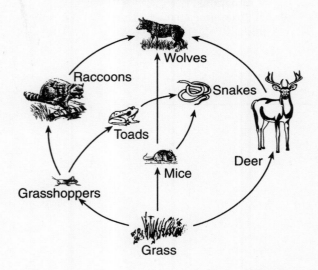

Two of the herbivores represented in this food web are

(1) toads and snakes
(2) deer and mice
(3) wolves and raccoons
(4) grasshoppers and toads

37 Compounds containing phosphorus that are dumped into the environment can upset ecosystems because phosphorus acts as a fertilizer. The graph below shows measurements of phosphorus concentrations taken during the month of June at two sites from 1991 to 1997.

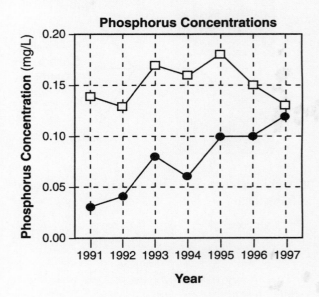

Phosphorus Concentrations

Key

Site 1 —□—
Site 2 —●—

Which statement represents a valid inference based on information in the graph?

(1) There was no decrease in the amount of compounds containing phosphorus dumped at site 2 during the period from 1991 to 1997.
(2) Pollution controls may have been put into operation at site 1 in 1995.
(3) There was most likely no vegetation present near site 2 from 1993 to 1994.
(4) There was a greater variation in phosphorous concentration at site 1 than there was at site 2.

Base your answers to questions 38 and 39 on the diagram below and on your knowledge of biology. The diagram illustrates a process by which energy is released in organisms.

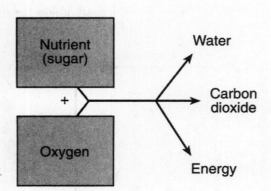

38 Cells usually transfer the energy that is released directly to

(1) glucose   (3) oxygen
(2) ATP   (4) enzymes

39 The energy released in this process was originally present in

(1) sunlight and then transferred to sugar
(2) sunlight and then transferred to oxygen
(3) the oxygen and then transferred to sugar
(4) the sugar and then transferred to oxygen

Base your answer to question 40 on the diagram below and on your knowledge of biology.

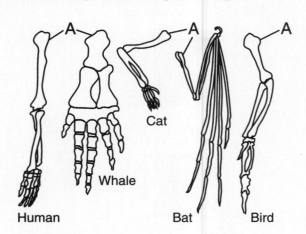

40 The similarities of the bones labeled A provide evidence that

(1) the organisms may have evolved from a common ancestor
(2) all species have one kind of bone structure
(3) the cells of the bones contain the same type of mutations
(4) all structural characteristics are the same in animals

Part B–2

Answer all questions in this part. [15]

Directions (41–55): For those questions that are followed by four choices, circle the *number* of the choice that best completes the statement or answers the question. For all other questions in this part, follow the directions given in the question.

Base your answers to questions 41 and 42 on the information below and on your knowledge of biology.

A biology student was given three unlabeled jars of pond water from the same source, each containing a different type of mobile unicellular organism: euglena, ameba, and paramecium. The only information the student has is that the ameba and paramecium are both heterotrophs and the euglena can be either heterotrophic or autotrophic, depending on its environment.

41 State *one* way the euglena's two methods of nutrition provide a survival advantage the other unicellular organisms do *not* have. [1]

_____

_____

_____   41 ☐

42 Which procedure and resulting observation would help identify the jar that contains the euglena?

(1) Expose only one side of each jar to light. After 24 hours, only in the jar containing euglena will most of organisms be seen on the darker side of the jar.

(2) Expose all sides of each jar to light. After 48 hours, the jar with the highest dissolved carbon dioxide content will contain the euglena.

(3) Over a period of one week, determine the method of reproduction used by each type of organism. If mitotic cell division is observed, the jar will contain euglena.

(4) Prepare a wet-mount slide of specimens from each jar and observe each slide with a compound light microscope. Only the euglena will have chloroplasts.   42 ☐

_____

Base your answers to questions 43 through 46 on the passage below and on your knowledge of biology.

Decline of the Salmon Population

Salmon are fish that hatch in a river and swim to the ocean where their body mass increases. When mature, they return to the river where they were hatched and swim up stream to reproduce and die. When there are large populations of salmon, the return of nutrients to the river ecosystem can be huge. It is estimated that during salmon runs in the Pacific Northwest in the 1800s, 500 million pounds of salmon returned to reproduce and die each year. Research estimates that in the Columbia River alone, salmon contributed hundreds of thousands of pounds of nitrogen and phosphorus compounds to the local ecosystem each year. Over the past 100 years, commercial ocean fishing has removed up to two-thirds of the salmon before they reach the river each year.

43 Identify the process that releases the nutrients from the bodies of the dead salmon, making the nutrients available for other organisms in the ecosystem. [1]

_____

43 ☐

44 Identify *one* organism, other than the salmon, that would be present in or near the river that would most likely be part of a food web in the river ecosystem. [1]

_____

44 ☐

45 Identify *two* nutrients that are returned to the ecosystem when the salmon die. [1]

_____

_____

45 ☐

46 State *one* impact, other than reducing the salmon population, that commercial ocean fishing has on the river ecosystem. [1]

_____

_____

_____

46 ☐

Base your answers to questions 47 through 51 on the information and data table below and on your knowledge of biology.

Biologists investigated the effect of the presence of aluminum ions on root tips of a variety of wheat. They removed 2-mm sections of the tips of roots. Half of the root tips were placed in a nutrient solution with aluminum ions, while the other half were placed in an identical nutrient solution without aluminum ions. The length of the root tips, in millimeters, was measured every hour for seven hours. The results are shown in the data table below.

Data Table

Time (hr)	Length of Root Tips in Solution With Aluminum Ions (mm)	Length of Root Tips in Solution Without Aluminum Ions (mm)
0	2.0	2.0
1	2.1	2.2
2	2.2	2.4
3	2.4	2.8
4	2.6	2.9
5	2.7	3.2
6	2.8	3.7
7	2.8	3.9

Directions (47–49): Using the information in the data table, construct a line graph on the grid *on the next page*, following the directions below.

47 Mark an appropriate scale on each labeled axis.   [1]

48 Plot the data for root tips in the solution with aluminum ions on the grid. Surround each point with a small circle and connect the points.   [1]

Example:

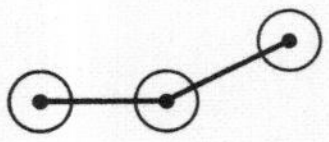

49 Plot the data for root tips in the solution without aluminum ions on the grid. Surround each point with a small triangle and connect the points.   [1]

Example:

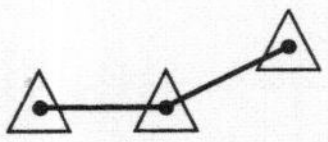

Growth of Wheat Root Tips

Length of Root Tips (mm)

Time (hr)

⊙ = Root tips in solution with aluminum ions

△ = Root tips in solution without aluminum ions

50 The aluminum ions most likely affected

(1) photosynthetic rate

(2) the union of gametes

(3) mitotic cell division

(4) starch absorption from the soil

51 Describe the effect of aluminum ions on the growth of the root tips of wheat. [1]

_____

_____

For Teacher Use Only

47 ☐

48 ☐

49 ☐

50 ☐

51 ☐

Base your answers to questions 52 and 53 on the information below and on your knowledge of biology.

A pond in the Adirondack Mountains of New York State was once a fishing spot visited by many people. It was several acres in size, and fishermen in boats were a common sight. Over time, the pond has become smaller in area and depth. Places where there was once open water are now covered by grasses and shrubs. Around the edges of the pond there are cattails and other wetland plants.

52 Identify the ecological process responsible for the changes to this pond. [1]

_____

53 Predict what will most likely happen to this pond area over the next hundred years if this process continues. [1]

_____

_____

Base your answers to questions 54 and 55 on the statement below and on your knowledge of biology.

The use of nuclear fuel can have positive and negative effects on an ecosystem.

54 State *one* positive effect on an ecosystem of using nuclear fuel to generate electricity. [1]

_____

_____

55 State *one* negative effect on an ecosystem of using nuclear fuel to generate electricity. [1]

_____

_____

For Teacher Use Only

52 ☐

53 ☐

54 ☐

55 ☐

Part C

Answer all questions in this part. [17]

Directions (56–65): Record your answers in the spaces provided in this examination booklet.

Base your answers to questions 56 and 57 on the statement below and on your knowledge of biology.

Selective breeding has been used to improve the racing ability of horses.

56 Define selective breeding and state how it would be used to improve the racing ability of horses. [2]

_____

_____

_____

_____

57 State *one disadvantage* of selective breeding. [1]

_____

_____

58 State *one* specific way the removal of trees from an area has had a *negative* impact on the environment. [1]

_____

_____

For Teacher Use Only

56 ☐

57 ☐

58 ☐

Base your answers to questions 59 through 61 on the information below and on your knowledge of biology.

It has been discovered that plants utilize chemical signals for communication. Some of these chemicals are released from leaves, fruits, and flowers and play various roles in plant development, survival, and gene expression. For example, bean plant leaves infested with spider mites release chemicals that result in an increase in the resistance to spider mites in uninfested leaves on the same plant and the expression of self-defense genes in uninfested bean plants nearby.

Plants can also communicate with insects. For example, corn, cotton, and tobacco under attack by caterpillars release chemical signals that simultaneously attract parasitic wasps to destroy the caterpillars and discourage moths from laying their eggs on the plants.

59 Identify the specialized structures in the cell membrane that are involved in communication. [1]

_____

59 ☐

60 Explain why chemicals released from one plant species may not cause a response in a different plant species. [1]

_____

_____

_____

60 ☐

61 State *two* advantages of relying on chemicals released by plants rather than using man-made chemicals for insect control. [2]

_____

_____

_____

_____

_____

61 ☐

Base your answers to questions 62 through 64 on the information below and on your knowledge of biology.

Cells of the immune system and the endocrine system of the human body contribute to the maintenance of homeostasis. The methods and materials these two systems use as they carry out this critical function are different.

62 State *two* ways cells of the immune system fight disease.   [2]

_____

_____

62 ☐

63 Identify the substance produced by the cells of all the endocrine glands that helps maintain homeostasis.   [1]

_____

63 ☐

64 Identify *one* specific product of one of the endocrine glands and state how it aids in the maintenance of homeostasis.   [1]

_____

_____

_____

64 ☐

65 A certain plant has white flower petals and it usually grows in soil that is slightly basic. Sometimes the plant produces flowers with red petals. A company that sells the plant wants to know if soil pH affects the color of the petals in this plant. Design a controlled experiment to determine if soil pH affects petal color. In your experimental design be sure to:

- state the hypothesis to be tested in the experiment [1]
- state *one* way the control group will be treated differently from the experimental group [1]
- identify *two* factors that must be kept the same in both the control group and the experimental group [1]
- identify the dependent variable in the experiment [1]
- state *one* result of the experiment that would support the hypothesis [1]

65 ☐

Part D

Answer all questions in this part. [13]

Directions (66–76): For those questions that are followed by four choices, circle the *number* of the choice that best completes the statement or answers the question. For all other questions in this part, follow the directions given in the question.

Base your answers to questions 66 and 67 on the information and data table below and on your knowledge of biology

For Teacher Use Only

Two students collected data on their pulse rates while performing different activities. Their average results are shown in the data table below.

Data Table

Activity	Average Pulse Rate (beats/min)
sitting quietly	70
walking	98
running	120

66 State the relationship between activity and pulse rate. [1]

____

____ 66

67 State *one* way that this investigation could be improved. [1]

____

____ 67

Base your answers to questions 68 through 71 on the information below and on your knowledge of biology.

To demonstrate techniques used in DNA analysis, a student was given two paper strip samples of DNA. The two DNA samples are shown below.

Sample 1: ATTCCGGTAATCCCGTAATGCCGGATAATACTCCGGTAATATC

Sample 2: ATTCCGGTAATCCCGTAATGCCGGATAATACTCCGGTAATATC

The student cut between the C and G in each of the shaded CCGG sequences in sample 1 and between the As in each of the shaded TAAT sequences in sample 2. Both sets of fragments were then arranged on a paper model of a gel.

68 The action of what kind of molecules was being demonstrated when the DNA samples were cut? [1]

_____

68 ☐

69 Identify the technique that was being demonstrated when the fragments were arranged on the gel model. [1]

_____

69 ☐

70 The results of this type of DNA analysis are often used to help determine

(1) the number of DNA molecules in an organism

(2) if two species are closely related

(3) the number of mRNA molecules in DNA

(4) if two organisms contain carbohydrate molecules

70 ☐

71 State *one* way that the arrangement of the two samples on the gel model would differ. [1]

_____

_____

71 ☐

Base your answers to questions 72 and 73 on the information below and on your knowledge of biology.

In birds, the ability to crush and eat seeds is related to the size, shape, and thickness of the beak. Birds with larger, thicker beaks are better adapted to crush and open seeds that are larger.

One species of bird found in the Galapagos Islands is the medium ground finch. It is easier for most of the medium ground finches to pick up and crack open smaller seeds rather than larger seeds. When food is scarce, some of the birds have been observed eating larger seeds.

72 Describe *one* change in beak characteristics that would most likely occur in the medium ground finch population after many generations when an environmental change results in a permanent shortage of small seeds. [1]

_____

_____

73 Explain this long-term change in beak characteristics using the concepts of:

- competition [1]
- survival of the fittest [1]
- inheritance [1]

_____

_____

_____

_____

_____

_____

72 ☐

73 ☐

Base your answers to questions 74 and 75 on the information and diagram below and on your knowledge of biology. The diagram represents some cells on a microscope slide before and after a substance was added to the slide.

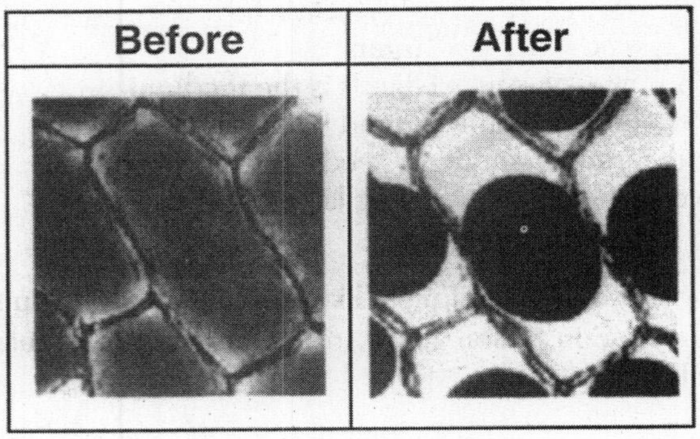

Before	After

74 Identify a substance that was most likely added to the slide to cause the change observed. [1]

_____

74 ☐

75 Describe a procedure that could be used to add this substance to the cells on the slide without removing the coverslip. [1]

_____

_____

_____

75 ☐

76 In the *Diffusion Through a Membrane* lab, the model cell membranes allowed certain substances to pass through based on which characteristic of the diffusing substance?

(1) size

(2) shape

(3) color

(4) temperature

76 ☐

LIVING ENVIRONMENT
JANUARY 2007

Part	Maximum Score	Student's Score
A	30	
B–1	10	
B–2	15	
C	17	
D	13	
Total Raw Score (maximum Raw Score: 85)		
Final Score (from conversion chart)		

Raters' Initials

Rater 1 . . . . Rater 2 . . . .

ANSWER SHEET

Student . . . .

Teacher . . . .

School . . . . Grade . . . .

Record your answers to Part A and Part B–1 on this answer sheet.

Part A

1	11	21 . . . .
2	12	22 . . . .
3	13	23 . . . .
4	14	24 . . . .
5	15	25 . . . .
6	16	26 . . . .
7	17	27 . . . .
8	18	28 . . . .
9	19	29 . . . .
10	20	30 . . . .

Part A Score

Part B–1

31	36 . . . .
32	37 . . . .
33	38 . . . .
34	39 . . . .
35	40 . . . .

Part B–1 Score

LIVING ENVIRONMENT

JUNE 2007

Part A

Answer all questions in this part. [30]

Directions (1–30): For *each* statement or question, write on your separate answer sheet the *number* of the word or expression that, of those given, best completes the statement or answers the question.

1 Which statement describes a role of fungi in an ecosystem?

(1) They transfer energy to decaying matter.
(2) They release oxygen into the ecosystem.
(3) They recycle chemicals from dead organisms.
(4) They synthesize organic nutrients from inorganic substances.

2 Which diagram best represents the levels of organization in the human body?

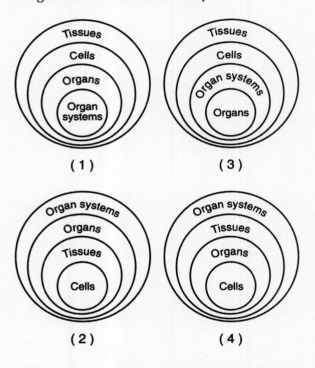

(1)　　　　　　(3)

(2)　　　　　　(4)

3 Which situation indicates that a disruption of homeostasis has taken place?

(1) the presence of hormones that keep the blood sugar level steady
(2) the maintenance of a constant body temperature
(3) cell division that is involved in normal growth
(4) a rapid rise in the number of red blood cells

4 A protein on the surface of HIV can attach to proteins on the surface of healthy human cells. These attachment sites on the surface of the cells are known as

(1) receptor molecules　　(3) molecular bases
(2) genetic codes　　　　(4) inorganic catalysts

5 Contractile vacuoles maintain water balance by pumping excess water out of some single-celled pond organisms. In humans, the kidney is chiefly involved in maintaining water balance. These facts best illustrate that

(1) tissues, organs, and organ systems work together to maintain homeostasis in all living things
(2) interference with nerve signals disrupts cellular communication and homeostasis within organisms
(3) a disruption in a body system may disrupt the homeostasis of a single-celled organism
(4) structures found in single-celled organisms can act in a manner similar to tissues and organs in multicellular organisms

6 Which statement best explains the observation that clones produced from the same organism may *not* be identical?

(1) Events in meiosis result in variation.
(2) Gene expression can be influenced by the environment.
(3) Differentiated cells have different genes.
(4) Half the genetic information in offspring comes from each parent.

7 A change in the base subunit sequence during DNA replication can result in

(1) variation within an organism
(2) rapid evolution of an organism
(3) synthesis of antigens to protect the cell
(4) recombination of genes within the cell

8 The diagram below represents a yeast cell that is in the process of budding, a form of asexual reproduction.

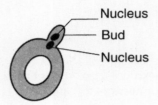

Nucleus
Bud
Nucleus

Which statement describes the outcome of this process?

(1) The bud will develop into a zygote.
(2) The two cells that result will each contain half the species number of chromosomes.
(3) The two cells that result will have identical DNA.
(4) The bud will start to divide by the process of meiotic cell division.

9 Two proteins in the same cell perform different functions. This is because the two proteins are composed of

(1) chains folded the same way and the same sequence of simple sugars
(2) chains folded the same way and the same sequence of amino acids
(3) chains folded differently and a different sequence of simple sugars
(4) chains folded differently and a different sequence of amino acids

10 Even though each body cell in an individual contains the same DNA, the functions of muscle cells and liver cells are *not* the same because

(1) mutations usually occur in genes when muscle cells divide
(2) liver tissue develops before muscle tissue
(3) liver cells produce more oxygen than muscle cells
(4) liver cells use different genes than muscle cells

11 The flounder is a species of fish that can live in very cold water. The fish produces an "antifreeze" protein that prevents ice crystals from forming in its blood. The DNA for this protein has been identified. An enzyme is used to cut and remove this section of flounder DNA that is then spliced into the DNA of a strawberry plant. As a result, the plant can now produce a protein that makes it more resistant to the damaging effects of frost. This process is known as

(1) sorting of genes
(2) genetic engineering
(3) recombination of chromosomes
(4) mutation by deletion of genetic material

12 Some human body structures are represented in the diagram below.

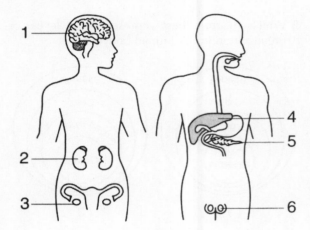

In which structures would the occurrence of mutations have the greatest effect on human evolution?

(1) 1 and 3   (3) 3 and 6
(2) 2 and 5   (4) 4 and 6

13 A single pair of goldfish in an aquarium produced a large number of offspring. These
offspring showed variations in body shape and
coloration. The most likely explanation for these
variations is that the

   (1) offspring were adapting to different environments
   (2) offspring were produced from different
   combinations of genes
   (3) parent fish had not been exposed to muta-
   genic agents
   (4) parent fish had not reproduced sexually

14 A certain species has little genetic variation. The
rapid extinction of this species would most likely
result from the effect of

   (1) successful cloning
   (2) gene manipulation
   (3) environmental change
   (4) genetic recombination

15 Which two structures of a frog would most
likely have the same chromosome number?

   (1) skin cell and fertilized egg cell
   (2) zygote and sperm cell
   (3) kidney cell and egg cell
   (4) liver cell and sperm cell

16 Tissues develop from a zygote as a direct result
of the processes of

   (1) fertilization and meiosis
   (2) fertilization and differentiation
   (3) mitosis and meiosis
   (4) mitosis and differentiation

17 The human female reproductive system is
adapted for

   (1) production of zygotes in ovaries
   (2) external fertilization of gametes
   (3) production of milk for a developing embryo
   (4) transport of oxygen through a placenta to a
   fetus

18 The letters in the diagram below represent
structures in a human female.

Estrogen and progesterone increase the chance
for successful fetal development by regulating
activities within structure

   (1) A   (3) C
   (2) B   (4) D

19 Which part of a molecule provides energy for life
processes?

   (1) carbon atoms   (3) chemical bonds
   (2) oxygen atoms   (4) inorganic nitrogen

20 Energy from organic molecules can be stored in
ATP molecules as a direct result of the process of

   (1) cellular respiration
   (2) cellular reproduction
   (3) diffusion
   (4) digestion

21 Which statement best describes how a vaccination can help protect the body against disease?

   (1) Vaccines directly kill the pathogen that
   causes the disease.
   (2) Vaccines act as a medicine that cures the
   disease.
   (3) Vaccines cause the production of specific
   molecules that will react with and destroy
   certain microbes.
   (4) Vaccines contain white blood cells that
   engulf harmful germs and prevent them
   from spreading throughout the body.

22 The diagram below represents four different species of wild birds. Each species has feet with different structural adaptations.

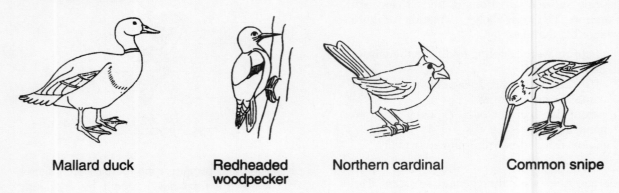

Mallard duck   Redheaded woodpecker   Northern cardinal   Common snipe

The development of these adaptations can best be explained by the concept of

(1) inheritance of resistance to diseases that affect all these species
(2) inheritance of characteristics acquired after the birds hatched from the egg
(3) natural selection
(4) selective breeding

23 The diagram below represents a nucleus containing the normal chromosome number for a species.

Which diagram bests illustrates the normal formation of a cell that contains all of the genetic information needed for growth, development, and future reproduction of this species?

(1)   (3)

(2)   (4)

24 The diagram below represents events associated with a biochemical process that occurs in some organisms.

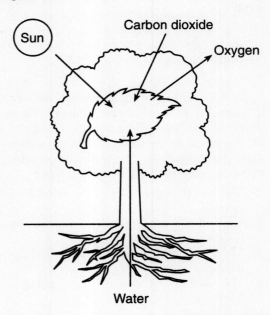

Which statement concerning this process is correct?

(1) The process represented is respiration and the primary source of energy for the process is the Sun.

(2) The process represented is photosynthesis and the primary source of energy for the process is the Sun.

(3) This process converts energy in organic compounds into solar energy which is released into the atmosphere.

(4) This process uses solar energy to convert oxygen into carbon dioxide.

25 In the transfer of energy from the Sun to ecosystems, which molecule is one of the first to store this energy?

(1) protein   (3) DNA
(2) fat   (4) glucose

26 The diagram below represents two molecules that can interact with each other to cause a biochemical process to occur in a cell.

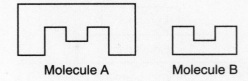

Molecules A and B most likely represent

(1) a protein and a chromosome
(2) a receptor and a hormone
(3) a carbohydrate and an amino acid
(4) an antibody and a hormone

27 The graph below represents the amount of available energy at successive nutrition levels in a particular food web.

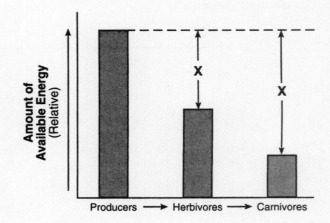

The Xs in the diagram represent the amount of energy that was most likely

(1) changed into inorganic compounds
(2) retained indefinitely by the herbivores
(3) recycled back to the producers
(4) lost as heat to the environment

28 The diagram below represents an energy pyramid constructed from data collected from an aquatic ecosystem.

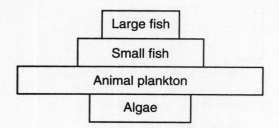

Which statement best describes this ecosystem?

(1) The ecosystem is most likely unstable.
(2) Long-term stability of this ecosystem will continue.
(3) The herbivore populations will continue to increase in size for many years.
(4) The producer organisms outnumber the consumer organisms.

29 In order to reduce consumption of nonrenewable resources, humans could

(1) burn coal to heat houses instead of using oil
(2) heat household water with solar radiation
(3) increase industrialization
(4) use a natural-gas grill to barbecue instead of using charcoal

30 In 1859, a small colony of 24 rabbits was brought to Australia. By 1928 it was estimated that there were 500 million rabbits in a 1-million square mile section of Australia. Which statement describes a condition that probably contributed to the increase in the rabbit population?

(1) The rabbits were affected by many limiting factors.
(2) The rabbits reproduced by asexual reproduction.
(3) The rabbits were unable to adapt to the environment.
(4) The rabbits had no natural predators in Australia.

Part B–1

Answer all questions in this part. [12]

Directions (31–42): For *each* statement or question, write on the separate answer sheet the *number* of the word or expression that, of those given, best completes the statement or answers the question.

31 What is the approximate length of the earthworm shown in the diagram below?

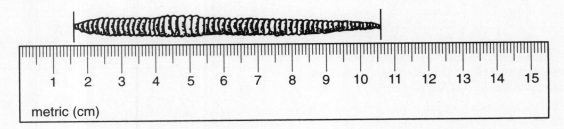

(1) 9 mm

(2) 90 mm

(3) 10.6 cm

(4) 106 cm

32 Information concerning the diet of crocodiles of different sizes is contained in the table below.

Percentage of Crocodiles of Different Lengths and Their Food Sources

Food Source	Group A 0.3–0.5 Meter	Group B 2.5–3.9 Meters	Group C 4.5–5.0 Meters
mammals	0	18	65
reptiles	0	17	48
fish	0	62	38
birds	0	17	0
snails	0	25	0
shellfish	0	5	0
spiders	20	0	0
frogs	35	0	0
insects	100	2	0

Which statement is *not* a valid conclusion based on the data?

(1) Overharvesting of fish could have a negative impact on group *C*.

(2) The smaller the crocodile is, the larger the prey.

(3) Group *B* has no preference between reptiles and birds.

(4) Spraying insecticides would have the most direct impact on group *A*.

33 The diagram below represents an incomplete section of a DNA molecule. The boxes represent unidentified bases.

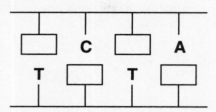

When the boxes are filled in, the total number of bases represented by the letter *A* (both inside and outside the boxes) will be

(1) 1   (3) 3
(2) 2   (4) 4

34 The graph below shows the growth of a population of bacteria over a period of 80 hours.

Growth of a Population of Bacteria

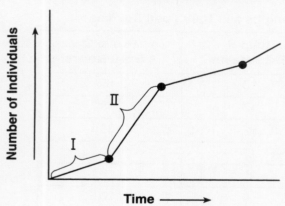

Which statement best describes section II of the graph?

(1) The population has reached the carrying capacity of the environment.
(2) The rate of reproduction is slower than in section I.
(3) The population is greater than the carrying capacity of the environment.
(4) The rate of reproduction exceeds the death rate.

35 A classification system is shown in the table below.

Classification	Examples
Kingdom — animal	△, ○, ◻, ☆, ▭, ◇, ℰ, ▽
Phylum — chordata	△, ◻, ℰ, ☆, ▭
Genus — *Felis*	▭, ℰ
Species — *domestica*	▭

This classification scheme indicates that ▭ is most closely related to

☆ △ ◻ ℰ
(1) (2) (3) (4)

36 Information concerning nests built in the same tree by two different bird species over a ten-year period is shown in the table below.

Distance of Nest Above Ground (m)	Total Number of Nests Built by Two Different Species	
	A	B
less than 1	5	0
1–5	10	0
6–10	5	0
over 10	0	20

What inference best describes these two bird species?

(1) They most likely do not compete for nesting sites because they occupy different niches.
(2) They do not compete for nesting sites because they have the same reproductive behavior.
(3) They compete for nesting sites because they build the same type of nest.
(4) They compete for nesting sites because they nest in the same tree at the same time.

37 The diagram below shows the effect of spraying a pesticide on a population of insects over three generations.

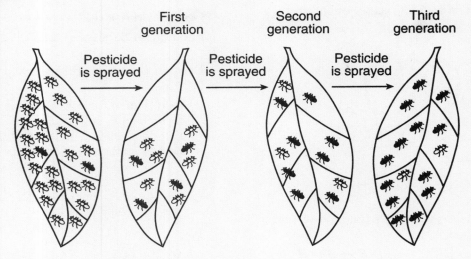

Which concept is represented in the diagram?

(1) survival of the fittest

(2) dynamic equilibrium

(3) succession

(4) extinction

38 In an ecosystem, the herring population was reduced by fishermen. As a result, the tuna, which feed on the herring, disappeared. The sand eels, which are eaten by herring, increased in number. The fishermen then overharvested the sand eel population. Cod and seabirds then decreased. Which food web best represents the feeding relationships in this ecosystem?

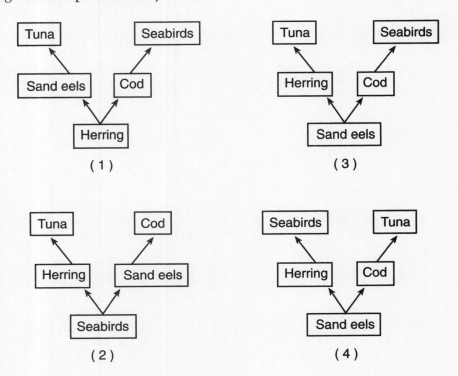

Base your answers to questions 39 through 41 on the diagram below, which represents systems in a human male and on your knowledge of biology.

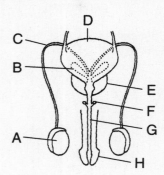

39 Which sequence represents the path of sperm leaving the body?

(1) $A \rightarrow C \rightarrow G$
(2) $A \rightarrow C \rightarrow B$
(3) $E \rightarrow F \rightarrow H$
(4) $D \rightarrow F \rightarrow G$

40 Which structures aid in the transport of sperm by secreting fluid?

(1) A and H
(2) B and E
(3) C and D
(4) D and H

41 Which structure has both reproductive and excretory functions?

(1) A
(2) G
(3) C
(4) D

42 Two food chains are represented below.

Food chain A: aquatic plant → insect → frog → hawk
Food chain B: grass → rabbit → hawk

Decomposers are important for supplying energy for

(1) food chain A, only
(2) food chain B, only
(3) both food chain A and food chain B
(4) neither food chain A nor food chain B

Part B–2

Answer all questions in this part. [13]

Directions (43–55): For those questions that are followed by four choices, circle the *number* of the choice that, of those given, best completes the statement or answers the question. For all other questions in this part, follow the directions given in the question and record your answers in the spaces provided.

Base your answers to questions 43 through 45 on the diagrams below and on your knowledge of biology. The diagrams represent two different cells and some of their parts. The diagrams are not drawn to scale.

Cell A   Cell B

For Teacher Use Only

43 Identify an organelle in cell *A* that is the site of autotrophic nutrition. [1]

_____

43 ☐

44 Identify the organelle labeled *X* in cell *B*. [1]

_____

44 ☐

45 Which statement best describes these cells?

(1) Cell *B* lacks vacuoles while cell *A* has them.

(2) DNA would not be found in either cell *A* or cell *B*.

(3) Both cell *A* and cell *B* use energy released from ATP.

(4) Both cell *A* and cell *B* produce antibiotics.

45 ☐

Base your answers to questions 46 through 48 on the diagram below and on your knowledge of biology.

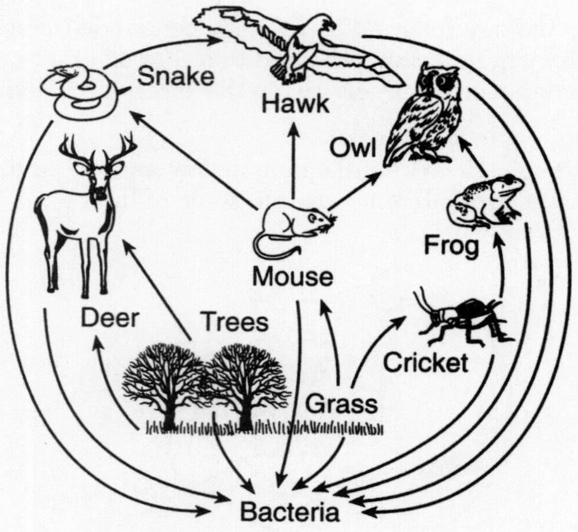

46 What is an appropriate title for this diagram?

(1) Energy Flow in a Community

(2) Ecological Succession

(3) Biological Evolution

(4) A Food Chain

46 ☐

47 Which organism carries out autotrophic nutrition?

(1) hawk

(2) cricket

(3) grass

(4) deer

47 ☐

48 State what would most likely happen to the cricket population if all of the grasses were removed. [1]

_____

_____

_____

48 ☐

Base your answers to questions 49 through 53 on the information and diagrams below and on your knowledge of biology.

The laboratory setups represented below were used to investigate the effect of temperature on cellular respiration in yeast (a single-celled organism). Each of two flasks containing equal amounts of a yeast-glucose solution was submerged in a water bath, one kept at 20°C and one kept at 35°C. The number of gas bubbles released from the glass tube in each setup was observed and the results were recorded every 5 minutes for a period of 25 minutes. The data are summarized in the table below.

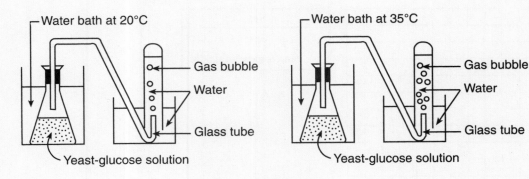

Data Table

Time (minutes)	Total Number of Bubbles Released	
	20°C	35°C
5	0	5
10	5	15
15	15	30
20	30	50
25	45	75

Directions (49–51): Using the information in the data table, construct a line graph on the grid on the next page, following the directions below.

49 Mark an appropriate scale on each axis. [1]

50 Plot the data for the total number of bubbles released at 20°C on the grid on the next page. Surround each point with a small circle and connect the points. [1]

Example:

51 Plot the data for the total number of bubbles released at 35°C on the grid. Surround each point with a small triangle and connect the points. [1]

Example:

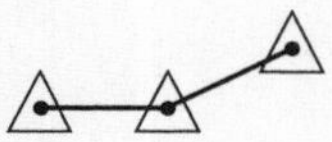

The Effect of Temperature on Respiration in Yeast

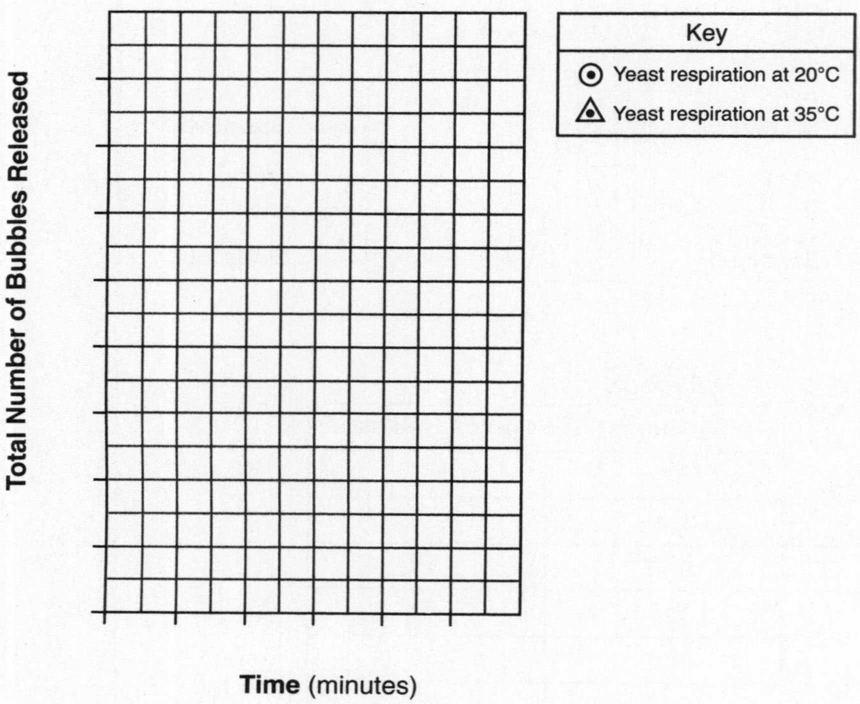

Key
⊙ Yeast respiration at 20°C
△ Yeast respiration at 35°C

Total Number of Bubbles Released

Time (minutes)

52 State *one* relationship between temperature and the rate of gas production in yeast. [1]

_____

_____

53 Identify the gas that would be produced by the process taking place in both laboratory setups. [1]

_____

_____

49 ☐

50 ☐

51 ☐

52 ☐

53 ☐

Base your answers to questions 54 and 55 on the diagram below and on your knowledge of biology.

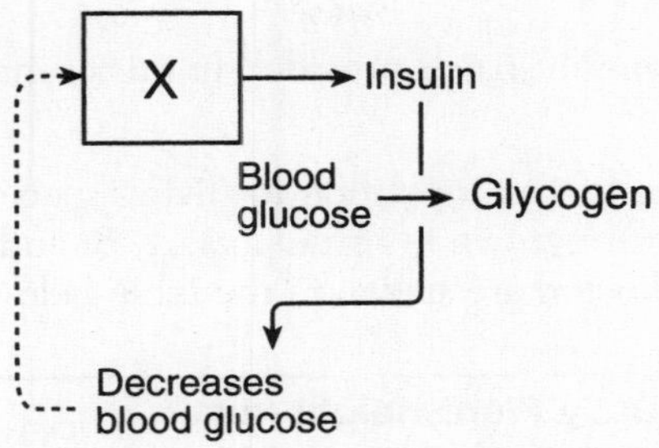

54 Identify the organ labeled *X*.   [1]

_____

54 ☐

55 The dashed line in the diagram represents

(1) a digestive process

(2) a feedback mechanism

(3) cellular differentiation

(4) recycling of organic chemicals

55 ☐

Part C

Answer all questions in this part. [17]

Directions (56–61): Record your answers in the spaces provided in this examination booklet.

56 An experiment was carried out to determine how competition for living space affects plant height. Different numbers of plants were grown in three pots, *A*, *B*, and *C*. All three pots were the same size. The data collected are shown in the table below.

	Average Daily Plant Height (mm)						
	Day 1	Day 2	Day 3	Day 4	Day 5	Day 6	Day 7
Pot A—5 plants	2	4	6	8	10	14	16
Pot B—10 plants	2	4	6	8	10	12	12
Pot C—20 plants	2	2	2	6	6	8	8

Analyze the experiment that produced the data shown in the table. In your answer be sure to:

- state a hypothesis for the experiment [1]
- identify *one* factor, other than pot size, that should have been kept the same in each experimental group [1]
- identify the dependent variable [1]
- state whether the data supports or fails to support your hypothesis and justify your answer [1]

_____

_____

_____

_____

_____

_____

_____

_____

_____

_____

_____

56

57 In many investigations, both in the laboratory and in natural environments, the pH of substances is measured. Explain why pH is important to living things. In your explanation be sure to:

- identify *one* example of a life process of an organism that could be affected by a pH change [1]
- state *one environmental* problem that is directly related to pH   [1]
- identify *one* possible cause of this environmental problem   [1]

_____

_____

_____

_____

_____

_____

_____

_____

_____

_____

_____

57

Base your answer to question 58 on the information below and on your knowledge of biology.

Cargo ships traveling to the Great Lakes from the Caspian Sea in Eurasia often carry water in tanks known as ballast tanks. This water helps the ships to be more stable while crossing the ocean. Upon arrival in the Great Lakes, this water is pumped out of the ships. Often this water contains species that are not native to the Great Lakes environment. The zebra mussel is one species that was introduced into the Great Lakes in this way.

Although large numbers of zebra mussels often clog water intake pipes of power plants and other industries, the mussels have a benefit. Each mussel filters about a quart of water per day, absorbing cancer-causing PCB's from lake water in the process.

The goby, a bottom-feeding fish from Europe, was introduced into the Great Lakes in a similar way a few years later. The gobies have become a dominant species in the Great Lakes, eating small zebra mussels and the eggs and young of other fish. Gobies are eaten by large sport fish. These sport fish have been tested and PCB's have been found in their tissues. Recommendations have been made that people limit the number of sport fish they eat.

58 Explain how the introduction of foreign species can often cause environmental problems. In your answer be sure to:

- state how the zebra mussels and gobies were introduced into the United States [1]
- state *one* way either the zebra mussels *or* gobies have become a problem in their new environment [1]
- describe how *both* zebra mussels and gobies contribute to increasing the concentration of PCB's in sport fish [2]

_____

_____

_____

_____

_____

_____

_____

_____

_____

_____

58 ☐

59 Knowledge of human genes gained from research on the structure and function of human genetic material has led to improvements in medicine and health care for humans.

- state *two* ways this knowledge has improved medicine and health care for humans [2]
- identify *one* specific concern that could result from the application of this knowledge [1]

_____

_____

_____

_____

_____

_____

59 ☐

Base your answers to questions 60 and 61 on the information below and on your knowledge of biology.

You are the owner of a chemical company. Many people in your community have been complaining that rabbits are getting into their gardens and eating the flowering plants and vegetables they have planted. Your company is developing a new chemical product called Bunny Hop-Away that repels rabbits. This product would be sprayed on the plants to prevent the rabbits from eating them. Certain concerns need to be considered before you make the product available for public use.

60 State *two* environmental concerns that should be considered before the product is sold and used by the public. [2]

_____

_____

_____

_____

60 ☐

61 State *one* safety procedure that should be followed when the product is sprayed on plants. [1]

_____

_____

61 ☐

Part D

Answer all questions in this part. [13]

Directions (62–73): For those questions that are followed by four choices, circle the *number* of the choice, that, of those given, best completes the statement or answers the question. For all other questions in this part, follow the directions given in the questions and record your answers in the spaces provided.

62 Students were asked to determine if they could squeeze a clothespin more times in a minute after resting than after exercising. An experiment that accurately tests this question should include all of the following *except*

(1) a hypothesis on which to base the design of the experiment

(2) a large number of students

(3) two sets of clothespins, one that is easy to open and one that is more difficult to open

(4) a control group and an experimental group with equal numbers of students of approximately the same age

62 ☐

63 Which statement best describes a controlled experiment?

(1) It eliminates the need for dependent variables.

(2) It shows the effect of a dependent variable on an independent variable.

(3) It avoids the use of variables.

(4) It tests the effect of a single independent variable.

63 ☐

64 Which statement best describes a change that usually takes place in the human body when the heart rate increases as a result of exercise?

(1) More oxygen is delivered to muscle cells.

(2) Blood cells are excreted at a faster rate.

(3) The rate of digestion increases.

(4) No hormones are produced.

64 ☐

Base your answers to questions 65 through 67 on the diagram below and on your knowledge of biology. The diagram shows the results of a technique used to analyze DNA.

DNA Samples

1   2   3   4   5   6   7

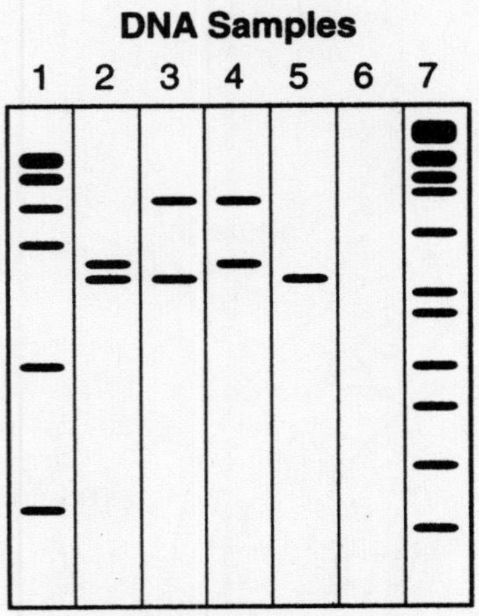

65 This technique used to analyze DNA directly results in

  (1) synthesizing large fragments of DNA

  (2) separating DNA fragments on the basis of size

  (3) producing genetically engineered DNA molecules

  (4) removing the larger DNA fragments from the samples

65 ☐

66 This laboratory technique is known as

  (1) gel electrophoresis

  (2) DNA replication

  (3) protein synthesis

  (4) genetic recombination

66 ☐

67 State *one* specific way the results of this laboratory technique could be used. [1]

_____

_____

_____

67 ☐

68 The cactus finch, warbler finch, and woodpecker finch all live on one island. Based on the information in the diagram below, which one of these finches is *least* likely to compete with the other two for food? Support your answer with an explanation. [1]

For Teacher Use Only

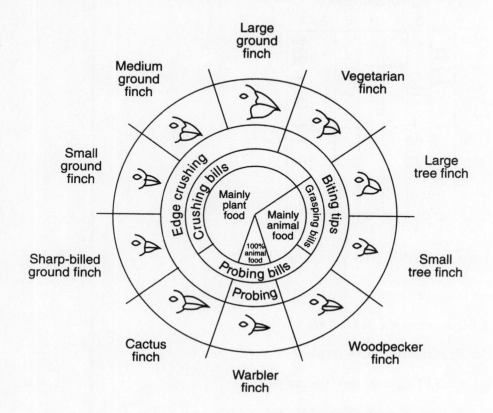

From: *Galapagos: A Natural History Guide*

Variations in Beaks of Galapagos Islands Finches

_____

_____

_____

68 ☐

Base your answers to questions 69 and 70 on the information below and on your knowledge of biology.

Evolutionary changes have been observed in beak size in a population of medium ground finches in the Galapagos Islands. Given a choice of small and large seeds, the medium ground finch eats mostly small seeds, which are easier to crush. However, during dry years, all seeds are in short supply. Small seeds are quickly consumed, so the birds are left with a diet of large seeds. Studies have shown that this change in diet may be related to an increase in the average size of the beak of the medium ground finch.

69 The most likely explanation for the increase in average beak size of the medium ground finch is that the

(1) trait is inherited and birds with larger beaks have greater reproductive success

(2) birds acquired larger beaks due to the added exercise of feeding on large seeds

(3) birds interbred with a larger-beaked species and passed on the trait

(4) lack of small seeds caused a mutation which resulted in a larger beak

69 ☐

70 In exceptionally dry years, what most likely happens in a population of medium ground finches?

(1) There is increased cooperation between the birds.

(2) Birds with large beaks prey on birds with small beaks.

(3) The finches develop parasitic relationships with mammals.

(4) There is increased competition for a limited number of small seeds.

70 ☐

Base your answers to questions 71 and 72 on the diagram below and on your knowledge of biology. The diagram shows the changes that occurred in a beaker after 30 minutes. The beaker contained water, food coloring, and a bag made from dialysis tubing membrane.

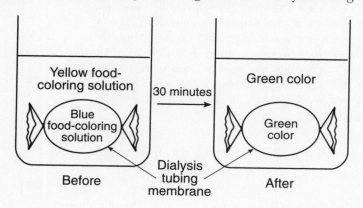

71 When the colors yellow and blue are combined, they produce a green color. Which statement most likely describes the relative sizes of the yellow and blue food-coloring molecules in the diagram?

(1) The yellow food-coloring molecules are small, while the blue food-coloring molecules are large.

(2) The yellow food-coloring molecules are large, while the blue food-coloring molecules are small.

(3) Both the yellow food-coloring molecules and the blue food-coloring molecules are large.

(4) Both the yellow food-coloring molecules and the blue food-coloring molecules are small.

71 ☐

72 Which statement best explains the changes shown?

(1) Molecular movement was aided by the presence of specific carbohydrate molecules on the surface of the membrane.

(2) Molecular movement was aided by the presence of specific enzyme molecules on the surface of the membrane.

(3) Molecules moved across the membrane without additional energy being supplied.

(4) Molecules moved across the membrane only when additional energy was supplied.

72 ☐

73 Cell *A* shown below is a typical red onion cell in water on a slide viewed with a compound light microscope.

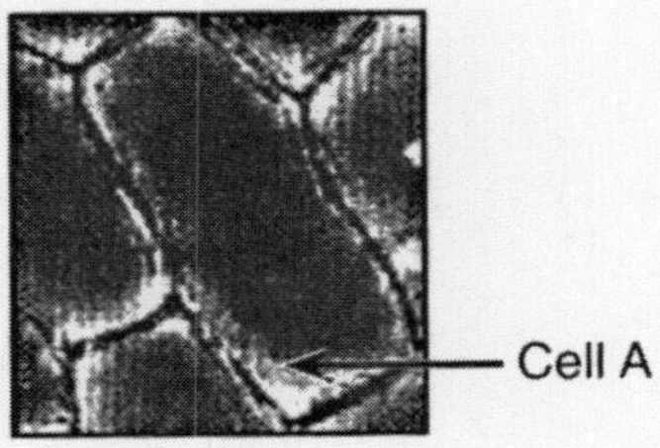

Cell A

Draw a diagram of how cell *A* would most likely look after salt water has been added to the slide and label the cell membrane in your diagram. [2]

73

LIVING ENVIRONMENT
JUNE 2007

ANSWER SHEET

Student . . . .

Teacher . . . .

School . . . . Grade . . . .

Part	Maximum Score	Student's Score
A	30	
B–1	12	
B–2	13	
C	17	
D	13	
Total Raw Score (maximum Raw Score: 85)		
Final Score (from conversion chart)		

Raters' Initials

Rater 1 . . . . Rater 2 . . . .

Record your answers to Part A and Part B–1 on this answer sheet.

Part A

1 . . . .   11 . . . .   21 . . . .

2 . . . .   12 . . . .   22 . . . .

3 . . . .   13 . . . .   23 . . . .

4 . . . .   14 . . . .   24 . . . .

5 . . . .   15 . . . .   25 . . . .

6 . . . .   16 . . . .   26 . . . .

7 . . . .   17 . . . .   27 . . . .

8 . . . .   18 . . . .   28 . . . .

9 . . . .   19 . . . .   29 . . . .

10 . . . .   20 . . . .   30 . . . .

Part A Score

Part B–1

31 . . . .   37 . . . .

32 . . . .   38 . . . .

33 . . . .   39 . . . .

34 . . . .   40 . . . .

35 . . . .   41 . . . .

36 . . . .   42 . . . .

Part B–1 Score